BIOCHEMICAL SOCIETY SYMPOSIA

No. 47

Messenger RNA and Ribosomes in Protein Synthesis

BIOCHEMICAL SOCIETY SYMPOSIUM No. 47
held at King's College, University of London, December 1981

Messenger RNA and Ribosomes in Protein Synthesis

ORGANIZED AND EDITED BY

C.F. PHELPS AND H.R.V. ARNSTEIN

1982

LONDON: THE BIOCHEMICAL SOCIETY

The Biochemical Society,
7 Warwick Court,
London WC1R 5DP, U.K.

ISBN: 0 904498 14 X
ISSN: 0067-8694

Printed in Great Britain by Spottiswoode Ballantyne Ltd., Colchester and London

List of Contributors

H.R.V. Arnstein (*Department of Biochemistry, King's College, Strand, London WC2R 2LS, U.K.*)

S. Blanquet (*Laboratoire de Biochimie, Ecole Polytechnique, 91128 Palaiseau-Cedex, France*)

G. Blobel (*Laboratory of Cell Biology, The Rockefeller University, New York, NY 10021, U.S.A.*)

W.A. Braell (*Department of Biology, Massachusetts Institute of Technology, Cambridge, MA 02139, U.S.A.*)

R. Brimacombe (*Max-Planck-Institut für Molekulare Genetik, Abteilung Wittmann, Ihnestrasse 63–73, D-1000 Berlin-Dahlem, Germany*)

M. Cannon (*Department of Biochemistry, King's College, University of London, Strand, London WC2R 2LS, U.K.*)

R.A. Cox (*National Institute for Medical Research, Mill Hill, London NW7 1AA, U.K.*)

E. Darzynkiewicz (*Institute of Experimental Physics, University of Warsaw, Warsaw, Poland*)

G. Fayat (*Laboratoire de Biochimie, Ecole Polytechnique, 91128 Palaiseau-Cedex, France*)

Y. Furuichi (*Roche Institute of Molecular Biology, Nutley, NJ 07110, U.S.A.*)

M. Grunberg-Manago (*Institut de Biologie Physico-Chimique, 13 rue Pierre et Marie Curie, 75005 Paris, France*)

M. Hümbelin (*Biocenter, University of Basel, CH-4056 Basel, Switzerland*)

J.M. Kelly (*National Institute for Medical Research, Mill Hill, London NW7 1AA, U.K.*)

M. Kozak (*Department of Biological Sciences, University of Pittsburgh, Pittsburgh, PA 15260, U.S.A.*)

H. Kroath (*Roche Institute of Molecular Biology, Nutley, NJ 07110, U.S.A.*)

C.G. Kurland (*Department of Molecular Biology, Biomedical Center, University of Uppsala, Box 590, S-75124 Uppsala, Sweden*)

H.F. Lodish (*Department of Biology, Massachusetts Institute of Technology, Cambridge, MA 02139, U.S.A.*)

M.A. Morgan (*Roche Institute of Molecular Biology, Nutley, NJ 07110, U.S.A.*)

J.A. Plumbridge (*Institut de Biologie Physico-Chimique, 13 rue Pierre et Marie Curie, 75005 Paris, France*)

C. Sacerdot (*Laboratoire de Biochemie, Ecole Polytechnique, 91128 Palaiseau-Cedex, France*)

A.J. Shatkin (*Roche Institute of Molecular Biology, Nutley, NJ 07110, U.S.A.*)

N. Sonenberg (*Department of Biochemistry, McGill University, Montreal, Canada*)

M. Springer (*Institut de Biologie Physico-Chimique, 13 rue Pierre et Marie Curie, 75005 Paris, France*)

C. Stähli (*Hoffmann-La Roche & Co. AG, Grenzacherstrasse, CH-4002 Basel, Switzerland*)

S.M. Tahara (*Roche Institute of Molecular Biology, Nutley, NJ 07110, U.S.A.*)

H. Trachsel (*Biocenter, University of Basel, CH-4056 Basel, Switzerland*)

C.P.H. Vary (*Department of Biology, Syracuse University, 130 College Place, Syracuse, NY 13210, U.S.A.*)

J.N. Vournakis (*Department of Biology, Syracuse University, 130 College Place, Syracuse, NY 13210, U.S.A.*)

P. Walter (*Laboratory of Cell Biology, The Rockefeller University, New York, NY 10021, U.S.A.*)

M. Yamakawa (*Roche Institute of Molecular Biology, Nutley, NJ 07110, U.S.A.*)

A. Zumbé (*Friedrich-Miescher Institut, Postfach 273, CH-4002 Basel, Switzerland*)

Preface

This is the first Biochemical Society Symposium with which I have been associated and I wish to thank all those contributors who accepted the Highwayman tactics of exchanging the manuscripts of their presentations for the reimbursement of their travelling expenses.

Many subjects vie for consideration for symposia. I have tried to seek advice as to three necessary prerequisites: timeliness for the airing of the subject matter for research workers in the field, need for state-of-the union resumé and distinctiveness in the material assembled. Because of excellent help from Dr. Ian Kerr, Dr. Tim Hunt and Professor H. R. V. Arnstein, these ends, I hope, have been met more realistically than if it had been left to my own devices.

Many generous contributions from learned societies and industry allowed this Symposium to invite a properly international representation of participants. I thank Ciba Laboratories, Imperial Chemical Industries Ltd., Beecham Research Laboratories, Glaxo Laboratories Ltd., Wellcome Research Laboratories and especially The Royal Society for their generosity.

'Work, finish, publish' was the phlegmatic advice of Faraday. This the excellent editorial office of the Biochemical Society has arranged with nary a hiccough. There remains that extra component that I hope readers will find in this and like symposia. It was well said by T. H. Huxley: 'It is a popular delusion that the scientific enquirer is under an obligation not to go beyond generalization of observed facts ... but anyone who is practically acquainted with scientific work is aware that those who refuse to go beyond the facts rarely get as far'.

CHARLES PHELPS

Chelsea College,
University of London,
Manresa Road,
London SW3 6LX, U.K.

Contents

Page

List of Contributors v

Preface vii

Prokaryotic Ribosome Structure: a Kinetic View
By C.G. Kurland 1

Structural Aspects of Eukaryotic Ribosomes
By R.A. Cox and J.M. Kelly 11

The Secondary Structure of Ribosomal RNA, and its Organization within the Ribosomal Subunits
By R. Brimacombe 49

Secondary Structure of Eukaryotic Messenger RNA
By C.P.H. Vary and J.N. Vournakis 61

Studies on the Structure and Biogenesis of Yeast Ribosomes
By M. Cannon 79

Translation Mechanism in Prokaryotes: Structure and Expression of *Escherichia coli* Initiation Factor IF3 Gene
By M. Grunberg-Manago, M. Springer, J. A. Plumbridge, S. Blanquet, G. Fayat and C. Sacerdot 95

How do Eukaryotic Ribosomes Recognize the Unique AUG Initiator Codon in Messenger RNA
By M. Kozak 113

5′-Terminal Caps, Cap-Binding Proteins and Eukaryotic mRNA Function
By A.J. Shatkin, E. Darzynkiewicz, Y. Furuichi, H. Kroath, M.A. Morgan, S.M. Tahara and M. Yamakawa 129

Association of an M_r 50000 Cap Binding Protein with the Cytoskeleton in BHK Cells
By H. Trachsel, A. Zumbé, C. Stähli, M. Hümbelin and N. Sonenberg . . 145

Messenger Ribonucleoprotein Complexes in Gene Expression
By H.R.V. Arnstein 163

Mechanism of Protein Translocation Across the Endoplasmic Reticulum
By P. Walter and G. Blobel. 183

Synthesis and Maturation of the Erythrocyte Anion Transport Protein – an Internal Sequence for Membrane Insertion
By H.F. Lodish and W.A. Braell 193

Subject Index 211

Abbreviations

BMV	Brome mosaic virus
CBP	Cap binding protein
CVR	Cobra venom ribonuclease
DNAase	Deoxyribonuclease
EF	Elongation factor
eIF	Eukaryotic initiation factor
EMC	Encephalomyocarditis virus
ER	Endoplasmic reticulum
FITC	Fluorescein isothiocyanate
Gm	2′-*O*-Methylguanosine
Hepes	4-(2-Hydroxyethyl)-1-piperazine-ethanesulphonic acid
hnRNA	Heterogeneous nuclear RNA
IF	Initiation factor
Ig	Immunoglobulin
m^7G	7-Methylguanosine
mRNA	Messenger RNA
Nm	2′-*O*-Methylnucleoside
N.m.r.	Nuclear magnetic resonance
PMSF	Phenylmethanesulphonyl fluoride
(p)ppGpp	Guanosine 2′-diphosphate 5′-di(tri)phosphate
RNAase	Ribonuclease
RNP	Ribonucleoprotein
rRNA	Ribosomal RNA
SDS	Sodium dodecyl sulphate
snRNA	Small nuclear RNA
SRP	Signal recognition protein
tRNA	Transfer RNA
U.v.	Ultraviolet
VSV	Vesicular stomatitis virus

Biochem. Soc. Symp. **47**, 1-9
Printed in Great Britain

Prokaryotic Ribosome Structure: a Kinetic View

C.G. KURLAND

Department of Molecular Biology, Biomedical Center, University of Uppsala, Box 590, S-75124 Uppsala, Sweden

Introduction

It is convenient to think of the ribosome as a machine. However, the use of this metaphor is not intended to justify a long discussion of the relevant parts list. Instead, we will be interested in the machine-like behaviour of the ribosome to the extent that we pose the following sorts of questions. What is the relationship between energy consumption and the speed of translation? How much energy must be consumed to translate with a certain accuracy? What design strategies are used to regulate the performance of the translation apparatus?

Critical to the present discussion is the notion that the translation apparatus is powered and regulated by the chemical potentials established by the metabolism of the bacterium or cell. Simple physical relationships that describe the interaction of the chemical potentials with the translation apparatus lead to a fundamental relationship between on the one hand the speed and accuracy of translation, and on the other, the energy consumed during translation. The details of these relationships are deeply influenced by those structural characteristics of the ribosome that determine the kinetic behaviour of the system. Accordingly, our concern with the energetics of translation is part of a more ambitious attempt to understand why the translation apparatus is constructed in its present form.

Unfortunately, our ambitions still exceed our capacities and we must content ourselves with a progress report. In this report we try to indicate some of the more obvious issues related to the larger themes as well as to describe two design strategies that are related to the mechanisms of translational accuracy. One of these is the strategy that our laboratory has used to develop systems that duplicate *in vitro* the activities of ribosomes in the living bacteria. The other concerns the strategy that the bacterium uses to limit the errors of translation when it is starved for substrates of protein biosynthesis.

External Constraints

The experimental situation in the 1970's could be summarized as follows. We were interested in understanding how a ribosome can synthesize polypeptides at rates between 15–20/s per ribosome (Kjeldgaard & Gausin, 1973) and with a mis-sense error frequency not greater than 1 in 3000 (Edelmann & Gallant, 1977). We pursued this interest with systems that translated poly(U) *in vitro* at less than 1% of the above rates and did so with a leucine mis-sense error frequency usually 100 times greater than that managed by bacteria. It was, therefore, not surprising that the mechanisms of protein biosynthesis seemed rather illusive.

The principal problem at that time was to account for the discrepancy between the discrimination we could expect from the codon–anticodon interaction, which was thought to necessitate mis-sense error frequencies in the range of 1% (Grosjean *et al.*, 1976), and the much lower mis-sense error frequency of protein biosynthesis. It seemed as though something was missing from our picture.

An extremely stimulating suggestion by Hopfield (1974) and Ninio (1975) provided both a theoretical solution to this problem, and, more importantly, the impetus to consider the problem from a broader perspective. Their suggestion provided a way to surpass a limitation in the capacity of an enzyme to distinguish two competing substrates on the basis of structurally determined binding interactions. The trick is to repeat the competition between substrate species one or more times, and thereby, multiply the discrimination obtained in successive competitions before forming the product. In order to mediate such a substrate proof-reading scheme, the enzyme must be constructed in such a way that the substrate flow can be branched: one pathway permitting the preferential flow of cognate substrate to product and the other pathway facilitating the preferential flow off the enzyme of noncognate substrate. We will have little more to say about proof-reading on ribosomes here, since we have other fish to fry. However, there is one aspect of the proof-reading notion that is of general relevance.

As pointed out in the original presentations of the proof-reading models, there is a thermodynamic constraint on chemical reactions with the weighty title of the 'principle of detailed balance', and this constraint must be broken in order for the discard branch of a proof-reading scheme to function in the desired way. Another way of saying this is that there must be a force to drive noncognate substrates out over the discard branch. Recognition of this requirement for driving forces to activate substrate flows was an important step for the following reasons.

The problem of maintaining the appropriate accuracy of translation can be restated as that of maintaining a high rate of flow into peptide of cognate amino acids and a low rate into peptide of noncognate amino acids. Therefore, we should be curious about the driving forces for these flows, not only in proof-reading schemes but for any arbitrarily chosen selection scheme. Thus, it is the effects of the thermodynamic forces on an enzyme that determine the accuracy of the product formation through the substrate flows. In what follows we will draw upon previous studies (Kurland, 1978; Ehrenberg & Blomberg, 1980; Blomberg *et al.*, 1980) to describe substrate flows. If through some wonder the reader should become curious about the details, these earlier studies can be consulted for a rigorous treatment.

We consider a steady state system in which the substrates and products of an enzymic reaction are held at constant levels. It is clear that the net flow over the enzyme is determined by the difference between the forward and backward flows. These in turn will be determined by the concentrations of substrates and products, respectively. In the particular case that the ratio of product to substrate concentrations is equivalent to the equilibrium constant, the flows in both directions will be evenly matched and there will be no net reaction. On the other hand, if the ratio of product to substrate concentrations is much smaller than that at equilibrium, there will be a net flow from substrate to product. We call the ratio of these concentrations, normalized to the equilibrium constant, the displacement from equilibrium (δ). Evidently, the smaller the value of δ, the faster the net forward flow to product.

Of course, we expect the structure of the enzyme to influence the flows, since in general the rates of reaction are catalysed by enzymes. It turns out that the kinetic contribution of the enzyme can be conveniently summarized in a rate factor (R) that is formally equivalent to the ratio of the more familiar $k_{cat.}$ to K_m for the enzyme–substrate interaction. We may describe the influence of the rate factor (R) for the enzyme (E), the displacement (δ), and the substrate concentration (S) on the flow to product (J) by the simple relationship:

$$J = R(1-\delta)(S)(E) \tag{1}$$

If we reflect on this equation we realize that with suitable biochemical manipulations all of the parameters except the rate factor can be manipulated by the experimentalist. In other words, the rate factor uniquely describes the functional properties of a given enzyme in a given reaction. This means that what is often referred to as an attempt to define the structure–function relationships for ribosomal components can be reduced to the more experimentally meaningful attempt to define structure–rate factor relations. This becomes quite obvious when we discuss the accuracy of translation.

For simplicity we consider a conventional enzymic reaction operating in the presence of two competing substrates, each of which can form product over the enzyme and for which the equilibrium constant with product is the same in both cases. In general, we expect the rate factors to be different for the two substrates with the ratio of the rate factor for cognate substrate to that of noncognate substrate defined as the discrimination factor (d). Here, the discrimination factor defines the capacity of the enzyme to distinguish effectively the two substrates. Next we ask how the accuracy of the reaction (A), defined as the ratio of the rate of formation of cognate to that of noncognate product, can be related to the cognate substrate displacement (δ) when the ratio of cognate to noncognate substrate species is S_r. Again we find a simple relationship for the example of substrate competition in the absence of such distractions as substrate allostery, hysteresis, etc:

$$A = dS_r\left[(1-\delta) + \frac{\delta}{d}\right] \tag{2}$$

We note that for a system that responds to substrate displacements as in eqns. (1) and (2), both the speed and the accuracy of function are monotonic increasing functions of the displacement. We call such systems balanced and, as we show below, the translation system seems to be balanced in this sense.

The significance of this balanced response to increasing displacements from equilibrium can be deepened by discussing the energetic costs of driving enzymic reactions. It is not infrequent that one comes across deeply confused discussions of the costs of enzymic reactions, either in conversation or in the literature. A good example of such confusion is the work of Freter & Savageau (1980). They have described a variety of enzymic proof-reading selection schemes that are required to obey only one physical constraint, the conservation of mass. The amusing aspect of this formalism is that by placing proof-reading schemes in a universe devoid of thermodynamic constraints, they have removed the original need to postulate the existence of proof-reading mechanisms. Hence, their cost calculations, which merely describe the fraction of the substrate flow that ends up in product, are at best incomplete.

In order to determine the cost of a reaction, we ask what is irreversibly lost when a mole of substrate is converted into a mole of product. We recall that, at equilibrium, the conversion of substrate to product is attended by no energetic loss because the molar free energies of substrate and product are the same. However, the further displaced from equilibrium the reactants are, the greater is the molar free energy difference between substrate and product. This free energy difference is the driving force for the reaction, and it is defined by two terms. One term is related to the standard free energies of the reactants and this is invariant. The other term describes the free energy loss due to the entropy of mixing, when substrate is diluted to the concentration level of product. It is this entropic term that increases with the displacement from equilibrium, and it is easily shown that the irreversible loss of free energy per mole of substrate converted to product is:

$$\mu_{\text{diss.}} = -RT \ln \delta \qquad (3)$$

Accordingly, the free energy cost per mole of substrate is uniquely determined by the displacement from equilibrium. Since the displacement also determines the speed and accuracy of the reactions, the balanced enzyme is one that in general responds most effectively to the free energy dissipation. Therefore, we might expect most enzymes to be balanced. This expectation is based on the assumption that the structural features of an enzyme that influence its rate factor will have evolved under a selective pressure to minimize the dissipation of free energy. Nevertheless, it is conceivable that such a minimization must be suspended when parameters other than the dissipative losses are limiting the performance of an enzyme system. Hence, the discovery that a particular enzyme system is unbalanced might indicate that the system is working near an extreme limit of the catalytic capacity of protein structure.

In summary, we identify the intrinsic structural properties of an enzyme system such as the ribosome and its factors with the rate factor for protein biosynthesis. The rate and accuracy outputs of such a system can be modulated by the external constraints imposed on the system in the form of the concentrations of substrates and products, summarized in the displacement from equilibrium. Since we began this discussion by emphasizing the unrealistically poor performance of conventional systems *in vitro*, it will not surprise the reader to learn that we next discuss data concerning the improvements of systems *in vitro* depending on the optimization of substrate displacements.

Translation *in vitro*

The experimental problem that has preoccupied us in recent years has been to learn how to apply the foregoing theoretical precepts to the practical task of characterizing the translation apparatus of *Escherichia coli*. This has meant searching for conditions *in vitro* under which the rate and discrimination factors of the system duplicate those expressed in the living bacterium. This effort initially centred around a study of the effects of the ionic environment on the rate factors and the effects of the displacements from equilibrium of the relevant substrates on the performance characteristics of the system.

It turns out that the Tris/Mg^{2+} buffers normally used for translation *in vitro* have bizarre effects on the rate factors of the translation system (Pettersson & Kurland, 1980). For example, the incorporation of phenylalanine with a poly(U)-primed system can be compared with the mis-sense incorporation of leucine as a function of Mg^{2+} concentration. It is observed that while the Mg^{2+} optimum for the rate of phenylalanine incorporation is greater than 10 mM, that for the minimum incorporation of leucine is below 5 mM. In other words, the poly(U) translation system in conventional buffers is unbalanced so that there is no Mg^{2+} concentration at which both the cognate rate factor and the discrimination factor can be simultaneously optimized.

This curious situation prompted us to study the effects of naturally occurring ions on the rate and discrimination factors for translation *in vitro*. A mixture of inorganic ions and polyamines, which is not very different from that found in *E. coli*, turned out to balance the rate factors of the poly(U) translation system so that maximum rate and accuracy could be obtained at comparable Mg^{2+} concentrations (Jelenc & Kurland, 1979; Pettersson & Kurland, 1980). This buffer, which we call polymix, then could be used to study systematically the influence of the substrate displacements on the system.

The substrates for protein biosynthesis are aminoacyl-tRNA and GTP. The displacement from equilibrium for these substrates can be coupled to that for ATP quite conveniently *in vitro*, as in fact it is done *in vivo*. Thus, by controlling the regeneration of aminoacyl-tRNA from deacylated tRNA and the regeneration of GTP from GDP with the aid of the appropriate enzymes and an ATP regenerating system, the relevant displacements become accessible experimental parameters. The response of the poly(U) translation system in polymix to increases in the absolute concentrations of substrates and substrate displacements is quite straightforward. The rates and accuracy of translation increase together, as would be expected for a balanced system (Jelenc & Kurland, 1979; Wagner & Kurland, 1980). Most important, the discrimination factor of the optimized system approaches that of the bacterium and yields a leucine mis-sense error frequency of 4×10^{-4}.

Nevertheless, the ribosomal rate factor for phenylalanine in this system seemed to be orders of magnitude slower than that expressed *in vivo*. Since the discrimination factor for the ribosome was reasonable, it seemed possible that the apparently low rate factor with phenylalanine was due to the fact that only a minor fraction of the ribosomes are active in our assays. Accordingly, we preincubated the ribosomes with poly(U) and *N*-acetyl-Phe-tRNA in order to both initiate and monitor the active ribosomes. Then all the purified components necessary for elongation were added, and the activities of the ribosomes were assayed at very short intervals (i.e. seconds after the start). The data show that approx. 10% of the ribosomes could be activated in this assay and that they elongate poly(Phe) at a rate of at least 8 peptide bonds/s per ribosome. During this 'burst', which lasts roughly 10 s, the missense error rate is 4×10^{-4} (Wagner *et al.*, 1982*a*).

Analysis of the 'burst' indicates that the elongation rate slows down as the ribosomes reach the 3′ end of the poly(U) messenger. This means that the lengths of poly(U) normally used restrict experiments with this system to unusually short times, but this is no problem, particularly if fast kinetic equipment is available. Furthermore, the stimulatory effect of the *N*-acetyl-Phe-tRNA suggests that in conventional systems it is the initiation step which limits the rate of polypeptide synthesis. Now that we can control these parameters, it is a routine matter to collect data with a system *in vitro* that elongates at a rate and with an accuracy comparable with those of *E. coli*.

There are a large variety of phenomena relevant to ribosome function that now can be analysed. These include the effects on accuracy of antibiotics, mutations, hypermodifications and the like, all of which are somewhat mysterious at this point. They are mysterious because not only do we not know how they influence the mechanism of translation, but until now we did not have the tools to find out. As an illustration of the changes in experimental possibilities, we consider a recent study of the effects of elongation factors Ts (EF-Ts) on the elongation rate.

It has long been supposed that the demonstrable stimulation *in vitro* of the rate of GDP release from the binary complex with EF-Tu represents the major function of EF-Ts in translation. However, the evidence for this supposition was derived from experiments done in the absence of most of the other components of the translation apparatus and under ionic conditions that, as we have indicated above, actually disturb the kinetic functions of the system. We, therefore, decided to try to measure the kinetic effects of EF-Ts on the elongation process in the optimized burst system.

The strategy of these experiments was that with the concentrations of all other components of the system arranged so that EF-Tu was the rate limiting factor, the catalytic effects of EF-Ts on the EF-Tu cycle could be measured through its effects on elongation rates (Ruusala *et al.*, 1982). The data showed that EF-Tu's cycle time in polypeptide synthesis is reduced by more than three orders of magnitude to 30 ms in the presence of EF-Ts. Separate measurements of the rate of release of GDP from the complex with EF-Tu in the presence of GTP and Phe-tRNA also indicate that the release time is reduced by EF-Ts from 90 s to 30 ms (Ruusala *et al.*, 1982).

These experiments represent the first kinetic measurements ever performed on individual factors during polypeptide synthesis. As such, they provide a paradigm for the analysis of the kinetic contributions of other components of the translation apparatus. They also foreshadow the types of experiments that will have to be done to identify the modes of action of antibiotics and mutations that effect protein synthesis. Most important, the results of these measurements define an intriguing problem.

Simply stated: why is the GDP–EF-Tu interaction so tight that it requires a separate factor to catalyse the regeneration of EF-Tu–GTP? We note that this is apparently not the case for the GDP–EF-G interaction. Stated another way, what is the problem solved by the peculiar arrangement of the EF-Tu, EF-Ts, guanine nucleotide cycle? It is our prejudice that answers to questions of this sort will provide a key to the design strategy of the translation apparatus.

The Stringent Response

As is evident from eqn. (2), which describes the accuracy of translation in terms of concentration terms, etc, that the ratio of cognate to noncognate substrates (S_r) should influence the error frequency of the process. That this is so *in vitro* has been verified in conventional assays as well as in the burst system (Wagner & Kurland, 1980; Wagner *et al.*, 1982*a*). In contrast, the response of bacteria to extreme changes of the substrate ratio, as for example during starvation, is more complex. If the bacterium is mutated so that it cannot express the stringent response, then the proteins synthesized during amino acid limitation or aminoacyl-tRNA synthetase inhibition are characterized by higher-than-normal mis-sense error frequencies (Edelmann & Gallant, 1977). If the bacteria are stringent, this effect is not observed. Hence, it seems that the stringent response protects the translation system against the errors induced by abnormal distortions of the substrate ratios. In other words, the wild type bacterium can regulate the error frequency of translation when it is exposed to a metabolic stress.

Since the guanine nucleotide analogues pppGpp (magic spot II) and ppGpp (magic spot I) suppress starvation-induced translation errors *in vitro*, it has been concluded that the magic spots themselves are the active agents of the stringent response in the bacterium. In addition, it has been observed *in vitro* that the addition of magic spots to unstarved systems has no effect on the accuracy, but produces a strong inhibition of the rate, of translation (Wagner & Kurland, 1980). This suggests that, whatever else they do, the magic spots do not change the intrinsic discrimination capacity of the ribosomes. The problem then is to determine how the magic spots can selectively influence the relative flows of cognate and noncognate substrates without influencing the selectivity of the ribosomes.

A very reasonable suggestion by O'Farrell (1978) to explain these effects focuses attention on the differences in absolute concentrations of the starved and unstarved aminoacyl-tRNA species. Thus, an inhibitor of translation such as the magic spots could inhibit the flow of the limiting amino acid sufficiently so that the concentration of the corresponding aminoacyl-tRNA would increase. This would tend to improve the substrate ratio at all of the codons cognate to the starved amino acid, and thereby improve the fidelity of translation.

Various aspects of this explanation have been verified in an amino acid-starved system *in vitro* (Wagner & Kurland, 1980). Nevertheless, data obtained by comparing stringent and relaxed bacteria during amino acid limitation contradicted the central prediction of the hypothesis. In fact, it was found that the aminoacyl-tRNA charging level of the starved species is unaffected by the presence or absence of the stringent response (Böck *et al.*, 1966; Yegian & Stent, 1969; Piepersberg *et al.*, 1979). We have, therefore, re-examined this problem with the aid of the new burst system.

Our account of the effects of magic spots on the accuracy of translation when the system is limited in one aminoacyl-tRNA species focuses attention on two points. First, the immediate substrate for polypeptide synthesis is not free aminoacyl-tRNA but the ternary complex of EF-Tu, aminoacyl-tRNA and GTP. Second, the magic spot I inhibition of EF-Tu is the dominant effect of magic spots during translation *in vitro* (S. Andersson, E.G.H. Wagner, M. Ehrenberg & C.G. Kurland, unpublished work). Therefore, we ask what happens when EF-Tu is inhibited by magic spot during aminoacyl-tRNA limitation in the steady state. Our answer is that magic spot I should selectively inhibit the flow of the high-concentration aminoacyl-tRNA species and have little effect on the low-concentration species. Therefore, it should improve the accuracy of translation at codons cognate to the starved species.

This effect can be deduced from the requirements for maintaining translation in the steady state. Thus, when the rate of cognate aminoacyl-tRNA conversion to peptide is limited by the rate of aminoacyl-tRNA formation, variations in the EF-Tu–GTP concentration will not affect this flow as long as the product of the concentration terms for aminoacyl-tRNA and EF-Tu–GTP binary complex is held constant. Therefore, the concentration of ternary complex containing the limiting species can be maintained at a low constant determined by the rate of aminoacyl-tRNA charging, even when EF-Tu function is inhibited.

In contrast, the high concentration aminoacyl-tRNA species will be very nearly fully charged. As a consequence, an agent such as magic spot, which inactivates EF-Tu, will make it impossible for the system to hold the product of the concentration terms for EF-Tu–GTP and these aminoacyl-tRNA species constant. Hence, the concentrations of the corresponding ternary complexes should decrease as the magic spot concentration increases. This will have the effect of decreasing the flow of aminoacyl-tRNA's that are noncognate to the codons corresponding to the limiting aminoacyl-tRNA species. In other words, the mis-sense error frequency will be decreased by magic spot at the starved codons.

This account of the effects of magic spot on the accuracy of translation under conditions of aminoacyl-tRNA limitation has been tested by comparing the effects of decreasing the concentrations of EF-G and EF-Tu in a poly(U) translation system limited on the Phe-tRNA synthetase (Wagner *et al.*, 1982*b*). The effects are straightforward. Decreasing the EF-G concentration yields a pure O'Farrell effect: the mis-sense error frequency decreases in parallel with an increase of the Phe-tRNA charging level. In contrast, decreasing the EF-Tu concentration over a relatively broad concentration range decreases the mis-sense error rate, while both the incorporation of phenylalanine and Phe-tRNA charging level remain relatively constant. In effect, decreasing the EF-Tu concentration *in vitro* mimics all the known physiological effects of the stringent response on translational accuracy *in vivo*. There remains one aspect of this response to explain. We postulate that the rate-limiting aminoacyl-tRNA species concentration will increase as the EF-Tu–GTP concentration decreases. However, the total limiting aminoacyl-tRNA concentration is observed both *in vivo* and *in vitro* to remain almost constant in response to magic spot. We explain away this apparent contradiction by suggesting that the total aminoacyl-tRNA concentration is dominated by the ternary complex species. When this larger concentration term is constant, the increase of the smaller free aminoacyl-tRNA concentration has a negligible effect on their sum.

In summary, the discussion of the stringent response hopefully has illustrated some of the major themes around which the present report is organized. First, it should be evident that it would be impossible to analyse the effects of the magic spots on accuracy without a well developed kinetic model for translation and without an experimental system that, at least approximately, duplicates the kinetic situation in the bacterium. Second, it should be clear that the particular problems that are solved by the translation system influence its design strategies. For example, part of the solution to the problem of regulating translational errors during metabolic stress depends on the separation of the nucleotide triphosphate pools for the aminoacyl-tRNA synthetase and ribosome components into adenosine and guanosine pools, respectively. Finally, there remains a great deal to do in the way of kinetic characterization before the design strategies of the system as a whole are understood.

References

Blomberg, C., Ehrenberg, M. & Kurland, C.G. (1980) *Q. Rev. Biophys.* **13,** 231–254

Böck, A., Fuiman, L.E. & Neidhardt, F.C. (1966) *J. Bacteriol.* **92,** 1076–1082

Edelmann, P. & Gallant, J. (1977) *Cell* **10,** 131–137

Ehrenberg, M. & Blomberg, C. (1980) *Biophys. J.* **31,** 333–358

Freter, R. & Savageau, M. (1980) *J. Theoret. Biol.* **85,** 99–123

Grosjean, H.J., Söll, D. & Crothers, D.M. (1976) *J. Mol. Biol.* **103,** 499–520

Hopfield, J.J. (1974) *Proc. Natl. Acad. Sci. U.S.A.* **71,** 4135–4139

Jelenc, P.C. & Kurland, C.G. (1979) *Proc. Natl. Acad. Sci. U.S.A.* **76,** 3174–3178

Kjeldgaard, N.O. & Gausing, K. (1974) in *Ribosomes* (Nomura, M., Tissieres, A. & Lengyel, P., eds.), pp. 369–392, Cold Spring Harbor Laboratory, Cold Spring Harbor, NY

Kurland, C.G. (1978) *Biophys. J.* **22,** 373–392

Ninio, J. (1975) *Biochimie* **57,** 587–595

O'Farrell, P.H. (1978) *Cell* **14,** 545–557

Piepersberg, W., Geyl, D., Buckel, P. & Böck, A. (1979) in *Regulation of Macro-Molecular Synthesis by Low Molecular Weight Mediators* (Koch, G. & Richter, D., eds.), pp. 39–52, Academic Press, New York

Pettersson, I. & Kurland, C.G. (1980) *Proc. Natl. Acad. Sci. U.S.A.* **77,** 4007–4010

Ruusala, T., Ehrenberg, M. & Kurland, C.G. (1982) *EMBO J.*, **1,** 75–78

Wagner, E.G.H. & Kurland, C.G. (1980) *Mol. Gen. Genet.* **180,** 139–145

Wagner, E.G.H., Jelenc, P.C., Ehrenberg, M. & Kurland, C.G. (1982*a*) *Eur. J. Biochem.* **122,** 193–197

Wagner, E.G.H., Ehrenberg, M. & Kurland, C.G. (1982*b*) *Mol. Gen. Genet.*, in press

Yegian, C.D. & Stent, G.S. (1969) *J. Mol. Biol.* **39,** 45–58

Biochem. Soc. Symp. **47**, 11–48
Printed in Great Britain

Structural Aspects of Eukaryotic Ribosomes

ROBERT A. COX and JOHN M. KELLY

National Institute for Medical Research, Mill Hill, London NW7 1AA, U.K.

Introduction

Recombinant DNA technology has made an impact on studies of cytoplasmic ribosomes in several ways. Firstly, the organization and structure of the genes for cytoplasmic and organelle rRNA are now known for a wide range of species (see below). Secondly, like other eukaryotic genes, active eukaryotic rRNA genes [e.g. those of *Physarum polycephalum* (Gubler *et al.*, 1979)] may be split by non-coding sequences (introns). Thirdly, the advent of rapid DNA sequencing techniques has led to the determination of the complete primary sequences of cytoplasmic and organelle S-rRNA (the RNA moiety of the smaller subparticle) and to partial sequences of L-rRNA (the major RNA component of the larger subparticle). These new data make it feasible to discuss in detail not only the relationship between cytoplasmic rRNA of different species but also the relationship between cytoplasmic rRNA and organelle rRNA, and the relationship between eukaryotic rRNA and bacterial rRNA. The aim of this report is to review the evidence relating to the evolution of rRNA and to summarize the current knowledge of the structure of cytoplasmic rRNA and its conformation within the functional ribosome or ribosomal subparticles.

We emphasize these aspects of the rRNA moiety of ribosomes because a review is timely and because of our own interest in the role of rRNA in protein biosynthesis. The first part of this report describes evolutionary conserved and divergent features of cytoplasmic rRNA. The appearance of 5.8 S rRNA as a separate species and the way in which L-rRNA has diverged during evolution have provided clues that suggest a scheme for the evolution of cytoplasmic rRNA genes and their relation with bacterial rRNA genes. The second part of the report concerns the conformation of rRNA within the ribosome based on both physiochemical methods and on the application of RNA sequencing methods to the study of rRNA conformation *in situ*.

Comparison of the Properties of Bacterial, Chloroplast, Mitochondrial and Cytoplasmic Ribosomes

Compared with prokaryotes, eukaryotic cells are not only larger but they are structurally more complex since they have, in addition to the nucleus, subcellular organelles, viz. chloroplasts (which are present in plant cells) and mitochondria (which are present in both plant and animal cells). In prokaryotes, the genetic message is translated as fast as it is transcribed (e.g. Miller & Hamkalo, 1972). In eukaryotes, transcription takes place in the nucleus and, after processing, mature mRNA is transported to the cytoplasm where it is translated (see Arnstein, 1982, this volume). There is still further specialization within eukaryotic cells because both chloroplasts and mitochondria not only have their own genome but they also have their own complement of ribosomes for the synthesis of specific proteins. However, most of the ribosomes (approx. 5×10^5) of a eukaryote are present in the cytoplasm.

Cytoplasmic, chloroplast and mitochondrial ribosomes each have their own distinctive structural features. These properties of eukaryotic ribosomes, together with those of bacterial ribosomes, are presented in Table 1. Despite their obvious differences, all of these species of ribosomes must have a very similar active centre because they all function in mRNA-directed protein biosynthesis, and they all use essentially the same genetic code. Thus an understanding of the similarities and differences in the structures of the difference species of ribosomes should prove valuable in defining the essential structure needed for the sequential formation of peptide bonds.

One feature of Table 1 is the variation in the mass of ribosomes, which ranges from the miniribosomes (approx. 1.6×10^6 daltons) of animal mitochondria to the larger cytoplasmic ribosomes (approx. 4.6×10^6 daltons) of the octopus. There is a correspondence between the size of the ribosome and its subparticles and the sizes of the major rRNA moieties (see Table 1). For example, S-rRNA (the RNA component of the smaller subparticle) ranges from 0.3×10^6 daltons (animal mitochondrial ribosomes) to 0.7×10^6 daltons (cytoplasmic ribosomes). There is even greater variation in the mass of L-rRNA (the major RNA component of the larger subparticle) which ranges from 0.5×10^6 daltons (animal mitochondrial ribosomes) to 1.9×10^6 daltons (octopus cytoplasmic ribosomes). Compared with bacterial ribosomes, both mitochondrial and cytoplasmic ribosomes have diverged more rapidly during evolution. Cytoplasmic ribosomes have become larger, whereas mitochondrial ribosomes have become smaller as they have evolved.

Cytoplasmic ribosomes are characterized by the size of their S-rRNA and L-rRNA components and also by the presence of 5.8 S rRNA within the larger subparticle. The absence of 5.8 S rRNA from prokaryotes has provoked speculation of the functions of 5 S rRNA in prokaryotes and of 5 S rRNA and 5.8 S rRNA in eukaryotes. It has been suggested (Erdmann, 1976; Wrede & Erdmann, 1977) that 5.8 S rRNA is the analogue of prokaryotic 5 S rRNA; and also that the roles of 5 S rRNA in prokaryotes have become specialized in eukaryotes so that two species (namely 5 S rRNA and 5.8 S rRNA) are needed to do the same job.

The appearance of 5.8 S rRNA as a separate species and the variations in the mass and nucleotide composition of L-rRNA can both be explained in the light of several recent developments. We present evidence that there are extensive homologies between mature L-rRNA of prokaryotes (e.g. *Escherichia coli* L-rRNA) and the precursor L-rRNA (pre L-rRNA) of eukaryotes (e.g. *Xenopus laevis* pre L-rRNA). These homologies suggest that the apparent evolutionary discontinuity separating prokaryotes and eukaryotes has arisen as a result of mutations altering the processing of pre L-rRNA.

Table 1. *Properties of prokaryotic and eukaryotic ribosomes*†

Ribosomes	Smaller subparticle $s_{20,w}$	Smaller subparticle Mass of S-rRNA	Smaller subparticle No. of protein subunits	Larger subparticle $s_{20,w}$	Larger subparticle Mass of L-rRNA	Larger subparticle 5 S rRNA	Larger subparticle 5.8 S rRNA	Larger subparticle No. of protein subunits
Bacterial (70 S)	30	0.55×10^6	21	50	1.1×10^6	Yes	–	32
Chloroplast (70 S)	30	0.55×10^6	21 (?)	50	1.1×10^6	Yes	–	32 (?)
Mitochondrial (50 S–80 S)	20–40	$(0.3–0.7) \times 10^6$	?	30–60	$(0.5–1.3) \times 10^6$	Yes*	–	?
Cytoplasmic (80 S)	40	0.7×10^6	29	60	$(1.3–1.9) \times 10^6$ ‡	Yes	Yes	42

* Presence of 5 S rRNA in mammalian mitochondrial ribosomes has not been confirmed.
† For review see Bielka & Stahl (1978); Cox (1977); Wool (1979).
‡ See Cammarano *et al.* (1980).

Cytoplasmic ribosomes have conserved features

It is feasible to regard cytoplasmic ribosomes as members of a closely related class, even though there is evidence for substantial evolutionary divergence. Cytoplasmic ribosomal subparticles appear to be mutually interchangeable (see Table 2). Inactive core particles may be generated by high-salt (2.75 M-NH_4Cl/65 mM-$MgCl_2$) treatment and activity in cell-free protein synthesis is restored by adding either homologous or heterologous split-protein fractions. Functional chimaera were constructed from rabbit S-subparticles, *X. laevis* core-particles and rabbit split-protein fraction (see Table 3). These data provide support for the view that the

Table 2. *Activity in polyphenylalanine synthesis of hybrid ribosomes formed from smaller and larger subparticles of different species (Cox & Greenwell, 1976)*

Activity in polyphenylalanine synthesis was measured by means of the standard assay. The radioactivity of the control (100%) amounted to 2688 c.p.m./assay (8.4 nCi/assay). These results suggest that the *N. crassa* S-subparticle fraction was partly inactivated during the isolation procedure, whereas the L-subparticle fraction was not impaired.

S-subparticle	L-subparticle	Activity in polyphenylalanine synthesis (% of rabbit control)
Rabbit	–	3
–	Rabbit	6
Rabbit	Rabbit	100
Rabbit	*X. laevis*	75
Rabbit	*N. crassa*	116
X. laevis	*X. laevis*	88
N. crassa	*N. crassa*	57

Table 3. *Activity in polyphenylalanine synthesis of L-subparticles reassembled from core-particles and either homologous or heterologous split-protein fractions (Cox & Greenwell, 1980)*

Activity in polyphenylalanine synthesis was measured by means of the standard assay. The control (100%) was 1472 c.p.m./assay for rabbit and 1024 c.p.m./assay for *X. laevis*.

		L-subparticle components		Activity in polyphenylalanine synthesis (% of control)	
S-subparticle	L-subparticle	Core-particle	Split-proteins	Rabbit	*X. laevis*
Rabbit	Rabbit	–	–	100	–
Rabbit	–	Rabbit	–	3	–
Rabbit	–	–	–	3	–
Rabbit	–	Rabbit	Rabbit	18	–
Rabbit	–	Rabbit	*X. laevis*	15	–
Rabbit	–	Rabbit	*N. crassa*	14	–
X laevis	*X. laevis*	–	–	–	100
X. laevis	–	*X. laevis*	–	–	5
X. laevis	–	–	–	–	4
X. laevis	–	*X. laevis*	*X. laevis*	–	21
X. laevis	–	*X. laevis*	Rabbit	–	16
Rabbit	–	*X. laevis*	Rabbit	–	32
Rabbit	–	*X. laevis*	*X. laevis*	–	22

architecture of the active centre of ribosomes is highly conserved among cytoplasmic ribosomes. Evidence that substantial features of the structure of cytoplasmic S-rRNA and L-rRNA have been conserved is presented below.

Conserved and divergent features of cytoplasmic rRNA

Studies of secondary structure. Historically, the study of the conformation of S-rRNA and L-rRNA preceded studies of primary structure. At the time, the possibility of obtaining nucleotide sequence data was remote and the study of physicochemical properties provided a method for gaining insight into rRNA structure and conformation. The partly bi-helical nature of rRNA was first suggested in 1959 [see for example Doty *et al*. (1959)], when it was recognized that the rRNA chain could fold back upon itself to form intramolecular bi-helical regions (see Fig. 1) stabilized mainly by A · U and G · C base-pairs (Cox, 1966). The formation and 'melting' of bi-helical structures can be monitored by optical methods. The asymmetry of the bi-helix confers optical activity on its constituent base-residues. This optical activity is most conveniently detected as circular dichroism (c.d.). The u.v. absorbance of the constituent base-residues of a bi-helix is also diminished over the range 230–300 nm as a result of the interaction between adjacent base-paired residues. When the temperature is raised a bi-helical region 'melts' over a narrow temperature range, the mid-point of which is termed T_m. The value $T_m(h_i)$ for helix h_i depends on the number (n_i) of base-pairs comprising the bi-helix and on the fraction (f_{GC}) of base-pairs that are G · C base-pairs (see eqn. 1).

$$T_m(h_i) = f(n_i, f_{GC}) \qquad (1)$$

The 'melting' of a bi-helix is manifest as a decrease in c.d., by an increase in u.v. absorbancy etc. The increase on 'melting' in u.v. absorbance over the range 220–300 nm reflects the nucleotide composition of the bi-helix. For example, the 'melting' of an A · U base-pair leads to an increase in A_{260} but A_{280} is unaffected, whereas the melting of a G · C base-pair leads to an increase in both A_{260} and A_{280} (see Fig. 2). Thus the ratio of the increments in A_{260} (ΔA_{260}) and in A_{280} (ΔA_{280}) measured over a particular temperature range yields the nucleotide composition of base-pairs 'melting' over that range. This relation is given in eqn. (2) where $\delta f_{(A \cdot U)}/\delta f_{(G \cdot C)}$ is the ratio of A · U and G · C base-pairs that melt over the temperature range being investigated (Cox, 1966, 1970)

$$\delta f_{(A \cdot U)}/\delta f_{(G \cdot C)} = 0.85\ (\Delta A_{260}/\Delta A_{280}) - 0.39 \qquad (2)$$

Since the temperature range over which the transition takes place provides a measure of $T_m(h_i)$, and since the nucleotide composition can be measured by means of eqn. (2), then n_i may be evaluated by using eqn. (3), where $T_{m(\infty)}$ is the value of T_m for a high-molecular-weight bi-helical RNA (cf. DNA) having the same G · C content (Cox, 1966).

$$T_m = T_{m(\infty)} - (172/n_i) \qquad (3)$$

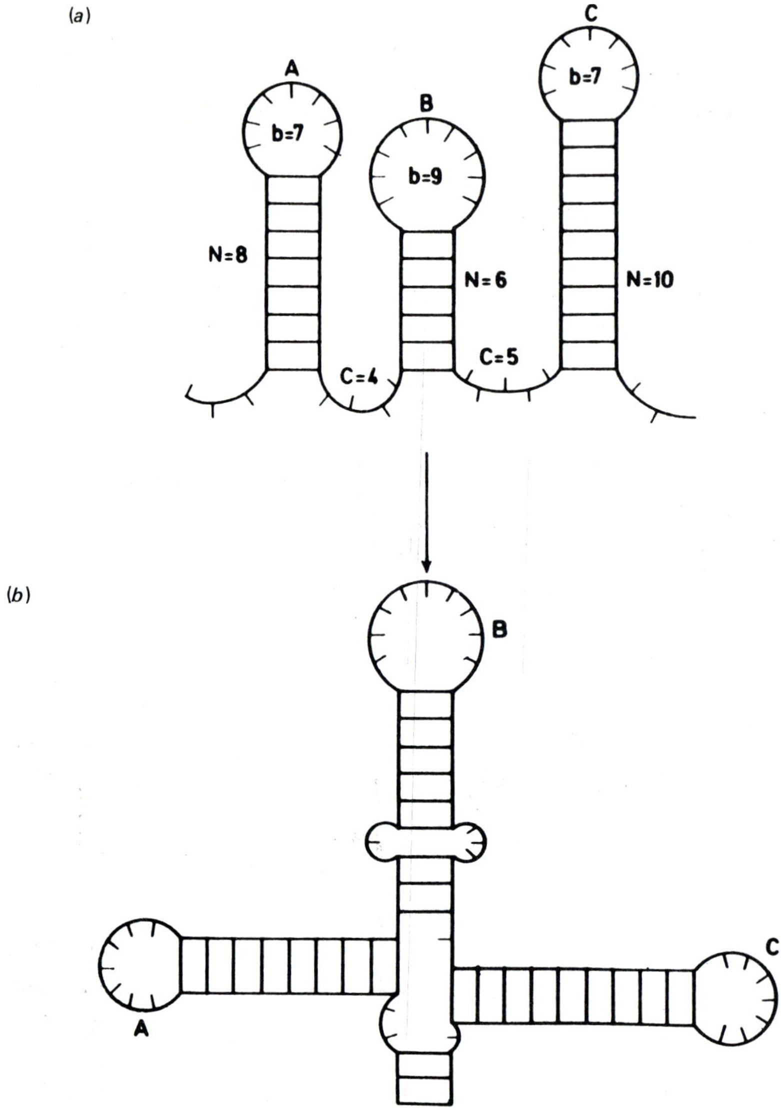

Fig. 1. *The 'simplest' scheme of secondary structure for RNA and more complicated structures (Cox & Godwin, 1975)*

(*a*) The simplest scheme comprising a series of hairpin loops. N is the number of base-pairs per loop; b is the number of unpaired residues per loop; c is the number of residues between one loop and the next. (*b*) The series of loops illustrated in (*a*) may further interact to yield more complicated structures. ∃ indicates base-residues that are 'stacked' but not base-paired, i.e. the base residues interact, the plane of one overlapping the plane of another, in much the same way as within the double helix.

In practice, the melting profile of an RNA species results from the 'melting' of a number of different bi-helices often with overlapping melting ranges. Thus, each RNA species has its own characteristic 'melting' profile.

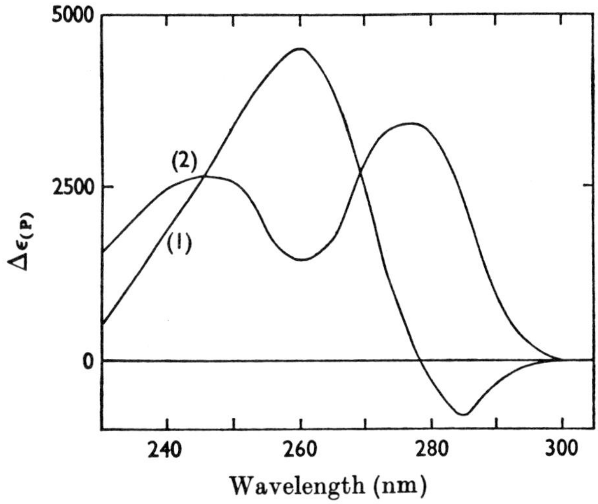

Fig. 2. *Reference denaturation spectra for (1) rA · rU and (2) rG · rC base pairs*

Each curve denotes the change in the spectrum obtained on 'melting' base-paired nucleotides (Cox, 1970) calculated for a 1 M solution of nucleotides.

Melting profile of cytoplasmic L-rRNA. Examination of the melting profiles of a range of L-rRNA species revealed that approx. 0.7×10^6 daltons had features of bi-helical secondary structure in common, comprising 50–55% G · C base-pairs on average, whereas other features of secondary structure were found to be highly divergent. Bi-helical regions of 25% G · C base-pairs were found in *Drosophila melanogaster* L-rRNA (Godwin *et al*., 1974; Cox *et al*., 1976*a*) and, in contrast, regions of 78% G · C base-pairs were identified in mammalian L-rRNA (Cox, 1966, 1970; Cox *et al*., 1973). These interpretations were confirmed by means of studies of partial nuclease digests of L-rRNA which led to the isolation of A + U-rich fragments from *Drosophila* L-rRNA (Cox *et al*., 1976*a*) and to the isolation of large (up to 0.4×10^6 daltons) fragments (78% G + C) from rabbit and amphibian (*X. laevis*) L-rRNA (Cox *et al*., 1973).

Drosophila L-rRNA and mammalian L-rRNA were investigated because they are representative of (A + U)-rich (42% G + C) and (G + C)-rich (67% G + C) L-rRNA species. Their melting profiles are unusual because they are biphasic. In the case of *Drosophila* L-rRNA bi-helical regions rich in A · U base-pairs 'melt' before the 'conserved' regions of secondary structure (see Fig. 3), whereas in (G + C)-rich L-rRNA species, regions of conserved secondary structure melt first and regions with G · C base-pairs 'melt' at higher temperatures (see Fig. 4). The conserved features of secondary structure were found to have many features in common with bacterial L-rRNA (see below and also Cox *et al*., 1973, 1976*a* for discussion). These experiments provided a basis for reconciling the evidence for conserved features and for evolutionary divergence in cytoplasmic rRNA. It was suggested that, e.g. L-rRNA comprises a conserved core comprising $(0.5\text{–}0.7) \times 10^6$ daltons that can form a characteristic secondary structure to which divergent sequences varying in nucleotide composition and in mass have been added. For summary see Figs. 5 and 6. It was also suggested that the divergent sequences could have arisen from mutations

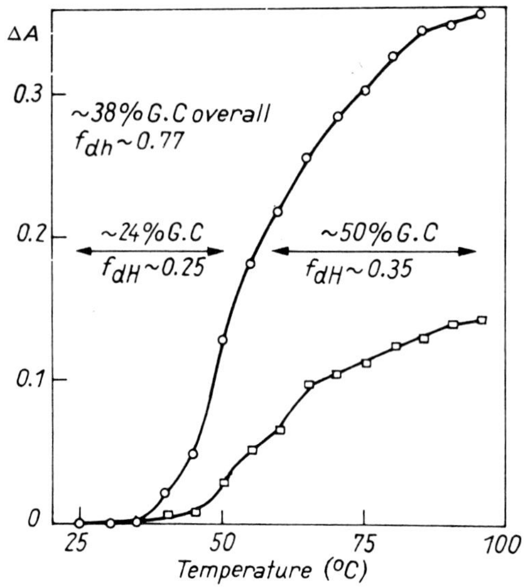

Fig. 3. *Melting profile of D. melanogaster L-rRNA*

The solvent was 0.01 M-Tris/HCl (pH 7.6)/0.1 mM-$MgCl_2$. ○, A_{260} (measuring the melting of both A·U and G·C base pairs); □, A_{280} (measuring the melting of G·C base pairs). f_{dh}, fraction of residues located in bi-helical regions melting over the temperature ranges indicated by the arrows (25–50°C and 60–95°C) or overall (i.e. over the range 25–95°C). The nucleotide composition of the population of bi-helical regions melting over these temperature ranges is also shown.

that affected the processing of precursor rRNA, i.e. regions of pre rRNA became incorporated in mature rRNA (Cox *et al.*, 1973; Godwin *et al.*, 1974). This possibility is discussed below.

Comparison of the properties of divergent sequences of L-rRNA with spacer sequences within rRNA genes

Within the rRNA gene cluster one repeating unit is separated from another by spacer sequences that are not transcribed (non-transcribed spacers). Mature rRNA is derived from a large precursor RNA (pre rRNA) that is cleaved to yield S-rRNA, 5.8 S rRNA and L-rRNA. Spacer sequences (transcribed spacers) that separate one gene from the next are excised from pre rRNA during processing (see Fig. 7).

The repeating unit of the *X. laevis* rRNA gene cluster comprises approx. 12000 base pairs. Approx. 6000 base pairs are accounted for by the sequences coding for S-rRNA and L-rRNA and the remaining 6000 base pairs comprise spacer sequences. The melting profile of the rRNA gene cluster of *X. laevis* (Birnstiel & Grunstein, 1972) is biphasic. The lower melting phase arises from sequences of 55% G+C comprising the regions coding for S-rRNA and L-rRNA. The second phase melts at a T_m that is 12°C higher than the T_m of the first phase and arises from sequences of 80% G+C. Thus, the sequences in *X. laevis* L-rRNA that melt to give rise to the second phase in the melting profile (see Fig. 8) have a G+C content similar to that of spacer sequences within the rRNA genes.

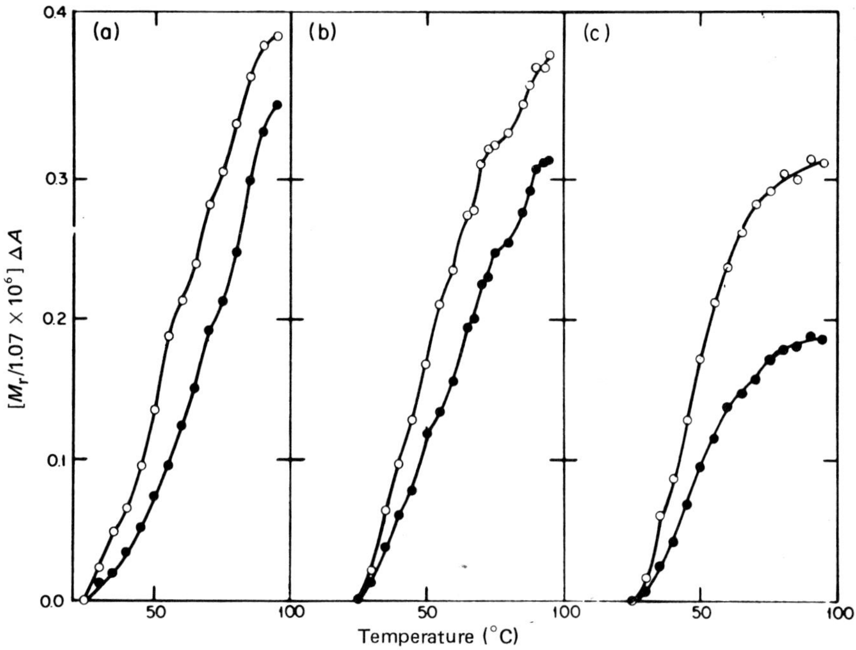

Fig. 4. *Melting profiles of L-RNA from rabbit reticulocytes, X. laevis and E. coli*

The solvent was 0.01 M-phosphate buffer, pH 7. A low ionic strength solvent is needed in order to achieve complete melting of the species containing G + C-rich double-helical regions, i.e. RNA of the eukaryotes; ○, A_{260} (measuring the melting of both A·U and G·C base pairs); ●, A_{280} (measuring the melting of G·C base pairs). (a) RNA of rabbit reticulocytes; (b) RNA of *X. laevis*; (c) RNA of *E. coli*. Values of *A* refer to a solution of A_{260} 1.0 at 25°C, i.e. for a concentration of 42 μg of RNA/ml ($A_{(p)260}$ 7750). To compare the profiles of the different species on a molar basis an allowance was made for differences in molecular weight. For this purpose *E. coli* RNA was taken as standard (Cox *et al.*, 1973).

The notion that the divergent (G + C-rich) sequences in mammalian L-rRNA and in *X. laevis* L-rRNA are related to spacer sequences within the rRNA gene repeat is supported by results obtained by denaturation mapping (Wellauer & Dawid, 1973*a*,*b*). Regions of secondary structure that withstand the spreading conditions are seen as characteristic looped structures. The high stability of these loops suggests that they are G + C-rich. Denaturation maps of the HeLa cell rRNA gene repeat and of *X. laevis* rRNA gene repeat are shown in Fig. 9. It is evident that parts of mature mammalian L-rRNA and *X. laevis* L-rRNA, especially regions near to the 5′-end, have the characteristic looped structures of spacer sequences. In the case of rabbit L-rRNA, the large tracts of 0.4×10^6 daltons of 78% G + C isolated after partial nuclease digestion of L-rRNA was shown (Godwin *et al.*, 1974) to have the same looped appearance when examined by denaturation mapping as the structure found at the 5′-end of L-rRNA. Thus the divergent regions of mammalian and amphibian L-rRNA closely resemble spacer sequences in their secondary structure as well as G + C content. Regions rich in A·U base-pairs are far more difficult to visualize. However, Greenberg (1969) has shown that in *Drosophila* the spacer sequences are A + U-rich. In this case too, there is a correspondence between the

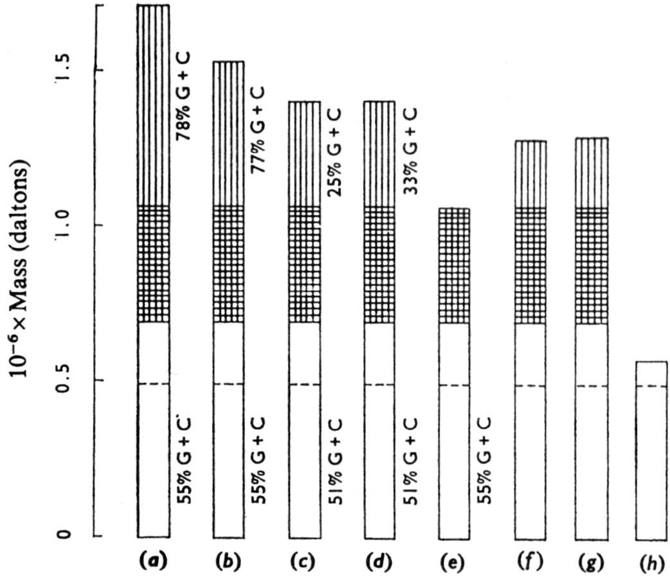

Fig. 5. *Summary of the properties of L-rRNA and their possible evolutionary significance (Cox et al., 1976a)*

(*a*) Schematic presentation of rabbit reticulocyte L-rRNA based on analysis of the 'melting' profile and on analysis of the products of partial digestion with T_1 ribonuclease digestion. (*b*) Data for *X. laevis* L-rRNA based on analysis of the 'melting' profile and on analysis of the products of partial T_1 ribonuclease digestion. The scheme does not conflict with the secondary-structure maps of Wellauer & Dawid (1973). (*c*) Scheme for *D. melanogaster* based on analysis of the 'melting' profile. (*d*) Scheme for *P. knowelsi* based on analysis of the 'melting' profile. (*e*) Scheme for *E. coli* L-rRNA based on the analysis of the 'melting' profile. □ (conserved) indicates the region of L-rRNA that apparently has features of secondary structure in common irrespective of its origin; ⊞ indicates the uncertainty in the estimation of the size of the conserved regions; ⊟ (non-conserved) denotes that part of the RNA moiety that may vary from one class to another, yet may be similar for species within a single class. (*f*) and (*g*) Proposed scheme for pea and yeast L-rRNA, based on the general hypothesis outlined above. (*h*) *X. laevis* mitochondrial L-rRNA. This species is included for comparison. According to our tentative hypothesis most of this species should form secondary structure corresponding to the conserved regions in eukaryotic cytoplasmic L-rRNA and prokaryotic L-rRNA, although there are no data relating to the secondary structure of this species.

nucleotide composition of spacer sequences and the divergent sequences of mature L-rRNA.

Distribution of conserved sequences within L-rRNA. Once rRNA genes were cloned it became feasible to probe for sequences conserved during evolution (see for example Cox & Thompson, 1980; Gourse & Gerbi, 1980). The genes were first fragmented with restriction endonucleases, the fragments were separated by agarose-gel electrophoresis, transferred to nitrocellulose film and fragments which contained conserved sequences were identified because they became radioactive after hybridization with heterologous probes (e.g. ^{125}I-labelled rRNA). Regions of high homology (90% or better) between *X. laevis* L-rRNA and *N. crassa* L-rRNA were found to comprise 1050 ± 200 nucleotides (Cox & Thompson, 1980). This estimate agrees fairly well with the estimate of (0.5–0.7) × 10^6 daltons of L-rRNA sequences having conserved conformational features (see above). The highly

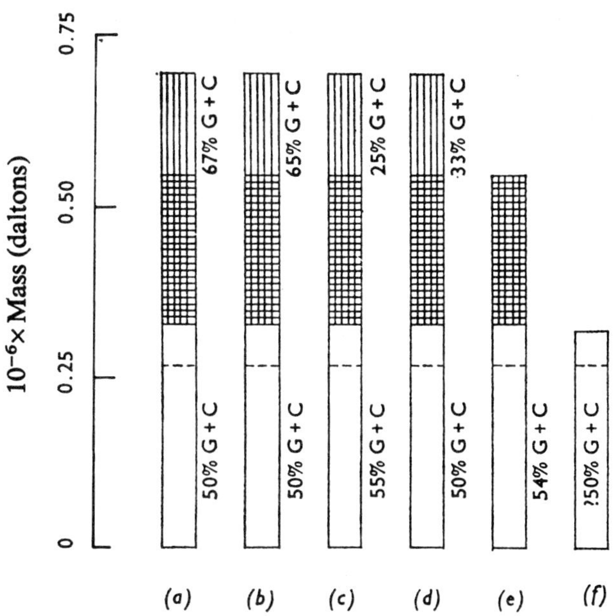

Fig. 6. *Summary of properties of S-rRNA and their possible evolutionary significance (Cox et al., 1976a)*

Schematic presentation based on analyses of 'melting' profiles for: (*a*) rabbit S-rRNA (e.g. Cox, 1970); (*b*) *X. laevis* S-rRNA; (*c*) *P. knowlesi* S-rRNA; (*d*) *D. melanogaster* S-rRNA; (*e*) *E. coli* S-rRNA (the analysis reveals that the percentage of G·C base-pairs of bi-helical region ranges from approx. 50–60% (cf. Cox, 1966); (*f*) *X. laevis* mitochondrial rRNA. □, Conserved region; ⊞ indicates uncertainty in the estimates of the conserved and non-conserved regions; ▤ non-conserved region.

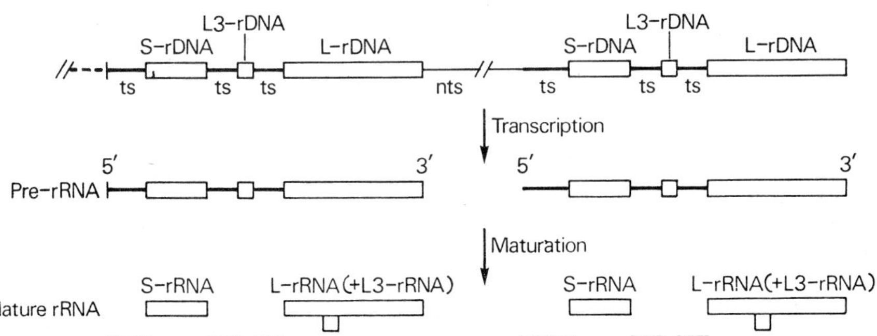

Fig. 7. *Scheme for the organization and transcription of rRNA genes and the processing of rRNA transcripts (see Hadjiolov & Nikolaev, 1976)*

Cytoplasmic rRNA genes are organized in clusters either as tandem repeats (e.g. *X. laevis*; Wellauer & Dawid, 1974) as shown, or as inverted repeats. ts indicates transcribed spacer sequences found in pre-rRNA and not in mature rRNA and nts denotes non-transcribed spacer sequences that separate one gene from the next.

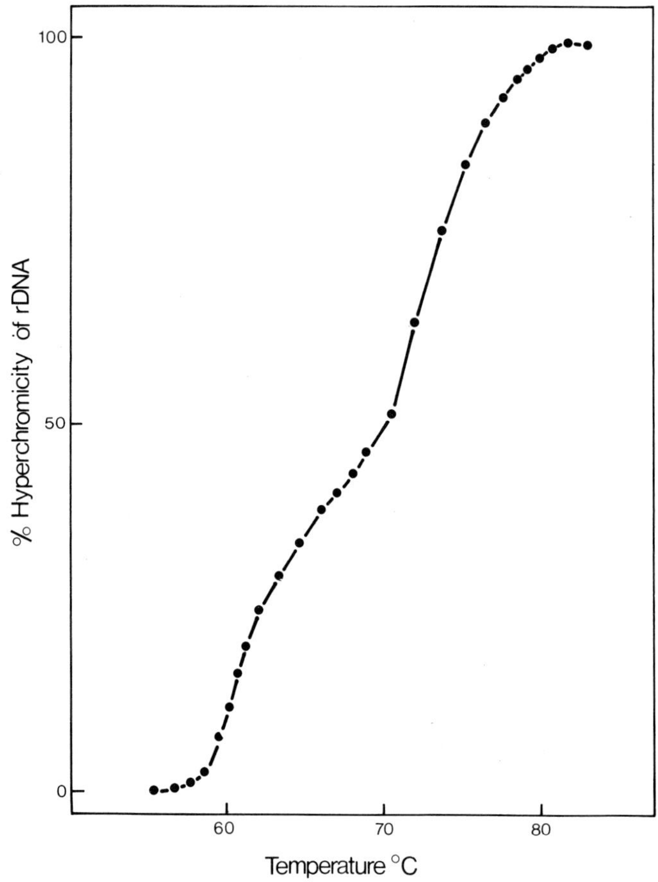

Fig. 8. *Melting profile of isolated rDNA of X. laevis (Birnstiel & Grunstein, 1972)*
The solvent was 50% (v/v) formamide/0.15 M-NaCl/0.015 M-sodium citrate, pH 7.0. The values for the two phases are approx. 62°C and 72.5°C, which correspond to DNA of 55% G + C and 80% G + C respectively.

conserved sequences are interspersed with regions of low or no homology and cover a range of approx. 3000 nucleotides extending from the 3′-end of *X. laevis* L-rRNA. However, few homologies were detected in the region of 1000 nucleotides from the 5′-end of *X. laevis* L-rRNA (Cox & Thompson, 1980). This result is in accord with the presence of G + C-rich sequences at the 5′-end of L-rRNA as shown by denaturation mapping (see Fig. 10). Within *N. crassa* L-rRNA the 1050 ± 200 nucleotides that comprise highly conserved sequences were found to span a region of not less than 1600 nucleotides and not more than 2700 nucleotides (Cox & Thompson, 1980). When *X. laevis* and *N. crassa* L-rRNA are compared (see Fig. 10) it is apparent that the increased mass of *X. laevis* L-rRNA is largely accounted for by the presence of G + C-rich sequences (approx. 1000 nucleotides) mainly at the 5′-end.

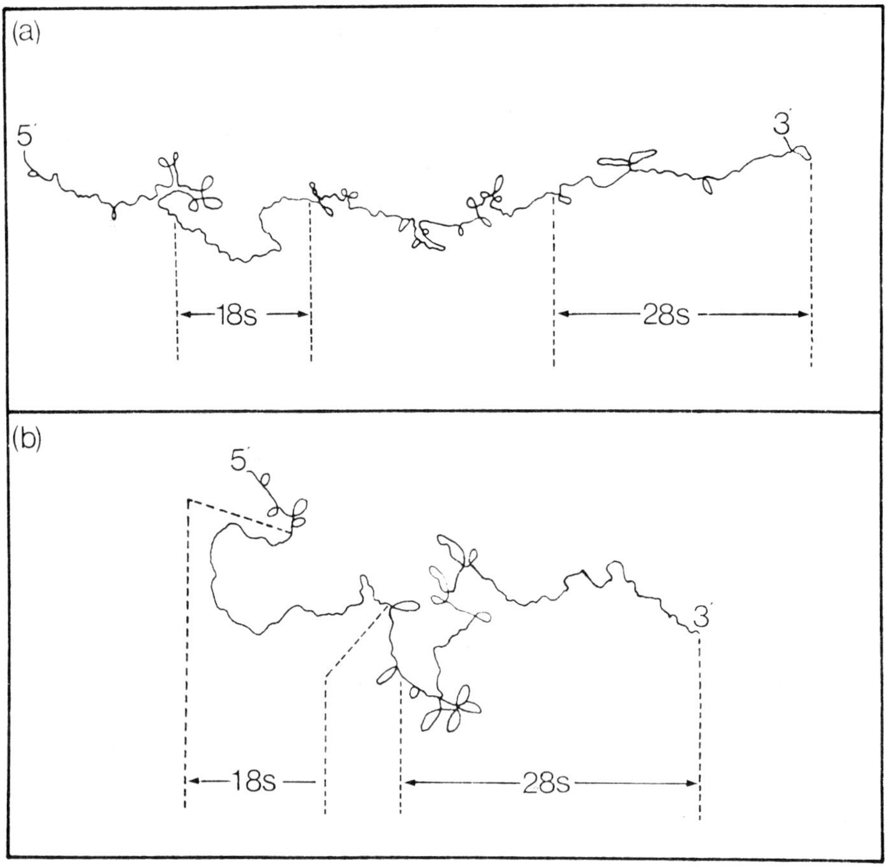

Fig. 9 *Secondary structure maps of (a) HeLa and (b) X. laevis rDNA*

Each figure is a tracing of an electron micrograph obtained by Wellauer & Dawid (1973*a*, *b*, 1974) after spreading rDNA in 80% (v/v) formamide/4 M-urea/1 mM-EDTA/0.03 M-Tris/HCl, pH 8.5.

Bacterial and cytoplasmic S-rRNA and L-rRNA have features of secondary structure in common

Cytoplasmic L-rRNA has sequences that have been conserved and sequences that have diverged during evolution. Although hybridization studies have revealed sequences of high homology (90% or more) that are common to cytoplasmic rRNA, few homologies were detected between bacterial L-rRNA and cytoplasmic L-rRNA (see also below). However, bacterial and cytoplasmic L-rRNA have extensive features of bi-helical secondary structure in common.

It was mentioned above that the melting profile of *E. coli* L-rRNA has features that correspond closely with the lower melting phase of rabbit and *X. laevis* L-rRNA and with higher melting phase of *D. melanogaster* (Cox *et al.*, 1976*a*). These homologies are further illustrated by the comparison of the melting profiles of rabbit L-rRNA and *Bacillus stearothermophilus* L-rRNA illustrated in

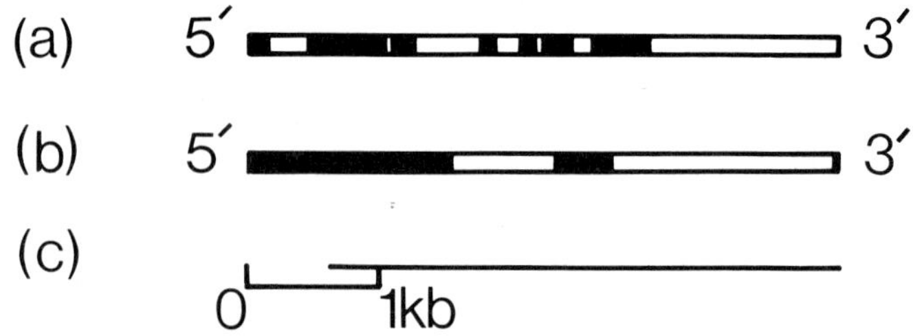

Fig. 10. *Comparison of the distribution within the L-rRNA gene of X. laevis of non-conserved sequences with the distribution within L-rRNA of regions of secondary structure of high thermal stability*

(a) Secondary-structure map of *X. laevis* L-rRNA (Wellauer & Dawid, 1974): ▪ denotes region of secondary structure detected by electron microscopy after spreading L-rRNA by means of a denaturing solvent. The length indicated corresponds to the estimated single-strand length of the hairpin loop. These regions of stable secondary structure are identified as G + C-rich on the basis of spectrophotometric analysis and partial nuclease digests (Cox *et al.*, 1973) and also electron microscopy (Godwin *et al.*, 1974). ▫, Regions appearing as a single-strand. (b) Distribution of non-conserved sequences identified in this study (see Fig. 4): ▪ indicates regions of very low homology with heterologous probes. (c) Size of *N. crassa* L-rRNA gene (Cox & Peden, 1979).

Fig. 11. Both *E. coli* L-rRNA and *B. stearothermophilus* L-rRNA are closely related and this is reflected in the similarity of their melting profiles. The congruence of the increments in A_{260}, A_{270} and A_{280} between the lower melting phase of rabbit L-rRNA and *B. stearothermophilus* L-rRNA demonstrates a high degree of homology extending over a substantial part (approx. 0.7×10^6 daltons) of the L-rRNA species (Cox *et al.*, 1973). For summary see also Figs. 5 and 6.

It was also concluded (Cox *et al.*, 1973, 1976*a*), that bacterial S-rRNA and cytoplasmic S-rRNA have features of secondary structure in common (see also Figs. 5 and 6). This conclusion is considerably strengthened by the availability of nucleotide sequence data for yeast S-rRNA (Rubstov *et al.*, 1980), *X. laevis* S-rRNA (Salim & Maden, 1981) and *E. coli* S-rRNA (Brosius *et al.*, 1978). The feasibility of these three S-rRNA species sharing a common secondary structure was discussed by Brimacombe (1982, this volume). The available evidence for sequence homologies found in bacterial L-rRNA and cytoplasmic L-rRNA is presented below.

Sequences common to bacterial L-rRNA and cytoplasmic 5.8 S rRNA and L-rRNA

One aspect of the evolutionary discontinuity separating bacterial and cytoplasmic ribosomes is the presence in the latter of 5.8 S rRNA. However, 5.8 S rRNA and L-rRNA are found as a hydrogen-bonded complex in cytoplasmic ribosomes. In this respect, the L-rRNA/5.8 S rRNA complex of cytoplasmic ribosomes is equivalent to L-rRNA in bacteria.

The 5.8 S rRNA gene of cytoplasmic ribosomes lies upstream from the 5′-end of the L-rRNA gene (Pellegrini *et al.*, 1977; Boseley *et al.*, 1978; Free *et al.*, 1979; Nath & Bollon, 1978; Sollner-Webb & Reeder, 1979; Veldman *et al.*, 1980) and 5.8 S rRNA appears as a separate entity only at a late stage in the processing of pre L-rRNA (Hadjiolov & Nikolaev, 1976). Nazar (1980) has suggested that sequences near to the 5′-terminus of L-rRNA (23 S rRNA) of *E. coli* are homologous with trout 5.8 S rRNA. We confirm this notion since sequences homologous with

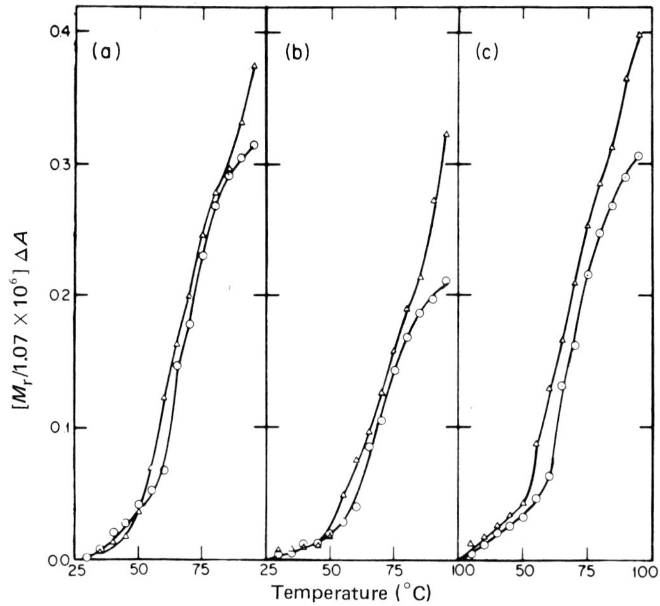

Fig. 11. *Comparison of the first phase of the melting profile of RNA from the larger ribosomal subparticle of a eukaryote (rabbit reticulocytes) with the melting profile of RNA from the larger ribosomal subparticle of a prokaryote (B. stearothermophilus)*

The solvent was 0.1 mM-$MgCl_2$/0.01 M-Tris/HCl, pH 7.6. The presence of Mg^{2+} stabilizes double-helical secondary structure so that the lower melting phase of eukaryotes (cf. Fig. 4) is more readily scrutinized. △, RNA from rabbit reticulocytes; ○, RNA from *B. stearothermophilus*; (a) ΔA_{260} (which reflects the melting of both A·U and G·C base-pairs); (*b*) ΔA_{280} (which reflects the melting of G·C base-pairs); (*c*) ΔA_{270} (which measures the melting of both A·U and G·C base-pairs; at 270 nm ΔA owing to the melting of an A·U base-pair is about the same as ΔA owing to the melting of a G·C base-pair). This increment provides a convenient index of the total base-pairs melting.

N. crassa (Selker & Yanofsky, 1979) and human 5.8 S rRNA (Maden & Robertson, 1974) can be identified close to the 5′-terminus of *E. coli* L-rRNA. Sequences of 5.8 S rRNA that have been most conserved during evolution, and so form a characteristic feature of this rRNA species, include a stretch of approx. 20 nucleotides located at approx. 40–60 nucleotides from the 5′-end (see Fig. 12). The homologous sequence is located within 50–70 nucleotides of the 5′-end of *E. coli* L-rRNA. A related phenomenon involving the 3′-end of L-rRNA has been described in e.g. tobacco chloroplasts. In some chloroplast ribosomes a 4.5 S rRNA component is found which is homologous to sequences at the 3′-end of *E. coli* L-rRNA. In the gene the 4.5 S rRNA sequences are found to be separated by spacer sequences from the 3′-end of L-rRNA (MacKay, 1981). As in the case of 5.8 S rRNA, the spacer sequences are excised but the fragments of mature L-rRNA are not ligated.

Presence at the 3′-end of E. coli L-rRNA of a binding site for 5.8 S rRNA-like sequences

Within the ribosome, cytoplasmic L-rRNA and 5.8 S rRNA form a specific complex. We have shown (Kelly & Cox, 1981*b*) that a binding site for 5.8 S rRNA

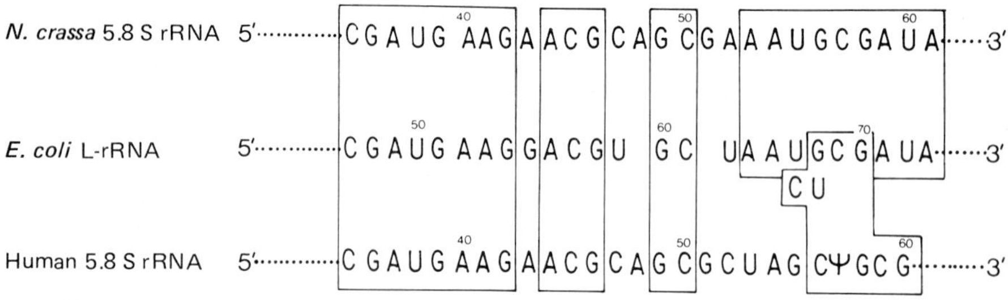

Fig. 12. *Sequences close to the 5′-end of E. coli L-rRNA (Brosius et al., 1980) are homologous with sequences of mould [N. crassa (Selker & Yanofsky, 1979)] and mammalian [human (Maden & Robertson, 1974)] 5.8 S rRNA*

is located in the region 30–80 nucleotides from the 3′-hydroxy end of *N. crassa* L-rRNA. This could be a general model for eukaryotic 5.8 S rRNA–L-rRNA interactions (see below).

By analogy, we suggest that within the ribosome the 5′- and 3′-terminal regions of *E. coli* L-rRNA are in close proximity forming a (5.8 S rRNA-like–L-rRNA) complex. Interaction between the 5′- and 3′-ends of *E. coli* L-rRNA has been proposed (Branlant *et al.*, 1981; Glotz *et al.*, 1981). We infer a (5.8 S rRNA-like–L-rRNA) complex in bacterial ribosomes homologous with the 5.8 S rRNA–L-rRNA complex in cytoplasmic ribosomes and that both complexes are also homologous in their functions. The finding by Wrede & Erdmann (1977) that certain *E. coli* L-subparticle proteins can bind yeast 5.8 S rRNA provides circumstantial evidence that these proteins interact with 5.8 S rRNA-like sequences in *E. coli* and that the binding sites of the proteins are sufficiently conserved to interact with yeast 5.8 S rRNA.

The region of E. coli L-rRNA (nucleotides 157–290) is homologous to sequences at the 5′-end of yeast and X. laevis L-rRNA

The hundred or so nucleotides at the 5′-end of yeast (Veldman *et al.*, 1981) and *X. laevis* (Hall & Maden, 1980) L-rRNA share a high degree of homology (see Fig. 13). The sequences common to both species have their counterpart in *E. coli* L-rRNA, but in the region of nucleotides 157–290 that follow immediately after the 5.8 S rRNA-like sequences (see Fig. 13).

The region of E. coli L-rRNA (nucleotides 1671–2520) is homologous with cytoplasmic L-rRNA sequences

Sequence studies of the exons flanking the two intervening sequences in *Physarum polycephalum* L-rRNA revealed that the introns were inserted into highly conserved regions of the molecule (Nomiyama *et al.*, 1981*a,b*). The corresponding region of *E. coli* L-rRNA (nucleotides 1671–2520) was shown to have sequences varying in length from about 20 to 90 nucleotides with high homology (75% or better) with cytoplasmic L-rRNA (including *X. laevis* L-rRNA as well as *P. polycephalum* L-rRNA). The conserved sequences were interspersed with regions of low homology. The sequences, nucleotides 2089–2199 of *E. coli* L-rRNA, comprise the binding site for protein L1 (Branlant *et al.*, 1981). A putative L1-like binding site is conserved in *Xenopus* L-rRNA and in *Dictyostelium* L-rRNA (Gourse *et al.*, 1981) and possibly also in *Physarum* (Nomiyama *et al.*, 1981*b*).

Bridging the Evolutionary Discontinuity

The organization of *E. coli* L-rRNA and *X. laevis* pre L-rRNA are compared in Fig. 14 and serve to illustrate a scheme for the evolution of L-rRNA. The two species differ by the presence in *X. laevis* pre L-rRNA of sequences analogous to intervening sequences in eukaryotic genes. We propose that these features arose

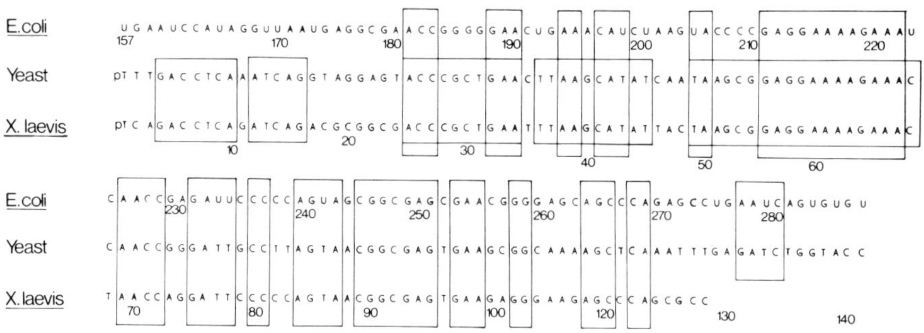

Fig. 13. *Sequences common to the 5′-ends of yeast and X. laevis L-rRNA and to residues 157-296 of E. coli L-rRNA*

The boxes indicate regions of homology. The sequences of *E. coli* refer to L-rRNA (Brosius *et al.*, 1980), whereas the sequences of the non-coding strand of the L-rRNA gene are presented for yeast (Veldman *et al.*, 1981) and *X. laevis* (Hall & Maden, 1980).

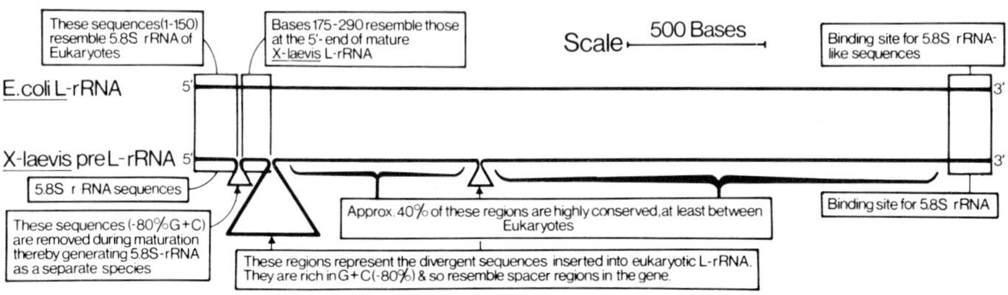

Fig. 14. *Comparison of the organization of E. coli L-rRNA and X. laevis pre-L-rRNA*

Known regions of sequence homology (see Figs. 12 and 13, Cox & Kelly, 1981) are boxed. Sequences amounting to approx. 0.4 $\times 10^6$ daltons are tightly conserved among eukaryotes, as shown by analysis with restriction endonucleases and hybridization with heterologous L-rRNA probes (Cox & Thompson, 1980; Gourse & Gerbi, 1980). Approx. 0.4 $\times 10^6$ daltons of L-rRNA of rabbit, *X. laevis*, *E. coli* and *B. stearothermophilus* have features of secondary structure in common (see Fig. 11). It is inferred that those sequences common to eukaryotes are also present, but less stringently conserved, in bacterial L-rRNA. This view is supported by evidence that nucleotides 1671–2520 of *E. coli* L-rRNA includes sequences that are homologous with cytoplasmic L-rRNA (see above). Tracts of eukaryotic L-rRNA have diverged very rapidly. In *X. laevis* L-rRNA gene (see Fi. 10). The position of the transcribed spacer region-2 that is spacer sequences (Hall & Maden, 1980). The positions of these tracts have been identified by electron microscopy using the technique of denaturation mapping (see Fig. 9 and Wellauer & Dawid, 1974). These locations have been related to the restriction endonuclease maps of *X. laevis* L-rRNA gene (see Fig. 10). The position of the transcribed spacer region-2 that is removed on processing to generate a separate 5.8 S rRNA species is indicated. This spacer region is G + C-rich (Hall & Maden, 1980). The approximate locations of major tracts of divergent sequences are also indicated and are designated spacer-like region A and spacer-like region B. They are considered as insertions into the L-rRNA, although they are not removed in processing. It is proposed that the evolutionary diversity apparent in eukaryotic L-rRNA (see Fig. 5) is generated by variations in the mass and nucleotide composition of the spacer-like insertions.

during evolution as the result of the insertion of additional sequences into L-rRNA genes and subsequent mutations at one or more processing sites leading to the incorporation of inserted sequences into mature L-rRNA and generation of a 5.8 S rRNA species. These divergent (non-conserved) sequences of cytoplasmic L-rRNA have no counterpart in bacterial L-rRNA, but they appear to resemble the spacer sequences of the rRNA gene cluster as was discussed above. In *X. laevis* and rabbit L-rRNA the divergent sequences are G + C-rich (Cox *et al.*, 1973) and so are the spacer sequences (Hall & Maden, 1980) and in *D. melanogaster* L-rRNA the divergent sequences are A + U rich (Cox *et al.*, 1976*a*) and apparently so are the spacer sequences (Greenberg, 1969; Jordan *et al.*, 1980).

In *X. laevis* pre L-rRNA, it appears as though 5.8 S rRNA and L-rRNA are separated by an inserted sequence (the so-called internal transcribed spacer-2 [ITS-2]). In this case, the ITS-2 sequence is excised, although the two L-rRNA fragments remain separated. The divergent sequences within L-rRNA may be regarded as insertions that are no longer excised.

The addition of non-conserved (possibly spacer-like) sequences to a conserved L-rRNA core (Cox *et al.*, 1973, 1976*a*; Godwin *et al.*, 1974) accounts for the elements for diversity (Cox, 1977; Cammarano *et al.*, 1980) that are found in this component of the ribosome. We also infer that as a consequence of the increase in the mass of L-rRNA additional binding sites for basic proteins may be created, thereby increasing the number of L-subparticle protein subunits. These additional protein subunits could be species-specific in view of the divergent properties of these particular RNA sequences.

Our suggestion that the evolutionary discontinuity separating prokaryotic and eukaryotic ribosomes is attributable to modifications in the structure of rRNA genes that alter processing of pre rRNA also implies that the basic partial reactions of mRNA-directed protein biosynthesis, especially those involving 5 S rRNA and 5.8 S rRNA, are the same in both prokaryotes and eukaryotes. Thus, we propose that 5.8 S rRNA in eukaryotes and the 5.8 S rRNA-like sequences in prokaryotes are not only homologous in structure but may also fulfil the same functional role. We also suggest that the 5 S rRNAs of eukaryotes, which are homologous in structure with 5 S rRNA of prokaryotes (Cox, 1977; Bielka & Stahl, 1978), are probably also homologous in function, contrary to previous suggestions (Erdmann, 1976; Wrede & Erdmann, 1977).

While it is evident that the putative alterations in rRNA gene structure and pre rRNA processing have become fixed, the advantages to the cell of a larger L-rRNA species and a separate 5.8 S rRNA entity are not yet apparent. The enlarged ribosome found in the cytoplasm of eukaryotes contrasts with the trend leading to the evolution of the mitochondrial ribosome that results in the miniribosome of amphibian and mammalian mitochondria (Chua & Luck, 1974).

Mechanisms of rRNA gene evolution

The origin of sequences inserted into the genes for rRNA (rDNA) including ITS-2 sequences that are excised from pre L-rRNA (see Fig. 14), is unknown. One possibility is that ITS-2, for example, is an intervening sequence that is excised but that mutations have prevented the ligation of the rRNA sequences. There is ample evidence for intervening sequences in L-rRNA. Chloroplast L-rDNA from *Chlamydomonas rheinhardii* contains an intervening sequence that is removed on processing and yields normal L-rRNA as a result of a ligation (Rochaix & Malnoe, 1978). In some cases, e.g. tobacco chloroplast L-rRNA, it appears that ligation is not normal so that a 4.5 S RNA fragment homologous with the 3′-terminus of *E. coli* L-RNA is generated after the intervening sequence is excised (MacKay, 1981). Intervening sequences are also known to be present in mitochondrial L-rRNA genes (Bos *et al.*, 1978; Heckman & Raj Bhandary, 1979; Lazarus *et al.*, 1980) and in cytoplasmic L-rRNA genes (Glover & Hogness, 1977; Cech & Rio, 1979; Din *et al.*, 1979; Wild & Gall, 1979; Gubler *et al.*, 1979). Therefore, the possibility that mature L-rRNA has retained additional sequences that resemble intervening sequences involves no new principles. The available sequence data for the exon/intron junction in rRNA genes (see Table 4) does not resolve the question of the origin of e.g. ITS-2. None of the processing sites within rRNA genes are similar to the consensus sequences found in mRNA. However, the sequences that act as signals during rRNA processing have not yet been identified. Until such sequences are known the relation between intervening sequences and e.g. ITS-2 remains obscure.

Structural Aspects of Eukaryotic Ribosomes

Role of the ionic environment

We seek to understand the structure of the functional ribosomes and so an essential step is to identify conditions that preserve their activity in protein biosynthesis. Ribosomes need the correct ionic environment to function efficiently, namely a bivalent cation such as Mg^{2+} and a monovalent cation such as K^+ or NH_4^+. Thus, monovalent and bivalent cations are an integral part of ribosomes. Ionic effects are important because of the polyelectrolyte properties of both the rRNA and protein moieties. The secondary and tertiary structures of RNA are dependent on cations because they screen the negative charges of the ribose-phosphate backbone; repulsions between phosphate residues oppose the interactions between purine and pyrimidine base residues. Because of its smaller size and higher charge density, Mg^{2+} screens the diesterified phosphate residues more efficiently than does K^+ (or NH_4^+) and so may allow interactions between nucleotides that are not permitted by K^+ (or NH_4^+). It was shown by Cox & Hirst (1976) that the addition of $MgCl_2$ to solutions of *E. coli* S-rRNA or L-rRNA or rabbit S-rRNA or L-rRNA in 0.35 M-KCl led to an increase in bi-helical secondary structure, as shown by u.v. and c.d. measurements, to an increase in the thermal stability of bi-helical secondary structure by approx. 5°C and to an increase in $s_{20,w}$ (e.g. for *E. coli* S-rRNA $s_{20,w}$ = 18 S in 0.35 M-KCl, 21 S in 0.35 M-KCl/15 mM-$MgCl_2$ and 24 S in 0.35 M-KCl/40 mM-$MgCl_2$).

The relation between the effects of K^+ (or NH_4^+) and Mg^{2+} on rRNA was analysed quantitatively (Cox & Hirst, 1976). Suppose that the conformation of an individual nucleotide, N, is changed according to eqn. (4) from state I (represented as N_I) to state II (shown as N_{II}) on the addition of Mg^{2+}:

$$K_m^+N_I + n\,Mg^{2+} \rightleftharpoons Mg_n^{2+}N_{II} + m\,K^+ \qquad (4)$$

Table 4. *Comparison of nucleotide sequences bordering ITS-2 with sequences at the borders of introns and exons of rRNA genes*

	Exon ↓ Intron ↓ Exon
Consensus sequences*	5′ A/C A G G A R A G · · · · · · · Y – Y Y Y – C A G · · · · · · · · · · 3′
X. laevis ITS-2†	T G T C T G A G G T C G C T C C G A C G T C C A T · · · C C C C A C G A C T C A G A C · · · ·
Physarum 1	A A G T T A C C A C A G G G A T T G A T A C A C A · · · G A G A A T C G G A A C T G G C T T G
Physarum 2‡	· · A A A T G C C T C G T C A T C G T A T A A C G · · · T G G G G G G T G T T A G T T T G T G
Yeast (mitochondrial)	A A G T T A C G C T A G G G A T A A T T T · · · · · · · A A A G A T T T G A A C A G G G T A A
Tetrahymena	T A A C T A T G A C T C T C T T A A A T T · · · · · · · G A C T A A T C G T A A G C T A G C C
Chlamydomonas (chloroplast)	C T G G G T T C A A A A A C G T A A A T A · · · · · · · A A A T T C A T G C G T G A G A C A G

* See Sharp (1981). The consensus sequences border introns inserted into genes coding for proteins.
† See Hall & Maden (1980).
‡ See Nomiyama *et al.* (1981*b*) and see Nomiyama *et al.* (1981*a*) for the other sequences shown.

The equilibrium constant, K, is given by eqn. (5)

$$K = [N_{II}]\,[K^+]^m/[N_I]\,[Mg^{2+}]^n \tag{5}$$

Eqn. (5) reveals that it is the ratio of the concentrations of K^+ (or NH_4^+) and Mg^{2+} that influence rRNA conformation and not the concentration of Mg^{2+} alone. In other words, in order to maintain the N_{II} conformation a certain level of Mg^{2+} must be bound by RNA; it is this level of bound Mg^{2+} that is governed by the ratio $[Mg^{2+}]:[K^+]$ (or NH_4^+). Weiss *et al.* (1973) showed that the activity in protein synthesis of *E. coli* S- and L-subparticles was preserved provided that the level of Mg^{2+} bound to the subparticle was maintained at 1 Mg^{2+}/5 nucleotides.

The equilibrium constant, K (see eqn. 5), was evaluated for both rRNA (K_{rRNA}) and for ribosomal subparticles ($K_{subparticles}$) and was found (Cox & Hirst, 1976) to have the same values (approx. 80) in each case (see eqn. 6).

$$K_{rRNA} = K_{subparticles} = 80 \tag{6}$$

Both m ($m = 1.3 \pm 0.1$) and n ($n = 1.16 \pm 0.1$) in eqn. (5) were found to have values close to unity (Cox & Hirst, 1976).

The significance of the observation that $K = 80$ is that the transition from N_I to N_{II} is half complete when there is 1 Mg^{2+} for every 80 K^+ (or NH_4^+). The transition is two-thirds complete when there is 1 Mg^{2+} per 40 K^+ (or NH_4^+). The results of Weiss & Morris (1973) and of Weiss *et al.* (1973), combined with the observations of Cox & Hirst (1976), suggest that ribosomal subparticles require that 1 Mg^{2+} is bound for every 5 nucleotides and that this level of Mg^{2+} binding is maintained over the range 0.1–3 M-KCl (or NH_4Cl) provided that Mg^{2+} is kept at a level of at least 1 Mg^{2+} ion for every 40 K^+ (or NH_4^+) ions. This interpretation was supported by experiment (Cox & Hirst, 1976).

Inactivation and reactivation of L-subparticles by NH_4Cl salt-shock treatment

L-subparticles are stable at 0°C in 0.1 M-NH_4Cl/5 mM-$MgCl_2$/0.05 M-Tris/HCl, pH 7.6. When the ratio $Mg^{2+}:NH_4^+$ was altered by raising the concentration of NH_4Cl up to 0.6 M (NH_4Cl salt-shock treatment), there was no loss of protein although there was a sharp decline in the activity of L-subparticles in both poly(U)-directed polyphenylalanine synthesis and in the puromycin reaction. Activity was halved in 0.35 M-NH_4Cl/5 mM-$MgCl_2$ (1 Mg^{2+} : 70 NH_4^+). The activity of salt-shocked L-subparticles was recovered when the level of Mg^{2+} was restored to at least 1 Mg^{2+}/40 NH_4^+ (e.g. 0.6 M-NH_4Cl/20 mM-$MgCl_2$) and the solution was kept at 37°C for approx. 20 min before assay. It appears that Mg^{2+} deficiency leads to a metastable inactive conformation of the L-subparticle. For summary see Fig. 15 and also Cox *et al.* (1976*b*).

Production of core-particles and split-protein fractions and the re-assembly of ribosomal subparticles

The concentration of salts such as KCl or NH_4Cl that release proteins from ribosomal subparticles depends on the concentration of $MgCl_2$. Proteins (20 ± 3%)

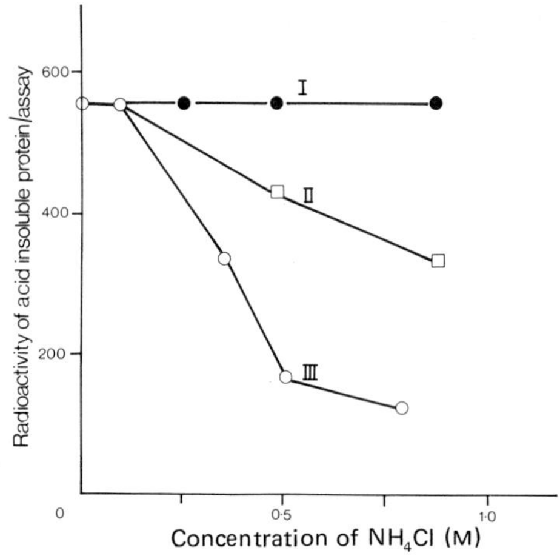

Fig. 15. *Effect of reducing the $MgCl_2$: NH_4Cl ratio on the activity of subparticles in poly(U)-directed polyphenylalanine synthesis*

Rabbit reticulocyte L-subparticles were stored in 0.1 M-NH_4Cl/5 mM-$MgCl_2$/20 mM-Tris/HCl, pH 7.6 (storage buffer). Curve I, $MgCl_2$ was first added to samples in storage buffer at 0°C to give a final concentration of 20 mM. The concentration of NH_4Cl was then increased to either 0.3 M, 0.5 M or 0.8 M. Thus, the ratio Mg^{2+} : NH_4^+ was kept throughout so that there was at least 1 Mg^{2+} for every 40 NH_4^+. Curve II, to samples of L-subparticles in storage buffer at 0°C, NH_4Cl was added to increase the concentration to 0.3 M, 0.5 M or 0.8 M so that the ratio Mg^{2+} : NH_4^+ ranged from 1:20 to 1:160. Curve III, an aliquot of each sample (see curve II) was further treated to give a final concentration of 20 mM-$MgCl_2$. The solution was kept at 37°C for 30 min then cooled in ice. In each case the final concentration of L-subparticles was 3 mg/ml and samples were kept at 0°C for 2 h. The capacity of L-subparticles to function in poly(U)-directed polyphenylalanine synthesis was assayed in a cell-free system.

are lost from L-subparticles in 1.2 M-NH_4Cl/2 mM-$MgCl_2$. However, provided that the level of Mg^{2+} is maintained at 1 Mg^{2+}/40 NH_4^+, rabbit subparticles may be exposed to solutions of up to 1.6 M-NH_4Cl at 0°C without the loss of either activity or of proteins. Even if Mg^{2+} is maintained at 1 Mg^{2+}/40 NH_4^+, raising the concentration of NH_4Cl above 1.6 M releases proteins from both rabbit S-subparticles and L-subparticles. For example, exposure to 2.5 M-NH_4Cl/61 mM-$MgCl_2$ at 0°C releases approx. 20% of proteins from rabbit L-subparticles (Cox & Greenwell, 1976) and approx. 30% of proteins from rabbit S-subparticles (Cox, 1980). In both cases the protein deficient core-particles were found to be inactive in cell-free protein biosynthesis. Activity was restored when the core-particles and split-protein fractions were allowed to recombine. Reconstruction of both subparticles was achieved either by diluting the concentrated salt solution to 0.35 M-NH_4Cl/10 mM-$MgCl_2$ before assaying the subparticles, diminishing the salt concentration by dialysis. Although reconstruction was easily achieved in this way, less success was obtained when protein-deficient core-particles were first isolated before the addition of the split-protein fraction. The variable results obtained by this method indicates that adventitious factors such as proteinases or nucleases might be responsible. The ease with which split-proteins recombine with core-particles, e.g. in dilution or dialysis of the concentrated salt solutions, is attributed to the rRNA moiety being kept in its correct conformation by maintaining Mg^{2+} at its optimum concentration. The influence of Mg^{2+} on ribosome conformation reflects the importance of rRNA conformation in binding ribosomal proteins and in maintaining the active conformation of the ribosome. A full complement of proteins is not sufficient to ensure full activity. A knowledge of the conformation of rRNA within functional subparticles is essential in order to understand ribosome function.

Conformation of rRNA within active ribosomal subparticles

The partly bi-helical structure of S-rRNA and L-rRNA within active subparticles can be investigated by c.d. measurements. The technique is useful because it is non-destructive, only small (μg) amounts of material are needed, and the c.d. of the

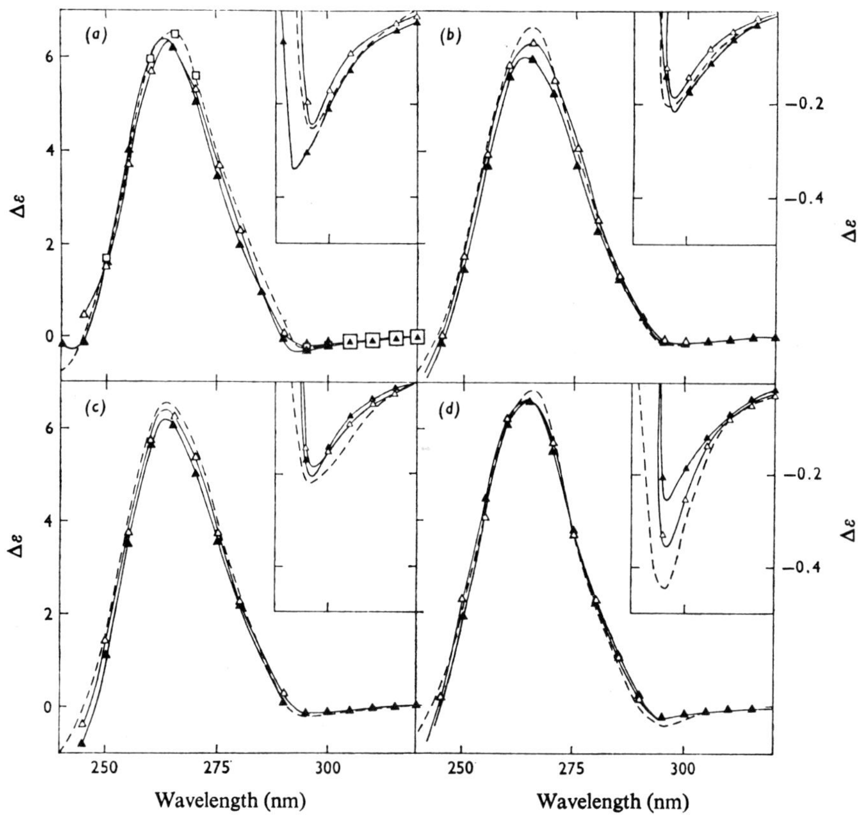

Fig. 16. *Comparison of the c.d. spectra of active ribosomal subparticles of E. coli and rabbit reticulocytes, ribosomal subparticles dissociated into rRNA and protein by treatment with 1% sodium dodecyl sulphate and purified rRNA*

The solvent in all experiments was 10 mM-NH_4Cl/0.1 mM-$MgCl_2$/10 mM-sodium phosphate buffer, pH 7. The spectrum was measured in 1 mm-path-length cells ($A_{260} = 1.2 \ell \cdot mol^{-1} \cdot mm^{-1}$), and, in addition, at wavelengths above 285 nm the spectrum was also measured in 1 cm-path-length cells ($A_{260} = 12 \ell \cdot mol^{-1} \cdot cm^{-1}$) to increase the accuracy in measuring the dip at 295 nm. (*a*) *E. coli* S-subparticle; (*b*) rabbit reticulocyte S-subparticle; (*c*) *E. coli* L-subparticles; (*d*) rabbit reticulocyte L-subparticles. ----, Free rRNA; ▲—▲, rRNA complexed in the intact subparticle; △—△, subparticles dissociated into rRNA and ribosomal proteins by 1% sodium dodecyl sulphate (Cox *et al.*, 1976*c*).

Table 5. *Properties of the rRNA moiety of functional ribosomal subparticles derived from c.d. measurements**

rRNA species	Size (nucleotides)	Number of base-pairs A + U	Number of base-pairs G + C	Fraction of residues forming base-pairs	Fraction of residues contributing to c.d.†
E. coli S-rRNA	1541	428 ± 28	513 ± 34	0.61 ± 0.04	0.83 ± 0.03
E. coli L-rRNA	2904	849 ± 26	1038 ± 32	0.65 ± 0.02	0.80 ± 0.03
Rabbit S-rRNA	1820	500 ± 49	611 ± 60	0.61 ± 0.06	0.75 ± 0.06
Rabbit L-rRNA	4400	845 ± 84	1840 ± 180	0.61 ± 0.06	0.90 ± 0.06

* See Cox *et al.* (1976*c*).
† Residues that are stacked (i.e. residues that interact to form part of a single-strand helix) as well as base-paired residues contribute to c.d.

protein moiety can be studied at the same time. The c.d. of both bacterial and cytoplasmic ribosomes (see Fig. 16) reveals that a high proportion of the nucleotide residues have structural asymmetry. Analysis of the c.d. spectra indicates (see Table 5) that approx. 60% of residues participate in base-pairing and that a further 15–20% of residues 'stuck' (i.e. interact to form part of a single-stranded helix). The c.d. of the rRNA moiety of ribosomal subparticles is close to (within 90%) but not necessarily identical with that of isolated RNA in the same buffer. Thus many of the features of the secondary structure of rRNA *in situ* are probably retained after the protein moiety is removed. However, it would be unwise to assume that all features of secondary structure identified in isolated rRNA are also present in the ribosome.

Dissociation of 5 S rRNA, 5.8 S rRNA and peptidyl-tRNA from ribosomal subparticles and polyribosomes

Both 5 S rRNA and 5.8 S rRNA are components of the larger ribosomal subparticle which also has a binding site for peptidyl-tRNA. The dissociation of these components from rabbit L-subparticles or from rabbit polyribosomes was examined. Whereas 5 S rRNA remains attached to protein-deficient core-particles produced by treating L-subparticles with high salt (2.5 M-NH_4Cl/61 mM-$MgCl_2$), it is readily dissociated by heat treatment. The effects of preheating either a rabbit polyribosome fraction or rabbit L-subparticles are summarized in Table 6. Activity in cell-free protein synthesis was retained after heating at 55°C for 3 min, whereas exposure to 60°C led to the loss of activity in protein biosynthesis and to the dissociation of 5 S rRNA but without an appreciable increase in the proportion of subparticle. However, distinct boundaries were still observed on sucrose-gradient centrifugation. Preheating to higher temperatures led to the formation of aggregated material. The dissociation of 5.8 S rRNA took place at 63–73°C (T_m = 67°C). The high T_m for the dissociation of 5.8 S rRNA from ribosomes is in contrast with the ease with which it can be dissociated from isolated L-rRNA. The basis for the increased stability of the 5.8 S rRNA complex within the L-subparticle is not known. The results are summarized in Table 7.

Peptidyl-tRNA was found to remain bound to the polyribosome fraction after heating at 60°C for 3 min. This stability is likely to be due to the nascent poly-

Table 6. *Comparison of the properties of the polyribosome fraction after heat-shock treatment at 55°C and 60°C*

	Polyribosome fraction after heating for 3 min	
	55°C	60°C
Activity in cell-free globin synthesis (% of non-heated control)	55	7
Activity in polyphenylalanine synthesis (% of non-heated control)	46	10
Presence of 55 S rRNA	Yes (> 90%)	No (< 10%)
Presence of 5.8 S rRNA	Yes	Yes
Presence of peptidyl-tRNA	Yes	Yes
$s_{20,w}$ values of principal species	80 S, 112 S, 140 S, 160 S	80 S, 112 S, 140 S, 160 S
$s_{20,w}$ values of heparin-derived subparticles†	40 S, 60 S	40 S, 50 S

* 5 S rRNA was released as an RNA species free of proteins.
† Heparin dissociates non-programmed ribosomes into subparticles.

Table 7. *Release of RNA components from L-subparticles or from the polyribosome fraction by heat-shock treatment*

Species	Ribosomal component	T_m (°C)*
L-subparticles	5 S rRNA	58 ± 2
	5.8 S rRNA	67 ± 2
Polyribosome fraction	Peptidyl-tRNA	>60
	5 S rRNA	58 ± 2
	5.8 S rRNA	67 ± 2

* T_m is the temperature at which 50% of 5 S or 5.8 S rRNA was released into solution. These components were released over a range of approx. 10°C.

peptide chain which is known to stabilize polyribosomes in conditions that would otherwise lead to dissociation into subparticles. The molecular basis of the interaction of the nascent chain with the ribosome is not understood.

The L-rRNA–5.8 S rRNA complex

The standard methods for the isolation of rRNA such as SDS/phenol used at 25°C or below lead to the isolation of a complex of L-rRNA and 5.8 S rRNA. The 5.8 S rRNA component is readily dissociated from L-rRNA by heating to 42°C (see for example King & Gould, 1970). Sequences near to the 3′- and 5′-ends of 5.8 S rRNA are implicated in binding to L-rRNA. Less is known about the sequences of L-rRNA that interact with 5.8 S rRNA. Sequences near to the 5′-end of mature yeast L-rRNA and *X. laevis* L-rRNA are complementary to sequences at the 3′-end of yeast 5.8 S rRNA and *X. laevis* 5.8 S rRNA respectively (Veldman *et al.*, 1981). There are also indications that 5.8 S rRNA can bind to sequences at the 3′-end of L-rRNA. Thus it is possible that 5.8 S rRNA serves to unite the 5′- and 3′-ends of L-rRNA and it is also possible that 5.8 S rRNA binds at different sites during the lifetime of the ribosome.

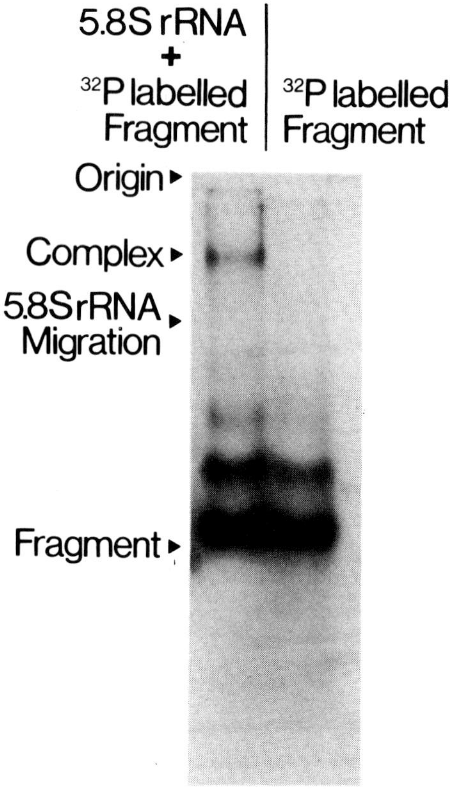

Fig. 17. *Interaction between rabbit 5.8S rRNA and a ^{32}P-labelled 3′-terminal fragment of rabbit L-rRNA of approx. 100 residues*

A ^{32}P-labelled 3′-terminal fragment of L-rRNA of approx. 100 nucleotides was extracted from a sequencing gel. The fragment was incubated with 5.8 S rRNA. The components of the mixture were separated by electrophoresis and the polyacrylamide gel was then subjected to radioautography.

The evidence that 5.8 S rRNA can bind to sequences at the 3′-end of L-rRNA is based on both experimental and on theoretical grounds. The experimental observation is that *N. crassa* 5.8 S rRNA binds to a fragment of 80 nucleotides from the 3′-end of *N. crassa* L-rRNA (Kelly & Cox, 1981*b*). Similarly, rabbit 5.8 S rRNA (but not *N. crassa* 5.8 S rRNA) binds to a fragment of approx. 100 nucleotides from the 3′-end of rabbit L-rRNA (see Fig. 17). The available sequence data reveals that homologous complexes of 5.8 S rRNA and L-rRNA are theoretically possible for yeast, *N. crassa*, *D. melanogaster* and *X. laevis* (see Fig. 18).

An attempt was made to investigate the native L-rRNA–5.8 S rRNA complex. The basis for this experiment was our ability to label independently the 3′-terminus of either the 5.8 S rRNA or the L-rRNA components of the complex to yield either L-rRNA-[^{32}P]pCp-labelled 5.8 S rRNA native complex or [^{32}P]pCp-labelled L-rRNA-5.8 S rRNA native complex (Kelly & Cox, 1981*a*). Very mild partial nuclease treatment of both complexes revealed that fragments of 600 nucleotides or more contained the 3′-ends of both L-rRNA and 5.8 S rRNA (Fig. 19). This observation is consistent with 5.8 S rRNA being bound to sequences within 450

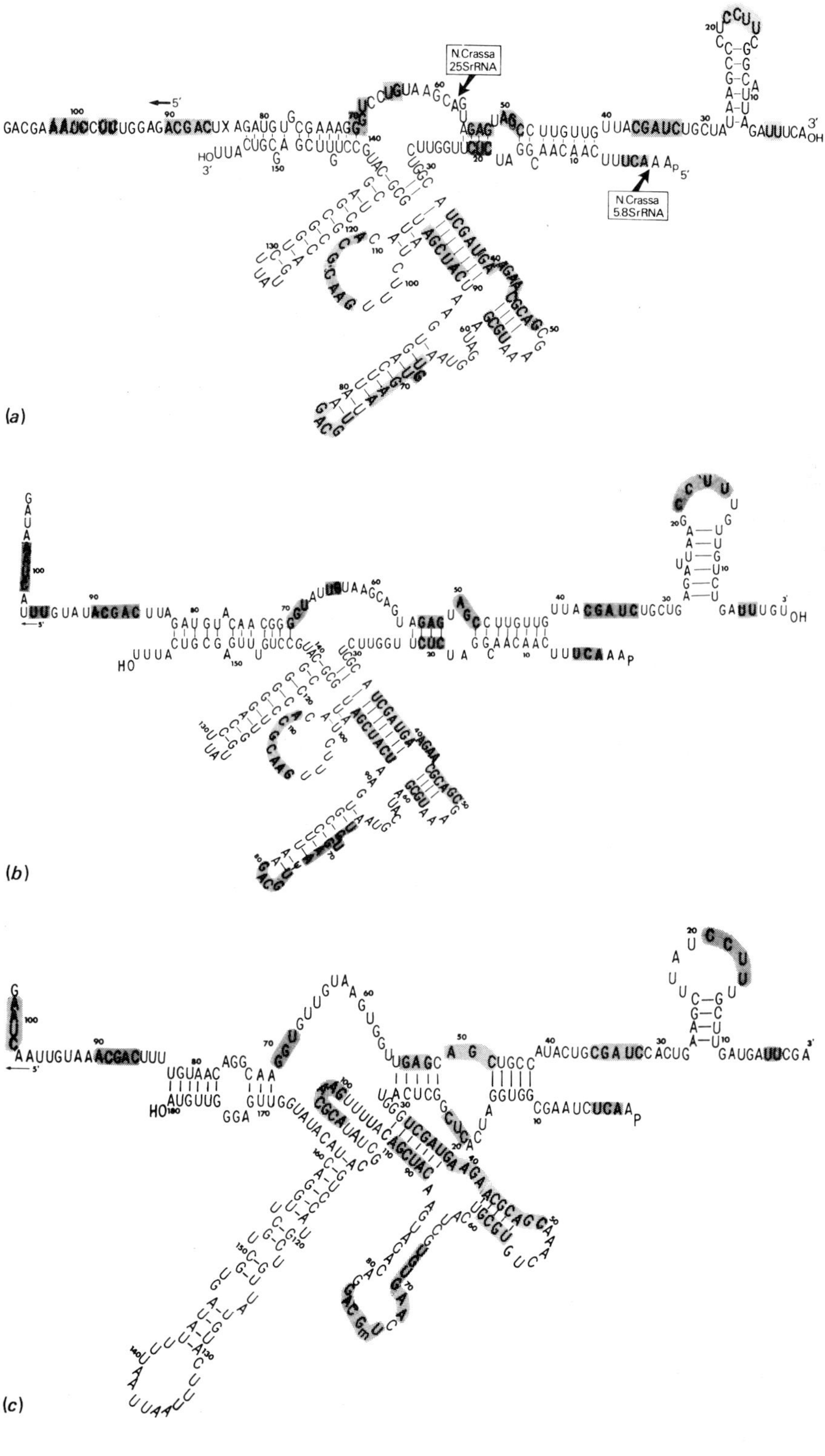
N.Crassa
25SrRNA
N.Crassa
5.8SrRNA
(a)
(b)
(c)

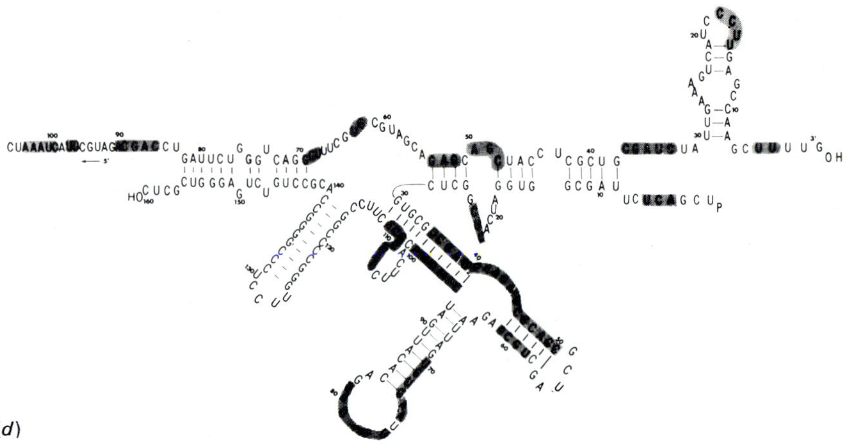

(*d*)

Fig. 18. *Possible schemes for the interaction of 5.8S rRNA with L-rRNA*

Interaction between 5.8S rRNA and a fragment of 80–100 nucleotides from the 3′-end of L-rRNA was shown for *N. crassa* (Kelly & Cox, 1981) and rabbit (see Fig. 17). Complexes similar to those proposed for (*a*) *N. crassa* (Kelly & Cox, 1981*b*) are theoretically possible for (*b*) yeast; (*c*) *D. melanogaster* and (*d*) *X. laevis*. Sequence data for L-rRNA were obtained from Veldman *et al.* (1981) or Georgiev *et al.* (1981) for yeast, from Mandal & Dawid (1981) for *D. melanogaster*, and Sollner-Webb & Reeder (1979) for *X. laevis*. Sequences for 5.8S rRNA were obtained from Rubin (1973) for yeast, Ford & Mathieson (1978) for *X. laevis*, and Pavlakis *et al.* (1979) for *D. melanogaster*, and Selker & Yanofsky (1979) for *N. crassa*. Sequences that are strictly conserved between all four species are shaded. An alternative scheme, based on sequence data alone, was proposed for the interaction of 5.8S rRNA with yeast L-rRNA (Veldman *et al.*, 1981).

nucleotides of the 3′-end of L-rRNA. Comparison of the products of partial nuclease digestion of [^{32}P]pCp-labelled L-rRNA and [^{32}P]pCp-labelled L-rRNA–5.8 S rRNA native complex indicate that the oligonucleotide fragments obtained from the region 40–80 nucleotides from the 3′-end of L-rRNA were not the same in both cases, therefore indicating a different conformational state in this region in the presence of 5.8 S rRNA. This result is in accord with the hypothetical complexes illustrated in Fig. 18. However, further study of the *in vivo* complex is needed to completely define those regions of L-rRNA that bind to 5.8 S rRNA. An alternative binding site for 5.8 S rRNA involving not only the residues 2–23 at the 5′-end of yeast L-rRNA but also residues in the region of 330–417 has been proposed (Veldman *et al.*, 1981) on the basis of primary sequence data.

The organization of rRNA within functional ribosomes

The complete nucleotide sequences of yeast and *X. laevis* 18 S rRNA and partial nucleotide sequences of 25–28 S rRNA from several species are now available. However, as yet little is known about the arrangement of the rRNA chains within the functional ribosome. It has been shown though, that the 3′-termini of 18 S and 28 S rRNA in functional 40 S and 60 S ribosomal subparticles are accessible to labelling with [5′-^{32}P]pCp by T_4 RNA ligase under conditions in which the 3′-termini of 5 S and 5.8 S rRNA are inaccessible (Kelly & Cox, 1981*a*). Further, on ribosome formation the access to the 3′-terminus of 18 S rRNA is reduced.

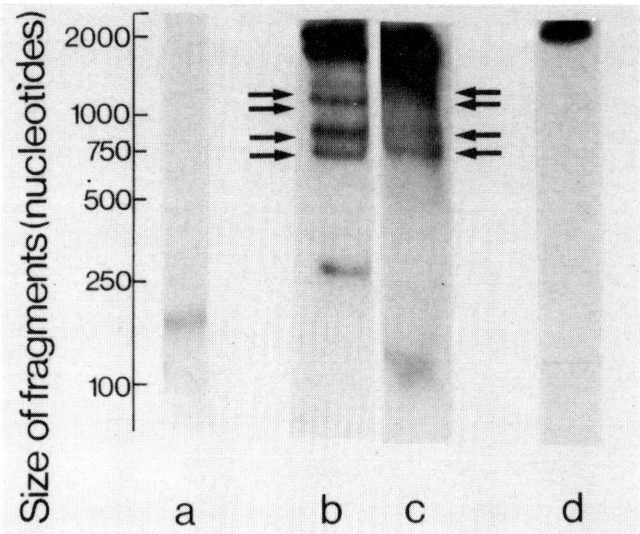

Fig. 19. *Comparison of the products of partial hydrolysis by pancreatic RNAase of* ^{32}*P-labelled 5.8 S rRNA–L-rRNA and 5.8 S rRNA–*^{32}*P-labelled L-rRNA native complexes*

Both complexes, suspended in 2 mM-$MgCl_2$/20 mM-Tris/HCl, pH 7.2, were treated with RNAase (0.2 ng/mg of RNA) for 30 min at 0°C. The products were separated by electrophoresis in non-denaturing conditions by electrophoresis in a 2.5% polyacrylamide gel. Fragments containing ^{32}P-labelled 3′-termini were visualized by autoradiography. (a) Control ^{32}P-labelled 5.8 S rRNA–L-rRNA complex heated to 90°C for 30 s; (b) partial RNAase digest of ^{32}P-labelled 5.8 S rRNA–L-rRNA complex (unheated); (c) partial RNAase digest of 5.8 S rRNA–^{32}P-labelled L-rRNA complex (unheated); (d) control 5.8 S rRNA–^{32}P-labelled L-rRNA complex heated to 90°C for 30 s.

The ability to attach a radioactive label to the 3′-termini of 18 S and 28 S rRNA within intact functional ribosomal subparticles provides a method to investigate the rRNA conformation *in situ*. Mild nuclease treatment of these labelled ribosomal subparticles, followed by separation of the digestion products of the RNA moiety on urea/acrylamide gels, reveals a spectrum of fragments. These fragments are generated as a result of a cleavage at sites where the rRNA is accessible to nuclease at the subparticle surface. Since only those oligonucleotide fragments containing a [^{32}P]pCp-labelled 3′-terminus are visualized by autoradiography, the accessible sites can be mapped from the 3′-end and their relative sensitivities calculated.

Cleavage of S-rRNA by nuclease treatment of S-subparticles. Rabbit S-subparticles labelled with [^{32}P]pCp were treated with either S_1 nuclease, T_1 ribonuclease or pancreatic ribonuclease under conditions in which ribosomal subparticles remain intact and functional. The resulting partially digested 18 S rRNA was fractionated on an 8 M-urea/2.5% (w/v) polyacrylamide gel and fragments containing the 3′-terminus were detected after autoradiography (Fig. 20). Similar results were obtained with *X. laevis* S-rRNA. These fragments are single polynucleotide chains free of hidden breaks generated by a single cleavage and therefore allow the accessible sites to be located. The sensitivity of these sites to nuclease digestion can be calculated from the absorbance of the autoradiograph at the points corresponding to the resultant 3′-end labelled fragment. The results are summarized in Fig. 21.

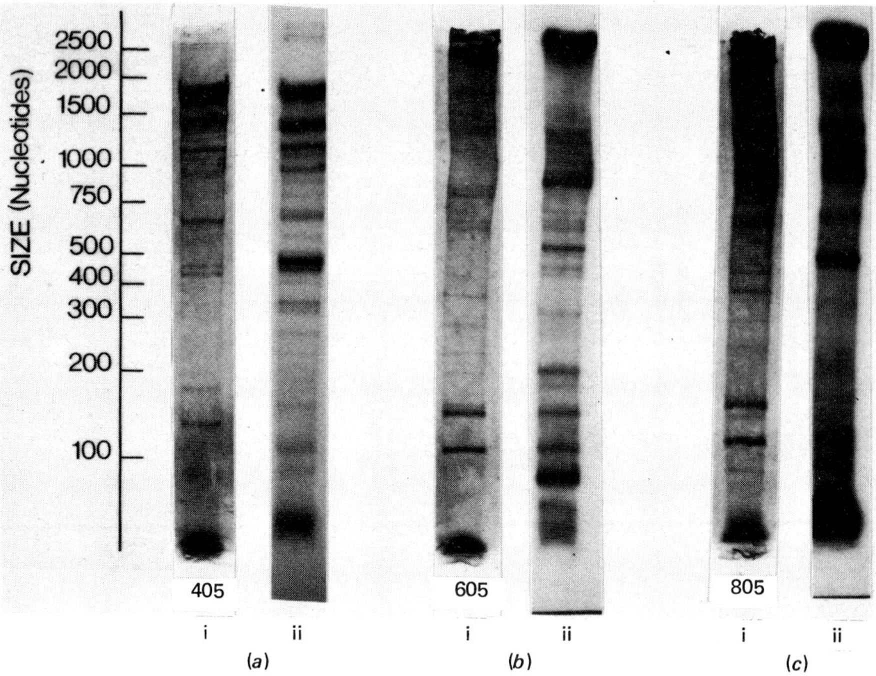

Fig. 20. *Distribution of* ^{32}*P-labelled 3'-terminal fragments of rRNA produced by nuclease treatment of functional subparticles of rabbit ribosomes*

Functional ribosomes and ribosomal subparticles of rabbit ribosomes were labelled at the 3′-ends of S-rRNA or L-rRNA with ^{32}P-labelled pCp by using T4 RNA ligase. The rRNA component was purified by extraction with phenol and fragments were separated by electrophoresis through 2.5% polyacrylamide gels under denaturing conditions. The gel was stained with Coomassie Blue, destained, dried and subjected to autoradiography. (*a*) rRNA from the S-subparticle; (*b*) rRNA from the L-subparticle; (*c*) rRNA from the ribosome fraction. In each case (i) indicates the stained gel, and (ii) indicates the autoradiograph.

Each enzyme gave rise to a distinctive pattern of radioactive fragments, although many features are common. The most sensitive regions were located 450–500 nucleotides, 1000–1200 nucleotides and 1350–1470 nucleotides measured from the 3′-end. In the region 100–500 nucleotides from the 3′-end sensitive sites are distributed at intervals of about 35 nucleotides. The sensitive sites along the 18 S rRNA chain must be accessible for cleavage by nucleases to occur and therefore they comprise regions that are exposed on the surface of the 40 S subparticle. Thus a substantial part of 18 S rRNA lies at the subparticle surface. Regions of 18 S rRNA unaffected by nuclease treatment extended over regions of no more than 150 nucleotides, e.g. nucleotides 580–650, nucleotides 720–820, nucleotides 1200–1300 and nucleotides 1470–1560 measured from the 3′-end.

The arrangement of sequences close to the 3′-end of 18 S and 28 S rRNA within ribosomal subparticles can be more precisely investigated by separating the digestion products on long polyacrylamide sequencing gels. For this investigation we used ribosomes from *X. laevis*, since the nucleotide sequences at the 3′-end of both 18 S and 28 S rRNA have been determined. The subparticles were subjected to more extensive nuclease treatment than before, since this reveals more clearly

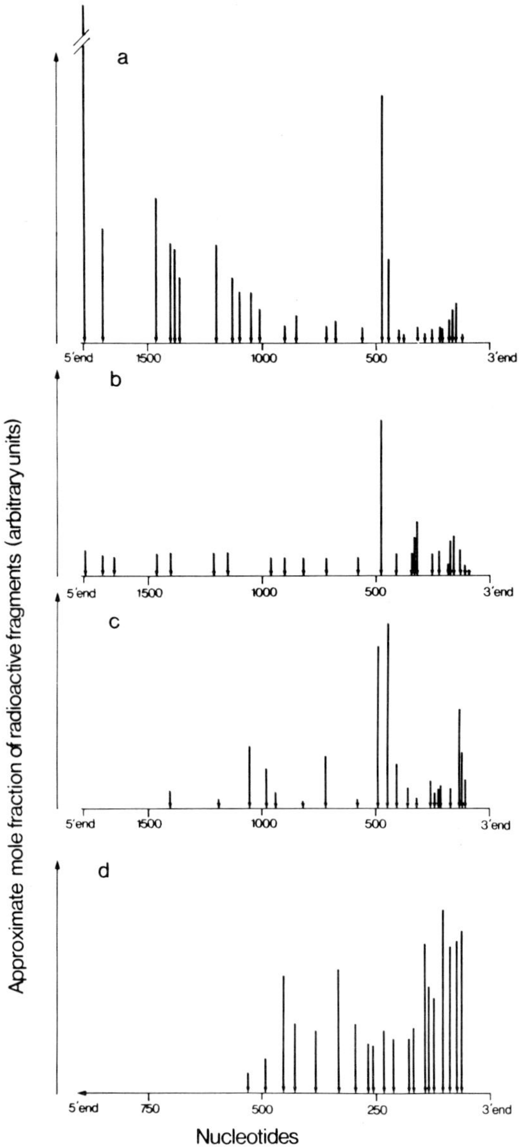

Fig. 21. *Summary of the location measured from the 3'-hydroxy ends of S-rRNA of internucleotide bonds cleaved by nuclease treatment of S-subparticles*

(*a*) Endogenous nuclease; (*b*) T_1 RNAase; (*c*) S_1 nuclease; (*d*) pancreatic RNAase (Cox & Kelly, 1981*b*). The length of the shaft of each arrow is proportional to the mole fraction of radioactive fragments produced as a result of the treatments, and the width of the arrowhead indicates the possible error in locating the cleavage site. Only those fragments that gave rise to distinct bands in the autoradiographs are included. The ordinate is directly proportional to the probability, p_n, of hydrolysing the nth bond during the course of the experiment. Inevitably the method highlights cleavage sites nearer to the 3'-end of S-rRNA.

cleavage sites close to the 3′-end. After partial nuclease digestion the oligonucleotide fragments were separated on 60 cm 8% (w/v) polyacrylamide gels and those fragments containing the 3′-terminus were visualized by autoradiography (Fig. 22). The precise size of the fragments can be determined by running the partially digested RNA in parallel with *N. crassa* 5 S rRNA or mammalian 5.8 S rRNA which had been partially degraded using one of the chemical sequencing reactions (usually A>G) of Peattie (1979).

Although these experiments are still at an early stage, several conclusions are already apparent.

(1) S-rRNA is packaged in the S-subparticle in such a way that regions are accessible to RNAase at intervals of no more than 150 nucleotides (see Fig. 23).

(2) The region of S-rRNA within 500 nucleotides of the 3′-end is cleaved at intervals of approx. 35 nucleotides upon RNAase treatment of S-subparticles.

(3) A scheme for the secondary structure of *X. laevis* S-rRNA was proposed by Brimacombe (1982, this volume) by analogy with that advanced for *E. coli* S-rRNA. The cleavage sites within 220 nucleotides of the 3′-end produced by nuclease S_1 treatment of *X. laevis* S-subparticles all map within sequences that are single-stranded in the model. Not all the proposed single-stranded nucleotide sequences were attacked by S_1 nuclease. If the model is correct then, within the S-subparticle, these sequences are not available to S_1 nuclease.

(4) Regions of S-rRNA that are protected in the S-subparticle from attack by nuclease can be mapped precisely.

(5) Direct comparison of regions susceptible to nuclease treatment *in situ* with sensitive regions in isolated S-rRNA is made difficult by the greater rate at which the latter is hydrolysed.

Cleavage of L-rRNA by nuclease treatment of L-subparticles. Using the same approach as before, we investigated the arrangement of L-rRNA within L-subparticles. This revealed that the 3′-half of L-rRNA is packed such that regions are accessible to nuclease at intervals of no more than 100 nucleotides (see Fig. 24). Closer study of the *X. laevis* L-rRNA within 260 nucleotides of the 3′-end (see Fig. 25) revealed that some regions are not extensively cleaved when L-subparticles are treated with nucleases. However, the region of 130–260 nucleotides from the 3′-end, which is G+C-rich and may be typical of non-conserved sequences (see above), was accessible to nucleases at several points.

Thus it appears that both S-rRNA and L-rRNA are packed with subparticles so that regions are accessible to nucleases at regular intervals of less than 100 nucleotides. In both cases the 3′-ends are accessible to T4 RNA ligase.

Conclusions

There is considerable evidence to show that sequences of both S-rRNA and L-rRNA are highly conserved, between bacterial and cytoplasmic ribosomes. Also, both chloroplast and mitochondrial rRNA can be shown to be related to bacterial RNA. It is now possible to begin to relate the genes for bacterial rRNA and for cytoplasmic rRNA and organelle rRNA through a series of additions or deletions (see also Noller *et al.*, 1981; Glotz *et al.*, 1981; Branlant *et al.*, 1981; Veldman

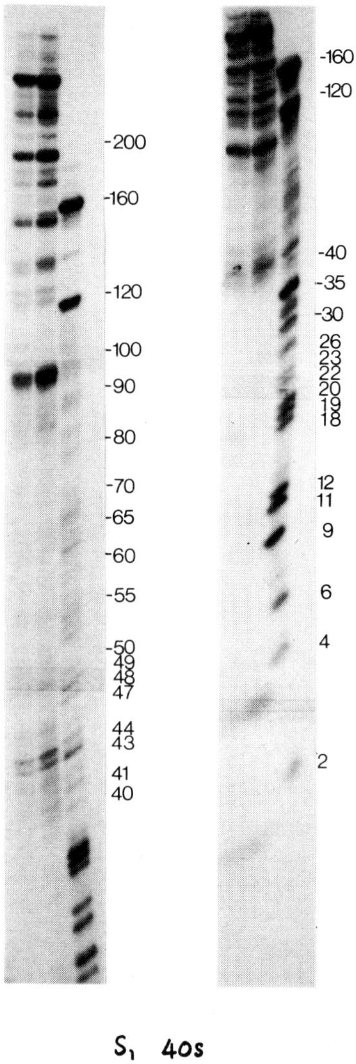

Fig. 22. *Autoradiographs of gels used to fractionate 3′-end labelled oligonucleotide products of X. laevis S-rRNA after treatment of S-subparticles with S_1 nuclease*

X. laevis S-subparticles, labelled with [^{32}P]pCp at the 3′-end of S-rRNA, were treated with S_1 nuclease (0.1–1 unit/μg of RNA) at 37°C for 30 min and layered on 12% (right) or 8% (left) 60 cm polyacrylamide gels. Electrophoresis was carried out at 2.0 kV for 2 h (right) and 6 h (left). The size of the oligonucleotide fragments were determined by running the digests in parallel with *N. crassa* 5 S rRNA (118 nucleotides) which had been chemically treated using the A > G sequencing reaction of Peattie (1979) and *N. crassa* 5.8 S rRNA (157 nucleotides) (right-hand track). The positions of the cleavage sites measured from the 3′-end of S-rRNA can be identified. In these regions, the RNA chain must be accessible to the nuclease at the subparticle surface.

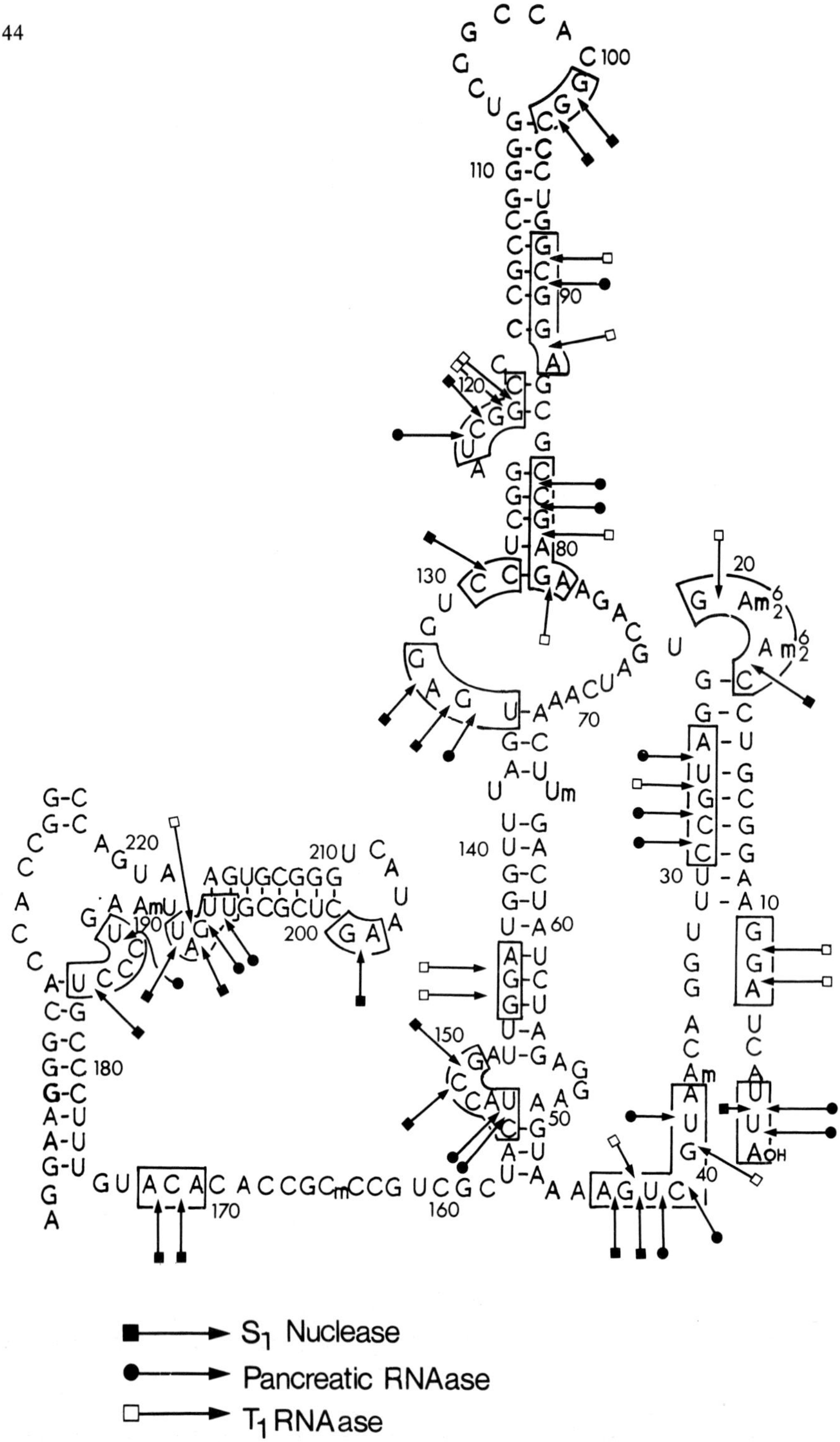

Fig. 23. *Cleavage sites at the 3′-end of X. laevis 18 S rRNA after nuclease treatment of S-subparticles*

The sequence at the 3′-end of *X. laevis* S-rRNA (Salim & Maden, 1981) is drawn in the conformation suggested by Glotz *et al.* (1981). Nucleotides are numbered from the 3′-end. 40 S-subparticles were treated with pancreatic ribonuclease, T_1 ribonuclease and S_1 nuclease. The cleavage sites identified from our preliminary data are indicated.

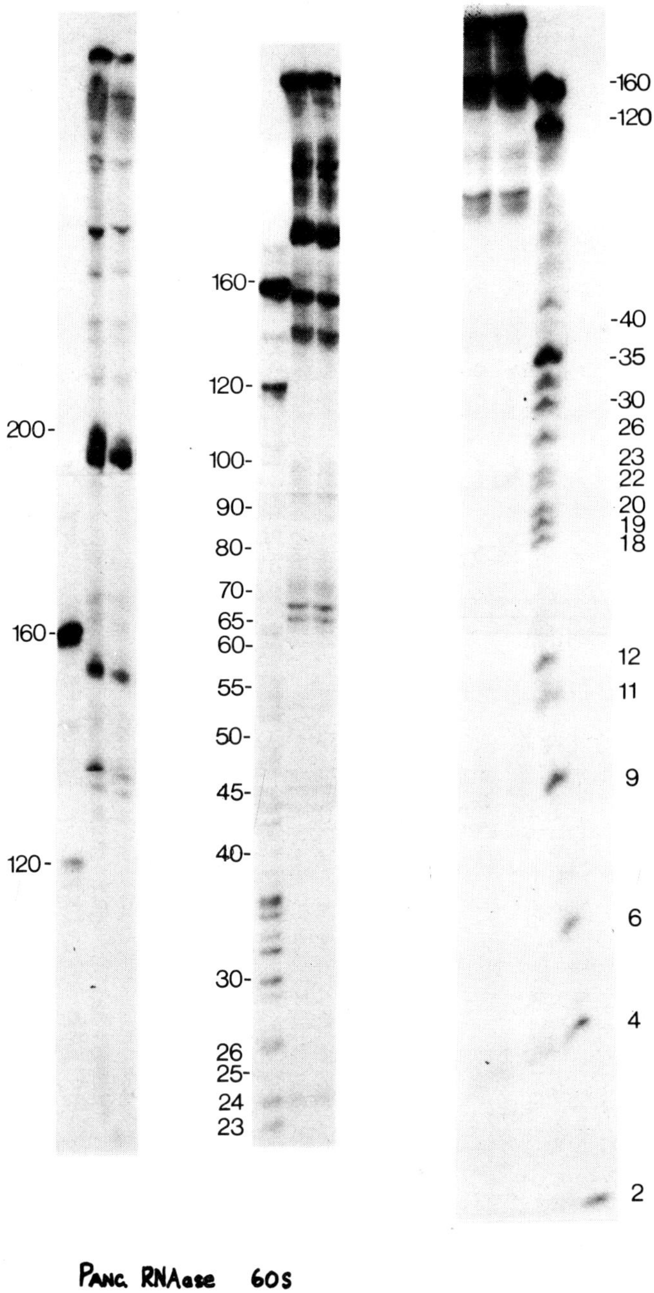

Fig. 24. *Autoradiograph of gels used to fractionate 3′-end labelled oligonucleotide products of X. laevis 28 S rRNA after treatment of L-subparticles with pancreatic ribonuclease*

X. laevis L-subparticles, labelled with [^{32}P]pCp at the 3′-end of 28 S rRNA, were treated with pancreatic ribonuclease (1 : 50000 and 1 : 500000 enzyme : substrate ratio) at 37°C for 30 min and layered on 12% (right) or 8% (centre and left) 60 cm polyacrylamide gels. Electrophoresis was carried out for 2 h (right), 6 h (centre) and 12 h (left). The location of cleavage sites was determined from the size of 3′-end labelled fragments as described above (see Fig. 22).

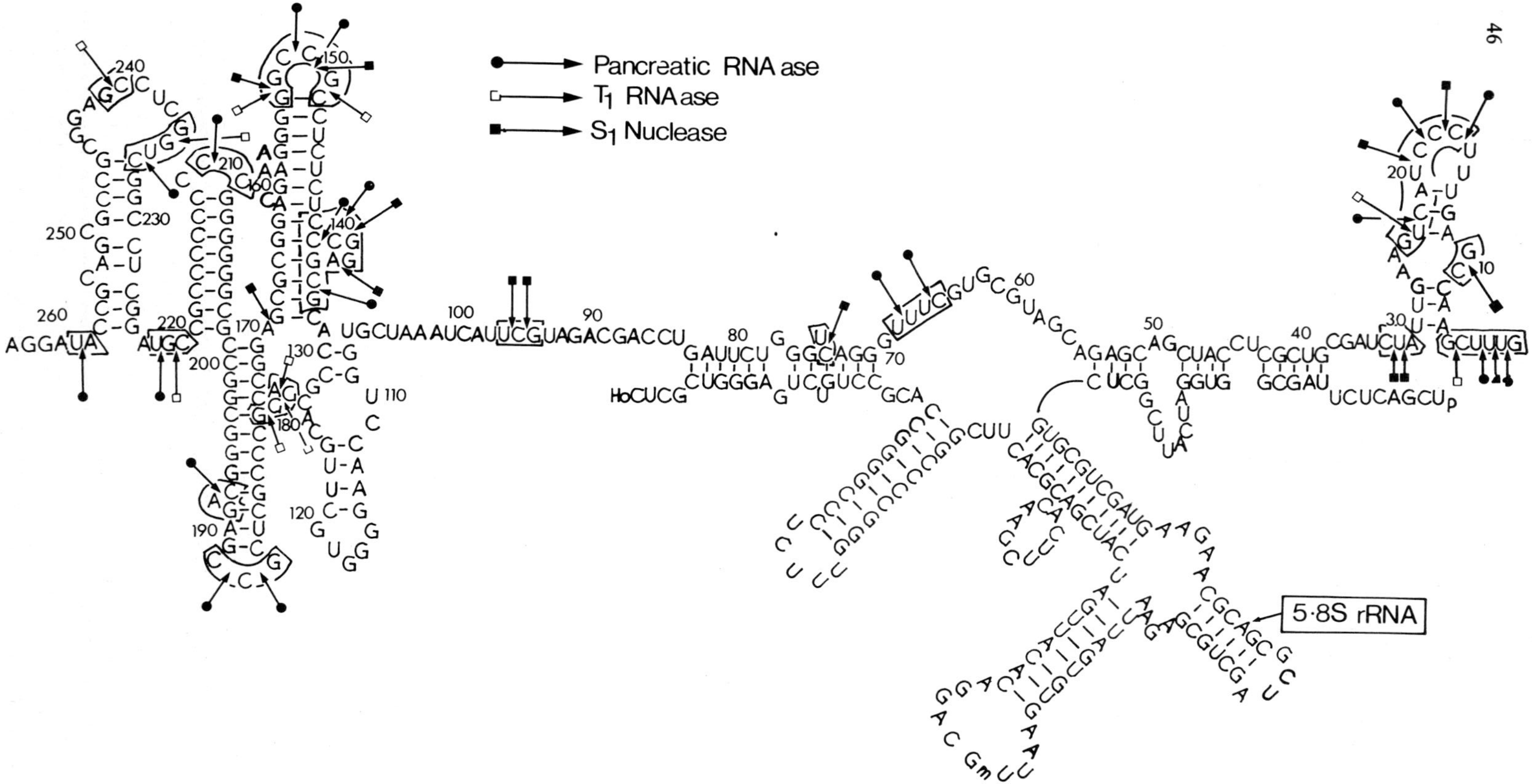

Fig. 25. *Cleavage sites at the 3′-end of X. laevis 28 S rRNA after treatment of L-subparticles with nucleases*

The sequence at the 3′-end of *X. laevis* 28 S rRNA (Sollner-Webb & Reeder, 1979) is shown with 5.8 S rRNA attached at a proposed binding site (Kelly & Cox, 1981). Base pairs in the region 100–260 nucleotides from the 3′-end have been maximized. L-subparticles were treated with pancreatic ribonuclease, T_1 ribonuclease and S_1 nuclease. The cleavage sites determined from our preliminary data are indicated.

et al., 1981). As more sequence data becomes available it is probable that precise mechanisms for rRNA evolution will emerge.

The availability of sequence data for cytoplasmic S-rRNA and L-rRNA has proved useful in delineating homologies. It is feasible to present the sequence data in the form of a scheme of secondary structure which can be related to schemes proposed for *E. coli* S-rRNA and L-rRNA. These schemes provide a useful starting point but the problem of establishing the secondary and tertiary structure of rRNA within the functional ribosome is the prime problem. Progress towards determining the way in which rRNA is packed within the ribosome is likely because it has been shown that rRNA lies at the surface at intervals of no more than approx. 100 nucleotides. A variety of approaches can be developed to identify these exposed sequences.

Although enormous progress has been made in establishing structural features of ribosomes, little is known about the mechanisms used by the ribosome in carrying out mRNA-directed protein synthesis by the stepwise incorporation of amino acids. Now that highly conserved rRNA sequences have been identified they provide a focus for investigation of the role of rRNA in ribosome function. It is heartening to see the convergence of information from both prokaryotic and eukaryotic ribosomes. We may need to accumulate still more structural data before we gain insight into ribosome function.

References

Arnstein, H.R.V. (1982) *Biochem. Soc. Symp.* **47**, 163–182
Bielka, H. & Stahl, J. (1978) *Int. Rev. Biochem.* **18**, 79–168
Birnstiel, M.L. & Grunstein, M. (1972) *FEBS Symp.* **23**, 349–364
Bos, J.L., Heyting, C., Borst, P., Arnberg, A.C. & Van Bruggen, E.F.J. (1978) *Nature (London)* **275**, 336–338
Boseley, P.G., Tuyns, A. & Birnstiel, M.L. (1978) *Nucleic Acids Res.* **5**, 1121–1137
Branlant, C., Knol, A., Machatt, M.A., Pouyet, J., Ebel, J.-P., Edwards, K. & Kossel, H. (1981) *Nucleic Acids Res.* **9**, 4303–4324
Brimacombe, R. (1982) *Biochem. Soc. Symp.* **47**, 49–60
Brosius, J., Palmer, M.L., Kennedy, P.J. & Noller, H.F. (1978) *Proc. Natl. Acad. Sci. U.S.A.* **75**, 4801–4805
Brosius, J., Dull, T.J. & Noller, H.F. (1980) *Proc. Natl. Acad. Sci. U.S.A.* **77**, 201–204
Cammarano, P., Londei, P. & Mazzei, F. (1980) *Biochem. J.* **189**, 313–335
Cech, T.R. & Rio, D.C. (1979) *Proc. Natl. Acad. Sci. U.S.A.* **76**, 5051–5055
Chua, N.H. & Luck, D. (1974) in *Ribosomes* (Nomura, M., Tissieres, A. & Lengyel, P., eds.), pp. 519–530, Cold Spring Harbor Laboratories
Cox, R.A. (1966) *Biochem. J.* **98**, 841–857
Cox, R.A. (1970) *Biochem. J.* **117**, 101–118
Cox, R.A. (1977) *Prog. Biophys. Mol. Biol.* **32**, 193–231
Cox, R.A. (1980) *Biochem. J.* **194**, 931–939
Cox, R.A. & Godwin, E.A. (1975) *MTP Int. Rev. Sci. Biochem. Ser. 1*, **7**, 179–253
Cox, R.A. & Greenwell, P. (1976) *Biochem. J.* **160**, 533–546
Cox, R.A. & Greenwell, P. (1980) *Biochem. J.* **186**, 861–872
Cox, R.A. & Hirst, W. (1976) *Biochem. J.* **160**, 505–519
Cox, R.A. & Kelly, J.M. (1981*a*) *FEBS Lett.* **130**, 1–6
Cox, R.A. & Kelly, J.M. (1981*b*) *FEBS Lett.* **136**, 306–310
Cox, R.A. & Peden, K. (1979) *Mol. Gen. Genet.* **174**, 17–24
Cox, R.A. & Thompson, R.D. (1980) *Biochem. J.* **187**, 75–90
Cox, R.A., Huvos, P. & Godwin, E.A. (1973) *Israel J. Chem.* **11**, 407–422
Cox, R.A., Godwin, E.A. & Hastings, J.R.B. (1976*a*) *Biochem. J.* **155**, 465–475
Cox, R.A., Greenwell, P. & Hirst, W. (1976*b*) *Biochem. J.* **160**, 521–533
Cox, R.A., Hirst, W., Godwin, E. & Kaiser, I. (1976*c*) *Biochem. J.* **155**, 279–295

Din, N., Engberg, J., Kaffenberger, W. & Eckert, W.A. (1979) *Cell* **18**, 525–532
Doty, P., Boedtker, H., Fresco, J.R., Haselkorn, R. & Litt, M. (1959) *Proc. Natl. Acad. Sci. U.S.A.* **45**, 482–487
Erdmann, V.A. (1976) *Progr. Nucleic Acids Res. Mol. Biol.* **18**, 45–90
Ford, P.J. & Mathieson, T. (1978) *Eur. J. Biochem.* **87**, 199–214
Free, S.J., Rice, P.W. & Metzenberg, R.L. (1979) *J. Bacteriol.* **137**, 1219–1226
Georgiev, O.L., Nikolaev, N., Hadjiolov, A.A., Skryabin, K.G., Zakharyev, V.M. & Bayev, A.A. (1981) *Nucleic Acids Res.* **9**, 6953–6974
Glotz, C., Zwieb, C., Brimacombe, R., Edwards, K. & Kossel, H. (1981) *Nucleic Acids Res.* **9**, 3287–3306
Glover, D.M. & Hogness, D.S. (1977) *Cell* **10**, 167–176
Godwin, E.A., Cox, R.A. & Huvos, P. (1974) *Acta Biol. Med. Germ.* **33**, 733–752
Gourse, R.L. & Gerbi, S.A. (1980) *J. Mol. Biol.* **140**, 321–340
Gourse, R.L., Thurlow, D.L., Gerbi, S.A. & Zimmerman, R.A. (1981) *Proc. Natl. Acad. Sci. U.S.A.* **78**, 2722–2726
Greenberg, J.R. (1969) *J. Mol. Biol.* **46**, 85–98
Gubler, U., Wyler, T. & Braun, R. (1979) *FEBS Lett.* **100**, 347–350
Hadjiolov, A.A. & Nikolaev, N. (1976) *Progr. Biophys. Mol. Biol.* **31**, 95–144
Hall, L.M.C. & Maden, B.E.H. (1980) *Nucleic Acids Res.* **8**, 5993–6005
Heckman, J.E. & Raj Bhandary, U.L. (1979) *Cell* **17**, 583–595
Jordan, B.R., Latil-Damotte, M. & Jourdan, R. (1980) *FEBS Lett.* **123**, 17–20
Kelly, J.M. & Cox, R.A. (1981*a*) *FEBS Lett.* **133**, 79–83
Kelly, J.M. & Cox, R.A. (1981*b*) *Nucleic Acids Res.* **2**, 1111–1121
King, H. & Gould, H.J. (1970) *J. Mol. Biol.* **51**, 687–692
Lazarus, C.M., Lunsdorf, H., Hahn, U., Stepien, P.P. & Kuntzel, H. (1980) *Mol. Gen. Genet.* **177**, 389–397
MacKay, R.M. (1981) *FEBS Lett.* **123**, 17–18
Maden, B.E.H. & Robertson, J.S. (1974) *J. Mol. Biol.* **87**, 227–255
Mandal, R.K. & Dawid, I.B. (1981) *Nucleic Acids Res.* **9**, 1801–1811
Miller, O.L. & Hamkalo, B.A. (1972) *FEBS Symp.* **23**, 367–378
Nath, K. & Bollon, A.P. (1977) *J. Biol. Chem.* **252**, 6562–6571
Nath, K. & Bollon, A.P. (1978) *Mol. Gen. Genet.* **160**, 235–245
Nazar, R.N. (1980) *FEBS Lett.* **119**, 212–214
Noller, H.F., Kop, J.A., Wheaton, V., Brosius, J., Gutell, R.R. & Kopylov, A.M. (1981) *Nucleic Acids Res.* **9**, 6167–6189
Nomiyama, H., Kuhara, S., Kukita, T., Otsuka, T. & Sakaki, Y. (1981*a*) *Nucleic Acids Res.* **9**, 5507–5520
Nomiyama, H., Sakaki, Y. & Takagi, Y. (1981*b*) *Proc. Natl. Acad. Sci. U.S.A.* **78**, 1376–1380
Pavlakis, G.N., Jordan, B.R., Wurst, R.M. & Vournakis, J.N. (1979) *Nucleic Acids Res.* **7**, 2213–2238
Peattie, D.A. (1979) *Proc. Natl. Acad. Sci. U.S.A.* **76**, 1760–1764
Pellegrini, M., Manning, J. & Davidson, N. (1977) *Cell* **10**, 213–224
Rochaix, J.D. & Malnoe, P. (1978) *Cell* **15**, 661–670
Rubin, G.M. (1973) *J. Biol. Chem.* **248**, 3860–3875
Rubstov, P.M., Musa Khanov, M.M., Zatharyev, V.M., Krayev, A.S., Skryabin, K.G. & Bayev, A.A. (1980) *Nucleic Acids Res.* **8**, 5779–5794
Selker, E. & Yanofsky, C. (1979) *Nucleic Acids Res.* **6**, 2561–2567
Salim, M. & Maden, B.E.H. (1981) *Nature (London)* **291**, 205–208
Sharp, P.A. (1981) *Cell* **23**, 643–646
Sollner-Webb, B. & Reeder, R.H. (1979) *Cell* **18**, 485–499
Stiegler, P., Carbon, P., Zuker, M., Ebel, J.-P. & Ehresmann, C. (1981) *Nucleic Acids Res.* **9**, 2153–2172
Veldman, G.M., Klootwijk, J., de Jonge, P., Leer, R.J. & Planta, R.J. (1980) *Nucleic Acids Res.* **8**, 5179–5192
Veldman, G.M., Klootwijk, J., de Regt, C.H.F. & Planta, R.J. (1981) *Nucleic Acids Res.* **9**, 6935–6952
Weiss, R.L. & Morris, D.R. (1973) *Biochemistry* **12**, 435–441
Weiss, R.L., Kimes, B.W. & Morris, D.R. (1973) *Biochemistry* **12**, 450–456
Wellauer, P.K. & Dawid, I. (1973*a*) *Cold Spring Harbor Symp. Quant. Biol.* **38**, 525–535
Wellauer, P.K. & Dawid, I. (1973*b*) *Proc. Natl. Acad. Sci. U.S.A.* **70**, 2827–2831
Wellauer, P. & Dawid, I. (1974) *J. Mol. Biol.* **89**, 379–395
Wild, M.A. & Gall, J.G. (1979) *Cell* **16**, 565–573
Wrede, P. & Erdmann, V.A. (1977) *Proc. Natl. Acad. Sci. U.S.A.* **74**, 2706–2709
Wool, I.G. (1979) *Annu. Rev. Biochem.* **48**, 719–754

Biochem. Soc. Symp. 47, 49–60
Printed in Great Britain

The Secondary Structure of Ribosomal RNA, and its Organization within the Ribosomal Subunits

RICHARD BRIMACOMBE

Max-Planck-Institut für Molekulare Genetik, Abteilung Wittmann, Ihnestrasse 63–73, D-1000 Berlin-Dahlem, Germany

Introduction

The ribosome is a complex array of protein and RNA molecules, and its structure can be investigated by a wide variety of methods at a number of different levels. In the past, most detailed studies have tended to concentrate on the ribosomal proteins, for the simple reason that the 20–40 individual proteins within each subunit can be separated from one another and examined as isolated entities. On the other hand, each ribosomal subunit contains only a single large rRNA molecule, and any correspondingly detailed study of particular regions of the RNA obviously requires sequence information. Until quite recently no complete base sequence was available for one of the large rRNA molecules, but during the last 2–3 years the dramatic advance in DNA sequencing technology has led to a veritable flood of DNA sequence determinations corresponding to ribosomal RNA molecules from many different sources.

The size of these rRNA molecules varies considerably from one organism to another. The smallest ribosomes are those from mammalian mitochondria, which have rRNA molecules of 12S and 16S in the small and large subunits (approx. 950 and 1550 nucleotides in length, respectively). Other types of mitochondria, and also chloroplasts, have larger rRNA molecules which are equivalent in size to the 16S and 23S species found in bacterial ribosomes (approx. 1550 and 3000 nucleotides), while still larger are the rRNA molecules from eukaryotic cytoplasmic ribosomes. These latter are 18S and 26–28S in the small and large subunits, respectively (approx. 1800, and up to 4000, nucleotides in length). The list of sequences currently completed or nearly completed is given in Table 1, and many other partial sequences, too numerous to be included, are also available. It can be seen from Table 1 that every common size class of ribosomal RNA is represented by sequences from two or more species, and with the help of this sequence information it is now possible not only to make detailed studies on the RNA moieties of the particular ribosomal species under investigation, but also to draw far-reaching conclusions concerning the secondary structure and evolutionary stability of ribosomal RNA in general.

The most widely-studied individual ribosome has for a long time been that of *Escherichia coli*, and in this article I shall describe firstly how detailed secondary structure models have been derived for both the 16S and 23S rRNA from this organism. Secondly, the question of the conservation or generality of this secondary

Table 1. *Known sequences of ribosomal DNA (or rRNA) molecules*

Known sequences are listed, showing the size (*s* value) of the corresponding rRNA molecule. ( ), the sequence concerned is not yet fully complete; –, the sequence is not available. All sequences were determined from ribosomal DNA, with the exception of those mentioned in the appropriate footnote giving the source of the sequence in the literature.

Organism	Small subunit rRNA	Large subunit rRNA
Human mitochondrion	12S[a]	16S[a]
Mouse mitochondrion	12S[b]	16S[b]
Rat mitochondrion	–	16S[c]
Saccharomyces cerevisiae mitochondrion	15S[d]	–
Aspergillus nidulans mitochondrion	15S[e]	–
Escherichia coli	16S[f]	23S[g]
Proteus vulgaris	16S[h]	–
Bacillus brevis	(16S)[i]	–
B. stearothermophilus	–	23S[j]
Zea mays chloroplast	16S[k]	23S[l]
S. cerevisiae cytoplasm	18S[m]	26S[n]
Xenopus laevis cytoplasm	18S[o]	(28S)[p]

[a] Eperon *et al.*, 1980. [b] van Etten *et al.*, 1980. [c] Saccone *et al.*, 1981. [d] Sor & Fukuhara, 1980. [e] Köchel & Küntzel, 1981. [f] Brosius *et al.*, 1978; Carbon *et al.*, 1978; the latter sequence was obtained direct from RNA. [g] Brosius *et al.*, 1980. [h] Carbon *et al.*, 1981; sequence determined from RNA. [i] Quoted in Noller & Woese, 1981. [j] Noller *et al.*, 1981. [k] Schwarz & Kössel, 1980. [l] Edwards & Kössel, 1981. [m] Rubtsov *et al.*, 1980. [n] Georgiev *et al.*, 1981; Veldmann *et al.*, 1981 (*S. carlsbergensis*). [o] Salim & Maden, 1981. [p] Preliminary report in Gourse & Gerbi, 1980.

structure will be discussed, both in relation to organisms with rRNA of the same size class (cf. Table 1) and also in relation to organisms having significantly longer or shorter rRNA molecules. Thirdly, I shall describe our attempts to gain an insight into the three-dimensional organization of the ribosomal RNA from *E. coli*, and the arrangement of the RNA relative to the proteins within the ribosomal subunits.

Secondary Structure Determination

The method which we have developed for the determination of secondary structure is the analysis of base-paired fragments generated by partial nuclease digestion either of ribosomal subunits or of isolated rRNA (Ross & Brimacombe, 1979; Glotz & Brimacombe, 1980). The principle of this method is illustrated in Fig. 1A, and involves a two-dimensional polyacrylamide gel electrophoresis. The first dimension of the gel system is run in SDS under non-denaturing conditions, so that base-paired interactions between fragments in the partially digested RNA sample remain largely intact. In the second dimension, which is run under denaturing conditions, the fragments dissociate, thereby giving rise to recognizable pairs or families of fragments which can then be subjected to 'fingerprint' analysis. The method gives direct information as to which sequences are base-paired together within the RNA, and has obvious advantages over the more commonly-used methods, such as determination of nuclease cutting points in single-stranded regions (e.g. Fiers *et al.*, 1976; Stiegler *et al.*, 1980) or points of attack by single-strand specific reagents (e.g. Woese *et al.*, 1980). These latter methods, while they are very useful for testing an established structure (e.g. Peattie & Gilbert, 1980), have a diminishing predictive value as the molecule under consideration becomes longer.

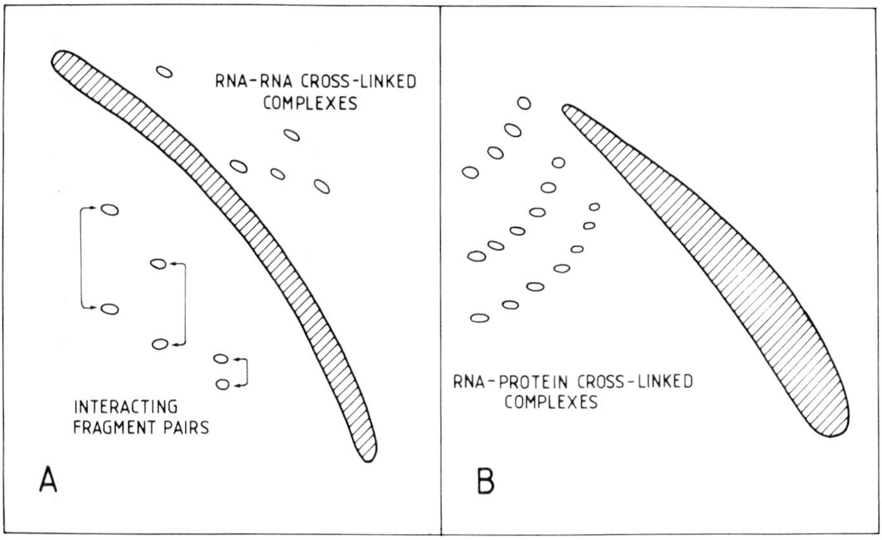

Fig. 1. *Two-dimensional gel systems used for separating interacting fragment pairs or RNA-RNA cross-linked complexes (A), or RNA-protein cross-linked complexes (B)*

First dimensions run from left to right, and second dimensions from top to bottom, in every case towards the anode. Cross-hatched regions denote 'diagonals' of free RNA fragments. See appropriate parts of the text for details.

Many hundreds of base-paired fragments have been analysed by means of the two-dimensional gel method just described, from both the 16S and 23S rRNA from *E. coli* (Glotz *et al.*, 1981; Zwieb *et al.*, 1981). The fragments are distributed fairly evenly throughout the sequences, and serve as a first basis for building models of the secondary structures. In particular, a number of 'long-range interactions', between RNA segments which are far apart in the primary sequence, could be clearly identified. Secondary structural information was also obtained from fragments containing intra-RNA cross-links (cf. Fig. 1A), and these cross-links will be discussed later.

The fragment data alone are of course not sufficient to build a secondary structure model. Often there are several base-pairing possibilities between even quite short fragments, and some regions of the sequences are not represented among those fragments so far obtained. The models therefore require a good deal of refinement, and the most powerful tool for this purpose is without doubt the sequence comparison method, which was first applied to the ribosome in the case of the small 5S rRNA by Fox & Woese (1975). This approach is in essence very simple. If similar but not identical sequences from different organisms are compared, then a base change in one strand of a putative helical region must be compensated by a complementary base change in the other strand. Thus, an A·U base-pair in one species could become a G·C pair at the corresponding positions in the second species, and so on. If the base changes do not compensate each other in this way, then the implication is that the proposed element of secondary structure is incorrect.

It is clear from Table 1 that a number of suitable sequences are available for comparison, and the question which immediately arises is: how similar are the various primary structures? The answer is that the sequences of rRNA molecules within the same size class have been conserved to a very large extent. *Zea mays* chloroplast 16S and 23S rRNA are for example both about 70% homologous with the corresponding *E. coli* molecules, and this degree of homology is ideal for secondary structural studies, since the two sets of sequences are similar enough to be compared directly with one another in detail, while at the same time the degree of divergence is sufficient to allow a large number of 'compensating base changes' to be found in the double-helical regions of the structures. The structural comparisons have generated a 'positive feed-back' situation, in which each new sequence studied not only gives support to the concept that secondary structure in rRNA has been very highly conserved between molecules of the same size class, but also serves to confirm and extend the original secondary structure model of the *E. coli* rRNA. To give just one example of the power of the sequence comparison method, a comparison of the structure derived for *E. coli* 23S RNA (see below) with the corresponding structure from *Z. mays* chloroplast ribosomes shows a total of over 450 compensating base changes.

Three research groups, in California, Strasbourg and here in Berlin, have been independently involved in the construction of these secondary structure models, each using different sets of experimental data derived from *E. coli* rRNA, and each using sequence comparisons to refine the models. Noller and his co-workers in Santa Cruz, California, have primarily exploited the analysis of sites of chemical modification by single-strand specific reagents in order to obtain their experimental data. Ehresmann, Branlant and colleagues in Strasbourg have made use of single- and double-strand specific nucleases to generate specific cuts in the RNA, and we, as already mentioned, have concentrated on the analysis of base-paired fragments. The latest versions of the secondary structures have all appeared in the literature quite recently for both 16S rRNA (Noller & Woese, 1981; Stiegler *et al.*, 1981*b*; Zwieb *et al.*, 1981) and for 23S rRNA (Glotz *et al.*, 1981; Branlant *et al.*, 1981; Noller *et al.*, 1981), and sketches of our own models for both the *E. coli* rRNA molecules are included in Figs. 2 and 3. These figures also include other structures as well as cross-linking data which will be discussed later.

There is no space here to discuss the models in detail, but it can be seen that the secondary structures of both 16S and 23S RNA are highly branched, and are organized into 'domains' which are clearly defined by the 'long-range interactions' already mentioned. In the case of the 23S RNA, the whole molecule appears to be a giant loop, closed by an interaction between the extreme 5′- and 3′-ends of the RNA strand. The models from the three different groups just cited are in agreement with one another to the extent of about 80%, and the main features of the structures can be regarded as firmly established. There are still some significant differences, however, and it will take some time before these are finally ironed out.

Universality of Secondary Structure

More interesting at the moment is the question as to whether the sequence comparison approach can be extended to other ribosomal RNA molecules, outside the 16S and 23S size class. For this purpose, both the smaller mitochondrial rRNA and the larger eukaryotic cytoplasmic rRNA species (Table 1) can be used for com-

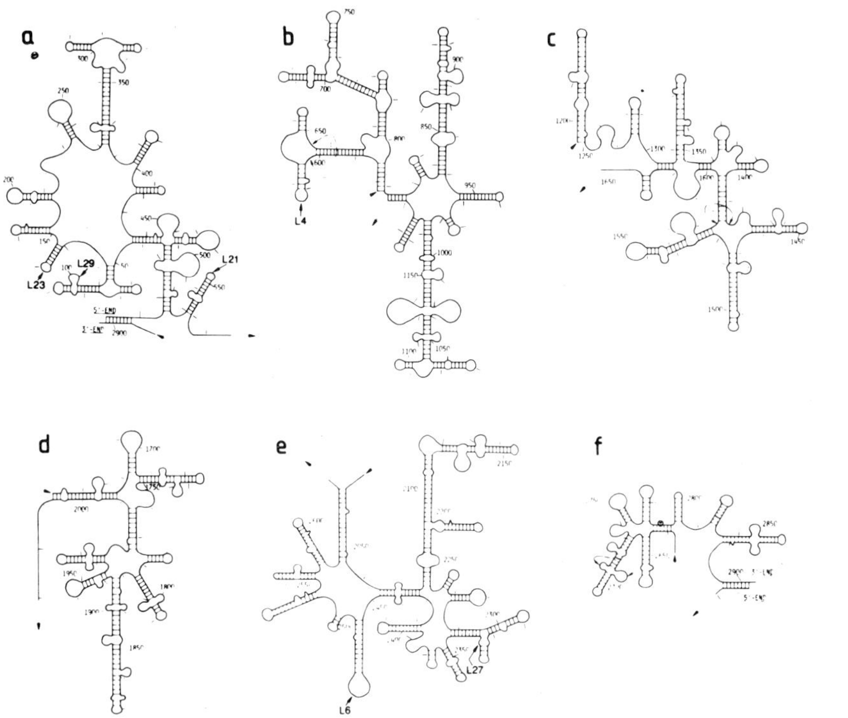

Fig. 2. *Sketch of the secondary structure of E. coli 23S rRNA*

The sequence is numbered from the 5′-end, and is divided into six domains (a–f) as in Glotz *et al.* (1981). Sites of intra-RNA cross-links are denoted by the arrowed loops (e.g. at positions 600–650), and sites of cross-linking to protein are also indicated (e.g. protein L4 at position 615).

parisons. In both cases the degree of primary sequence homology with respect to *E. coli* rRNA is at first sight not very high, but nevertheless there are significant regions of conserved sequence, and secondary structures can be built up around such conserved regions, using the *E. coli* secondary structure models as a basis. The results of these comparisons are very striking indeed.

The small mitochondrial 12S and 16S rRNA species, although they are only about half the size of the corresponding *E. coli* molecules, contain many elements of secondary structure which are precisely equivalent to their 16S and 23S counterparts. The reduction in size is achieved by simple amputation of secondary structural loops, or by erosion of whole domains; sketches of the structures derived for human mitochondrial 12S and 16S rRNA (Glotz *et al.*, 1981; Zwieb *et al.*, 1981) are shown in Fig. 4. If the mitochondrial 16S RNA (Fig. 4) is compared with the *E. coli* 23S RNA (Fig. 2), it can be seen that in domains ‘b’, ‘d’ and ‘e’ there is strong conservation of the secondary structure, with a certain number of loops being cleanly amputated. In domains ‘a’, ‘c’ and ‘f’, on the other hand, the mitochondrial domains are considerably reduced in size. A similar situation pertains in the mitochondrial 12S RNA (Fig. 4), as compared with the *E. coli* 16S RNA (Fig. 3).

If the *E. coli* structures are compared with the larger eukaryotic molecules, then a precisely analogous pattern is observed, except that here of course there are

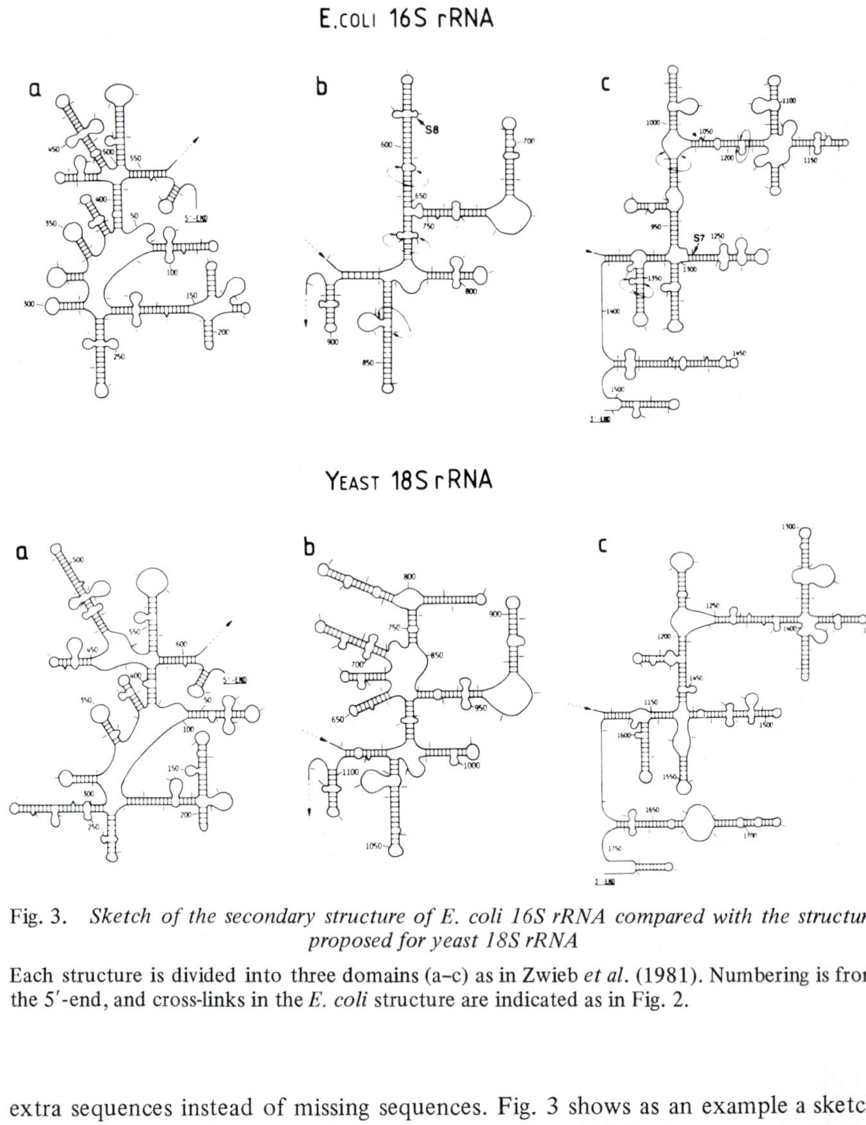

Fig. 3. *Sketch of the secondary structure of E. coli 16S rRNA compared with the structure proposed for yeast 18S rRNA*

Each structure is divided into three domains (a–c) as in Zwieb *et al.* (1981). Numbering is from the 5′-end, and cross-links in the *E. coli* structure are indicated as in Fig. 2.

extra sequences instead of missing sequences. Fig. 3 shows as an example a sketch of the 18S rRNA from yeast (Zwieb *et al.*, 1981), and if this is compared with the *E. coli* 16S RNA, it can be seen that the extra sequences in the yeast largely occur in the region from bases 640–850 (equivalent to 590–650 in *E. coli*), and that precisely this region is entirely missing in the mitochondrial 12S RNA (Fig. 4). Other regions of the yeast structure (e.g. domain ‘c’) are virtually superimposable on the corresponding *E. coli* structure. Furthermore, if the sequences are compared within these secondary structure models, then it becomes immediately clear that the degree of sequence homology between the various species is in fact a good deal higher than was at first supposed. Many short stretches of four to five base homology, which would not be regarded as significant in a simple comparison of the primary sequences, appear at identical positions in the various secondary structures, and appear to constitute a ‘core’ of conservation running right through the mole-

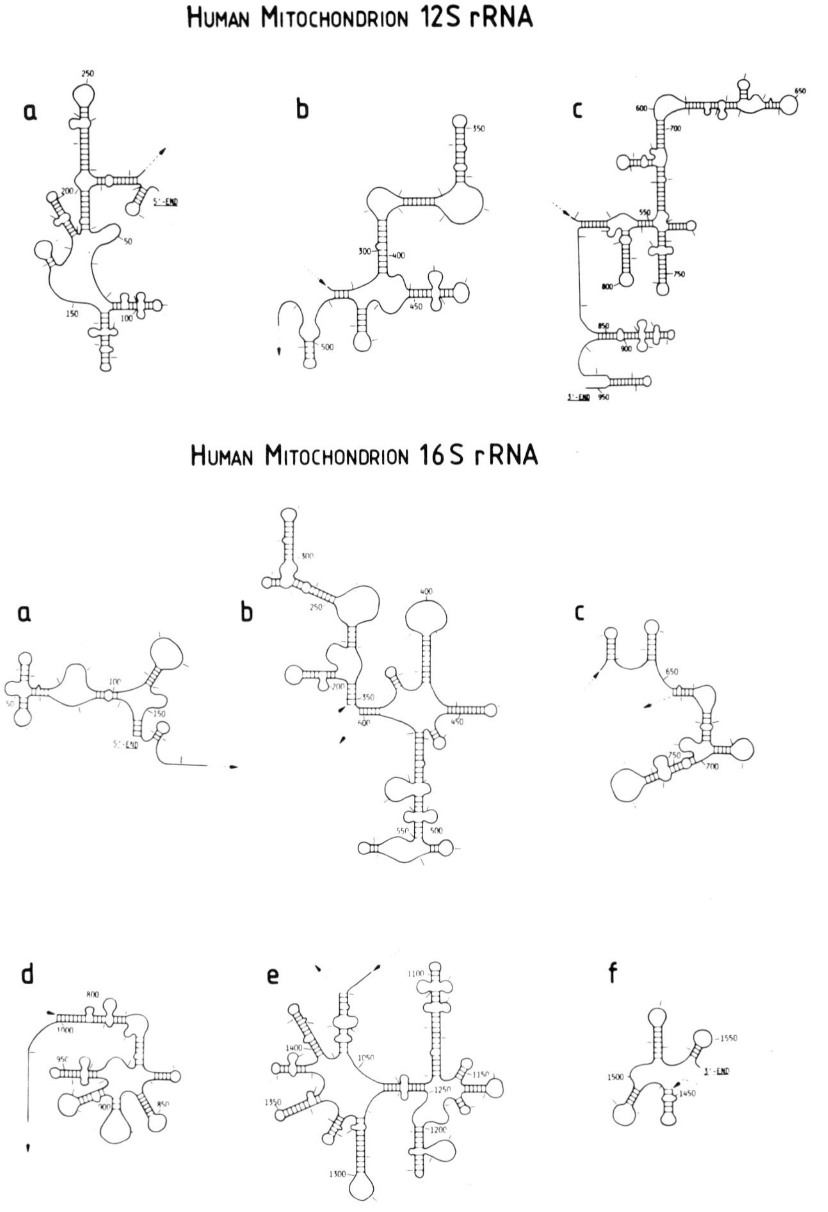

Fig. 4. *Sketch of the secondary structure proposed for human mitochondrial 12S rRNA (as in Zwieb et al., 1981) and 16S rRNA (as in Glotz et al., 1981)*

The structures should be compared with Figs. 3 and 2, respectively, and are arranged so that equivalent secondary structural elements lie in the same orientation to facilitate this comparison.

cules. The longer stretches of sequence homology tend to be in single-stranded regions, but there are also significant sequence homologies in double-helical regions. The secondary structure comparisons also support the concepts, based on observations of sequence homology, that eukaryotic 5.8S rRNA is equivalent to the 5′-end of prokaryotic 23S rRNA (Nazar & Roy, 1978; Nazar, 1980; Pace *et al.*, 1977; cf. also Kelly & Cox, 1981), and the 4.5S rRNA found in plant chloroplast ribosomes is equivalent to the 3′-end of bacterial 23S rRNA (MacKay, 1981; Machatt *et al.*, 1981; Edwards *et al.*, 1981).

Taken together, these studies demonstrate that despite differences in size, base composition and base sequence, all ribosomal RNA molecules so far studied fall into the same overall pattern of structure. This immediately raises the question as to why the high level of conservation has been necessary, and here it is difficult to avoid the conclusion that the ribosomal RNA must be intimately involved in the protein biosynthetic function. This in turn leads to the question as to whether the secondary structures are more or less rigid, or whether gross changes in conformation take place during the ribosomal cycle. There is some fairly convincing evidence that the latter possibility is the correct one, and this evidence comes from the analysis of base-paired fragments (cf. Fig. 1A). In our two-dimensional gels, a few RNA sequences from both *E. coli* 16S and 23S RNA were found with different partners in the various experiments (Glotz & Brimacombe, 1980; Glotz *et al.*, 1981; cf. Noller, 1980). In particular, a fragment of 5S rRNA was identified in association with the 23S sequence region 1760–1770 which is base-paired to the 1980–1990 region in Fig. 2 (domain 'd'); in the 16S RNA, the sequence 1050–1070, which was normally found associated with the sequence 1190–1210 (domain 'c', Fig. 3), also occurred in association with the sequence 385–400 (domain 'a', Fig. 3). In such 'switch' situations, the sequence comparison approach can again be invoked, to test whether the results are experimental artefacts, or whether they represent genuine multiple conformations of the RNA.

In the case of the 5S–23S RNA interaction, a similar but not entirely satisfactory structure can be drawn for the corresponding sequences in *Z. mays* chloroplasts, with some compensating base changes. In the case of the 16S 'switch', the base-pairing concerned is not only conserved from *E. coli* to *Z. mays* chloroplast, but a precisely similar switch structure can also be drawn in both yeast and *Xenopus* 18S rRNA (Brimacombe, 1980; Zwieb *et al.*, 1981), i.e. between sequences 1250–1280 (domain 'c' in yeast RNA, Fig. 3) and sequences 390–410 (domain 'a'). This structure is in every case a strong helix composed of 14 consecutive base-pairs. On the other hand, however, there is no trace of a corresponding 'switch' interaction in the 12S mitochondrial rRNA, and it should also be mentioned in this context that the secondary structure of the 18S rRNA in the region between bases 200 and 500 (encompassing the 'switch' sequence) is currently in dispute. The controversy arises because two distinct sets of primary sequence homology between yeast and *E. coli* can be found in this region; one set of homology leads to the structure shown for yeast in Fig. 3 (domain 'a'), whereas the other set leads to a different structure (Mankin *et al.*, 1981; Stiegler *et al.*, 1981*a*). The former shows the greatest similarity to the *E. coli* secondary structure in the region between bases 400 and 500, whereas the latter structure shows better primary sequence homology, but less secondary structural similarity.

I have gone into some detail here, in order to illustrate the sort of problem which can arise in these studies, and it is clear that the important question is not yet finally settled as to whether or not the 'switch' situation under discussion exists. There are two obvious ways to gain more insight into the question. One is to undertake a base-paired fragment analysis of yeast or *Xenopus* 18S rRNA, in order to obtain direct experimental evidence in the disputed region of the structure; such experiments are currently in progress (in collaboration with Dr. B.E.H. Maden). The other, and more general, approach is to study the three-dimensional organization of the RNA; sequences suspected of being involved in switches must be located in the three-dimensional structure in a manner which is topographically compatible with both conformations of the switch. A detailed examination of the three-dimensional structure of the RNA is in any case the next logical development in this area of research, and our experiments in this direction are based on cross-linking studies, involving both intra-RNA and RNA–protein cross-links.

Intra-RNA Cross-Linking

Our purpose is to introduce cross-links into the intact ribosomal subunits, and then to pinpoint the sites of cross-linking in order to obtain detailed topographical information. For this reason our efforts have been divided between a search for suitable cross-linking agents on the one hand, and the development of suitable methodology for cross-link site analyses on the other. In the case of intra-RNA cross-linking, we were able to show some time ago that mild u.v. irradiation of *E. coli* ribosomal subunits causes formation of cross-links within the RNA (Zwieb *et al.*, 1978), but it was not until quite recently that it became possible to localize the sites of reaction. The answer to the problem has turned out to be a two-dimensional polyacrylamide gel system very similar to the one already described for the analysis of base-paired fragments. If a partial digest of RNA containing intra-RNA cross-links is applied to such a gel system, the cross-linked fragments run above the 'diagonal' of free RNA fragments (Fig. 1A; Zwieb & Brimacombe, 1980). These fragments have the expected properties, namely they consist in general of two non-contiguous sequence regions in the RNA, and each sequence shows a 'gap' in its oligonucleotide fingerprint, which represents the cross-link site. The mobility of the fragments in the gel is not affected by treatment with proteinase K, indicating that RNA–protein cross-links (which are also formed under these cross-linking conditions; see below) are not involved.

One would expect that intra-RNA cross-links could be formed either within an element of the secondary structure, or between two separate elements of the secondary structure which lie close together in the three-dimensional structure. Several cross-links have already been identified by the procedure just outlined within both 16S (Zwieb & Brimacombe, 1980) and 23S rRNA (Glotz *et al.*, 1981) from *E. coli* (see Figs. 2 and 3), and so far all of them belong to the first of these two categories. Since cross-links of the second category are known to exist, although none has as yet been localized in detail (Thammana *et al.*, 1979; Wollenzien *et al.*, 1979), this result almost certainly reflects the conditions used for the partial digestion of the RNA, which usually involves a mild treatment with ribonuclease T1. Such treatment leads to a spectrum of fragments in which stable elements of secondary structure predominate. Since the cross-linking yield is itself not high, any cross-links occurring outside of such stable structural elements will only be represented in fragments appearing in very low yield, and will in consequence simply vanish out of sight. Recently we have been able to show that under different digestion conditions, using the double-strand-specific nuclease from cobra venom (Vassilenko & Ryte, 1975), very specific fragments can be obtained in high yield (cf Stiegler *et al.*, 1981*b*). The yield of fragments containing cross-links is thereby increased 10–100-fold, and a totally different spectrum of cross-linked fragments is obtained. These are currently being analysed, and it is to be expected that at least some of the cross-links will provide information directly relevant to the detailed three-dimensional folding of the rRNA. Several bi-functional chemical reagents are also availble for inducing the intra-RNA cross-links.

RNA–Protein Cross-Linking

As already mentioned above, u.v. irradiation of *E. coli* ribosomal subunits also causes formation of RNA–protein cross-links (cf. Gorelic, 1975). Under suitably mild conditions, only protein S7 in the 30S subunit and proteins L2 and L4 in the 50S subunit become cross-linked to their respective RNA molecules (Möller & Brimacombe, 1975), and the sites of cross-linking on protein and RNA have been precisely identified both for S7 (Möller *et al.*, 1978; Zwieb & Brimacombe, 1979) and for L4 (Maly *et al.*, 1980). In the latter case the methodology used was developed with a view to the simultaneous analyses of multiple cross-links involving several proteins, the situation usually pertaining when a chemical RNA–protein cross-linking agent is used. The principle of the method for analysis of cross-link sites on the RNA is illustrated in Fig. 1B, and again involves partial nuclease digestion and two-dimensional gel electrophoresis. In this instance the cross-linked sample is first separated from non-cross-linked protein by sucrose density gradient centrifugation in SDS, and the isolated RNA together with RNA cross-linked to protein is digested and applied to the gel system. The first dimension of the gel is run in a non-ionic detergent, so that the proteins exert their positive charge and retard the RNA fragments to which they are attached. The second dimension is run in SDS, so that the separation is on a simple size basis. Under these conditions, the RNA–protein cross-linked species appear as rows of spots lying roughly at right-angles to the 'diagonal' of free RNA fragments (Fig. 1B). These spots can be extracted from the gel, and subjected to both protein analysis and oligonucleotide fingerprinting, the cross-link site appearing as a 'gap' in the RNA sequence.

A number of RNA–protein cross-linking agents have been tested (e.g. Ulmer *et al.*, 1978; Oste & Brimacombe, 1979; cf. Bäumert *et al.*, 1978), the most successful being those carrying an imido function to react with protein, plus a photo-activatable group to react non-specifically with the RNA. The first such reagent to be developed was *p*-azidophenylacetic imidoester ('APAI', Rinke *et al.*, 1980; cf. Millon *et al.*, 1980), which generates a useful number of RNA–protein cross-links in the ribosomal subunits. More recently we have used the well-known protein–protein cross-linking reagent 2-iminothiolane (Traut *et al.*, 1973; Sommer & Traut, 1976), which under non-oxidizing conditions can be used to generate photo-activated RNA–protein cross-links in precisely the same way as APAI. With this reagent, a series of cross-link sites on *E. coli* 23S RNA has already been identified, for six different proteins (Wower *et al.*, 1981). Analysis of the corresponding sites of cross-linking within the individual proteins is still proving troublesome, but nevertheless the location of the cross-link sites on the RNA provides vital information for integrating the RNA structure into the mass of data which is already available on the spatial arrangement of the ribosomal proteins (see e.g. Chambliss *et al.*, 1980). All the RNA-protein cross-link sites which we have found so far are included in Figs. 2 and 3.

In summary, the ribosomal RNA field has made enormous progress since the first complete ribosomal RNA sequence became available 3 years ago. We now have convincing secondary structure models for the RNA from both ribosomal subunits, and it is clear that these structures have been conserved to a remarkable degree throughout evolution. In addition, the advances in cross-linking technology described here now offer realistic possibilities for both converting the 'two-dimensional' secondary structure maps into three dimensional models, and of fitting the ribosomal proteins to these models. It is not overly optimistic to predict that very soon we will see the appearance of the first detailed approximation to the structure of the ribosomal RNA within the subunits.

None of this work would have been possible without the energetic and skilful co-operation of Carola Glotz, Peter Maly, Marlis Meinke, Klaus Möller, Christian Oste, Jutta Rinke, Alex Ross, Wolfgang Stiege, Edith Ulmer, Iwona and Jacek Wower, and Christian Zwieb. I am also very grateful to Dr. H.G. Wittmann for his continued advice and encouragement, and to the Deutsche Forschungsgemeinschaft for financial support.

References

Bäumert, H.G., Sköld, S.E. & Kurland, C.G. (1978) *Eur. J. Biochem.* **89**, 353–359

Branlant, C., Krol, A., Machatt, M.A., Pouyet, J., Ebel, J.P., Edwards, K. & Kössel, H. (1981) *Nucleic Acids Res.* **9**, 4303–4324

Brimacombe, R. (1980) *Biochem. Int.* **1**, 162–171

Brosius, J., Palmer, M.L., Kennedy, P.J. & Noller, H.F. (1978) *Proc. Natl. Acad. Sci. U.S.A.* **75**, 4801–4805

Brosius, J., Dull, T.J. & Noller, H.F. (1980) *Proc. Natl. Acad. Sci. U.S.A.* **77**, 201–204

Carbon, P., Ehresmann, C., Ehresmann, B. & Ebel, J.P. (1978) *FEBS Lett.* **94**, 152–156.

Carbon, P., Ebel, J.P. & Ehresmann, C. (1981) *Nucleic Acids Res.* **9**, 2325–2333

Chambliss, G.R., Davies, J., Davis, K., Kahan, L. & Nomura, M. (eds.) (1980) *Ribosomes*, pp. 1–984, University Park Press, Baltimore

Edwards, K. & Kössel, H. (1981) *Nucleic Acids Res.* **9**, 2853–2869

Edwards, K., Bedbrook, J., Dyer, T. & Kössel, H. (1981) *Biochem. Int.* **2**, 533–538

Eperon, I.C., Anderson, S. & Nierlich, D.P. (1980) *Nature (London)* **286**, 460–467

Fiers, W., Contreras, R., Duerinck, F., Haegeman, G., Iserentant, D., Merregaert, J., Min Jou, W., Molemans, F., Raeymaekers, A., Van den Berghe, A., Volckaert, G. & Ysebaert, M. (1976) *Nature (London)* **260**, 500–507

Fox, G. & Woese, C.R. (1975) *Nature (London)* **256**, 505–507

Georgiev, O.I. Nikolaev, N., Hadjiolov, A.A., Skryabin, K.G., Zakharyev, V.M. & Bayev, A.A. (1981) *Nucleic Acids Res.* **9**, 6953–6958

Glotz, C. & Brimacombe, R. (1980) *Nucleic Acids Res.* **8**, 2377–2395

Glotz, C., Zwieb, C., Brimacombe, R., Edwards, K. & Kössel, H. (1981) *Nucleic Acids Res.* **9**, 3287–3306

Gorelic, L. (1975) *Biochim. Biophys. Acta* **390**, 209–225

Gourse, R. & Gerbi, S. (1980) *Nucleic Acids Res.* **8**, 3623–3637

Kelly, J.M. & Cox, R.A. (1981) *Nucleic Acids Res.* **9**, 1111–1121

Köchel, H.G. & Küntzel, H. (1981) *Nucleic Acids Res.* **9**, 5689–5696

Machatt, M.A., Ebel, J.P. & Branlant, C. (1981) *Nucleic Acids Res.* **9**, 1533–1549

MacKay, R.M. (1981) *FEBS Lett.* **123**, 17–18

Maly, P., Rinke, J., Ulmer, E., Zwieb, C. & Brimacombe, R. (1980) *Biochemistry* **19**, 4179–4188
Mankin, A.C., Kopylov, A.M., Rubtsov, P.M. & Skryabin, K.G. (1981) *Dokl. Acad. Nauk S.S.S.R.* **256**, 1006–1010
Millon, R., Olomucki, M., LeGall, J.Y., Golinska, B., Ebel, J.P. & Ehresmann, B. (1980) *Eur. J. Biochem.* **110**, 485–492
Möller, K. & Brimacombe, R. (1975) *Mol. Gen. Genet.* **141**, 343–355
Möller, K., Zwieb, C. & Brimacombe, R. (1978) *J. Mol. Biol.* **126**, 489–506
Nazar, R.N. (1980) *FEBS Lett.* **119**, 212–214
Nazar, R.N. & Roy, K.L. (1978) *J. Biol. Chem.* **253**, 395–399
Noller, H.F. (1980) in *Ribosomes* (Chambliss, G.R., Davies, J., Davis, K., Kahan, L. & Nomura, M., eds.), pp. 3–22, University Park Press, Baltimore
Noller, H.F. & Woese, C.R. (1981) *Science* **212** 403–411
Noller, H.F., Kop, J.A., Wheaton, V., Brosius, J., Gutell, R.R., Kopylov, A.M., Dohme, F., Herr, W., Stahl, D.A., Gupta, R. & Woese, C.R. (1981) *Nucleic Acids Res.* **9**, 6167–6189
Oste, C. & Brimacombe, R. (1979) *Mol. Gen. Genet.* **168**, 81–86
Pace, N.R., Walker, T.A. & Schroeder, E. (1977) *Biochemistry* **16**, 5321–5328
Peattie, D. & Gilbert, W. (1980) *Proc. Natl. Acad. Sci. U.S.A.* **77**, 4679–4682
Rinke, J., Meinke, M., Brimacombe, R., Fink, G., Rommel, W. & Fasold, H. (1980) *J. Mol. Biol.* **137**, 301–314
Ross, A. & Brimacombe, R. (1979) *Nature (London)* **281**, 271–276
Rubtsov, P.M., Musakhanov, M.M., Zakharyev, V.M., Krayev, A.S., Skryabin, K.G. & Bayev, A.A. (1980) *Nucleic Acids Res.* **8**, 5779–5794
Saccone, C., Cantatore, P., Gadaleta, G., Gallerani, R., Lanave, C., Pepe, G. & Kroon, A.M. (1981) *Nucleic Acids Res.* **9**, 4139–4148
Salim, M. & Maden, B.E.H. (1981) *Nature (London)* **291**, 205–208
Schwarz, Zs. & Kössel, H. (1980) *Nature (London)* **283**, 739–742
Sommer, A. & Traut, R.R. (1976) *J. Mol. Biol.* **106**, 995–1015
Sor, F. & Fukuhara, H. (1980) *C.R. Acad. Sci. Paris Ser. D* **291**, 933–936
Stiegler, P., Carbon, P., Zuker, M., Ebel, J.P. & Ehresmann, C. (1980) *C.R. Acad. Sci. Paris Ser. D* **291**, 937–940
Stiegler, P., Carbon, P., Ebel, J.P. & Ehresmann, C. (1981*a*) *Eur. J. Biochem.* **120**, 487–495
Stiegler, P., Carbon, P., Zuker, M., Ebel, J.P. & Ehresmann, C. (1981*b*) *Nucleic Acids Res.* **9**, 2153–2172
Thammana, P., Cantor, C.R., Wollenzien, P.L. & Hearst, J.E. (1979) *J. Mol. Biol.* **135**, 271–283
Traut, R.R., Bollen, A., Sun, T.T., Hershey, J.W.B., Sundberg, J. & Pierce, L.R. (1973) *Biochemistry* **12**, 3266–3272
Ulmer, E., Meinke, M., Ross, A., Fink, G. & Brimacombe, R. (1978) *Mol. Gen. Genet.* **160**, 183–193
van Etten, R.A., Walberg, M.W. & Clayton, D.A. (1980) *Cell* **22**, 157–170
Vassilenko, S.K. & Ryte, V.C. (1975) *Biokhimia* **40**, 578–583
Veldmann, G.M., Klootwijk, J., de Regt, V.C.H.F., Planta, R.J., Branlant, C., Krol, A. & Ebel, J.P. (1981) *Nucleic Acids Res.* **9**, 6935–6952
Woese, C.R., Magrum, L.J., Gupta, R., Siegel, R.B., Stahl, D., Kop, J., Crawford, N., Brosius, J., Gutell, R., Hogan, J.J. & Noller, H.F. (1980) *Nucleic Acids Res.* **8**, 2275–2293
Wollenzien, P., Hearst, J.E., Thammana, P. & Cantor, C.R. (1979) *J. Mol. Biol.* **135**, 255–269
Wower, I., Wower, J., Meinke, M. & Brimacombe, R. (1981) *Nucleic Acids Res.* **9**, 4285–4302
Zwieb, C. & Brimacombe, R. (1979) *Nucleic Acids Res.* **6**, 1775–1790
Zwieb, C. & Brimacombe, R. (1980) *Nucleic Acids Res.* **8**, 2397–2412
Zwieb, C., Ross, A., Rinke, J., Meinke, M. & Brimacombe, R. (1978) *Nucleic Acids Res.* **5**, 2705–2720
Zwieb, C., Glotz, C. & Brimacombe, R. (1981) *Nucleic Acids Res.* **9**, 3621–3640

Biochem. Soc. Symp. **47**, 61–78
Printed in Great Britain

Secondary Structure of Eukaryotic Messenger RNA

CALVIN P.H. VARY and JOHN N. VOURNAKIS

Department of Biology, Syracuse University, 130 College Place, Syracuse, NY 13210, U.S.A.

Introduction

Importance of mRNA structure

A considerable literature exists that documents evidence for the relevance of secondary structure in mRNA function. The early work of Lodish [1], Steitz [2] and Fiers *et al.* [3] strongly suggests that secondary structure acts to regulate protein synthesis in the bacteriophage f2, R17 and MS2 systems, respectively. Studies from Roth's laboratory on the regulation of the histidine operon in *Salmonella* [4] which combines genetic and translation analysis, the work of Yanofsky and co-workers on the *Escherichia coli* tryptophan operon [5], and the model of enteric bacteria [6], all invoke the existence of specific secondary structures in the respective mRNAs as a key regulatory feature. Isertant & Fiers [7] used a series of recombinant plasmids containing the *cro* gene of λ phage to show a correlation between the level of expression of the *cro* protein and secondary structure models of the 5′-regions of the mRNAs coded for by the respective plasmids. They argue that initiation of translation in prokaryotes is influenced by the position of the initiator AUG in the secondary structure, i.e. whether or not the AUG is base-paired is a more important factor than the accessibility of the Shine & Dalgarno sequence for ribosome binding [8]. They also argue that the 5′-region must be folded into a specific ribosome binding site following transcription, prior to initiation.

Considerable recent work in eukaryotic systems is relevant to the role of secondary structure in mRNA function. The observation initially by Protzel & Morris [9], and more recently by Lizardi *et al.* [10] that protein synthesis can proceed in a discontinuous fashion *in vivo*, and the studies of C.P.H.V. & A.J. Morris (unpublished observations; [11]) of the accumulation patterns of rabbit α- and β-globin nascent chain polypeptides *in vitro*, suggest that ribosome transit is discontinuous along the mRNA.

Work on ribosome binding sites in BMV mRNA [12], ovalbumin mRNA [13], alfalfa mosaic virus mRNA [14], and globin mRNA [15] all invoke base-pairing schemes to rationalize data on the size and sequence of 5′-region nucleotides protected by 40 S and 80 S ribosome initiation complexes. Kozak [16] has demonstrated that disruption of secondary structure by chemical means can alter the formation of 40 S ribosomal initiation complexes in reovirus mRNA. Theoretical models for the translation of mRNA always involve secondary structure as a parameter [17]. Translation of mRNA can be affected by factors that affect RNA structure such as ionic conditions [18]. Direct measurements by physical techniques [19-22], and enzymic means [23, 24] demonstrate that eukaryotic mRNA has considerable specific secondary structure.

A recent publication by Rosa [26] describes experiments in which the enzymic mapping technique developed in this laboratory [27, 28] was used to study the structures of the protein synthesis initiator region in the three T7 late mRNA species IIIb, IV and V. Although it is known, as mentioned earlier, that prokaryotic ribosomes locate the correct initiator through base-pairing between the 3′-end of 16 S rRNA and a region preceding the AUG [8], those factors that influence the translational efficiency of an mRNA are unclear. Minor changes in sequence around the initiator region seem to have a dramatic effect on the level of expression of proteins in prokaryotes [29, 30]. It is plausible that secondary and tertiary structure changes caused by alterations in sequence can affect ribosome binding and thereby translational efficiency. The conclusions reached by Rosa [26] are consistent with this notion. In addition, secondary-structure models proposed by Rosa for ribosome binding regions in the three T7 mRNAs, which were arrived at by combining nuclease mapping data and computer-assisted thermodynamic secondary structure predictions using the strategy of Pavlakis *et al.* [24], indicate that the initiator AUG may in some cases be base-paired (mRNAs IIIb and IV) and in other cases unpaired (mRNA V). The relative efficiency of ribosome binding may, therefore, involve the extent to which stable hairpin regions must be disrupted (melted-out) during ribosome loading at the initiation site. A similar conclusion is reached by Pinck *et al.* [14] for the binding of ribosomes to alfalfa mosaic virus mRNA 3.

It is suggested, therefore, that knowledge of the structure of mRNA molecules may be key in attempting to begin to understand the subtleties of the regulation and modulation of protein synthesis in both prokaryotic and eukaryotic systems.

Methods for RNA structure analysis

The development of rapid methods for DNA sequence analysis [31] resulted in a general strategy that could be applied to RNA sequence and structure analysis [27]. Both enzymic [27, 28] and chemical [32] techniques for structure mapping were developed that involve the partial cleavage of specifically end-labelled RNAs followed by high resolution acrylamide-gel electrophoresis and autoradiography to identify nucleotides that are either base-paired or unpaired. The enzymic technique has the advantages that both single-strand specific (S1, *Neurospora crassa* and mung-bean nucleases, etc.) and base-pair specific (cobra venom RNAase) enzymic probes are available and have been developed as effective tools [28], and that either 5′- or 3′-end labelled RNA can be studied. It has the major disadvantage that the enzyme probes are large molecules and are unable, often, to penetrate regions of dense conformational packing within RNA molecules. Therefore, enzymic structure-mapping provides data on the most exposed regions only. The enzymes used do not have significant base-specificity, so that all four major nucleotides can be studied, although the detailed structural specificities of certain of the enzymes are not known. The chemical technique has the great advantage that the probes are small and can penetrate throughout, thus providing nearly complete information on base-pairing, base-stacking, and on shielding of reactive sites by tertiary structure, bound ions or bound proteins [32, 33]. The major disadvantages are that the method is limited to only three of the four major nucleotide bases (A, G and C), and that it is applicable only to 3′-end labelled RNA. The second limitation is particularly serious for proposed studies of poly(A)-containing mRNA, requiring that an effective method for specifically deadenylylating mRNA molecules be available.

Computer aided studies of RNA secondary structure

Computer methods for predicting RNA secondary structure have been developed by Pipas & McMahon [34], Studnicka *et al.* [35], and Zuker & Stiegler [36], all of which impose thermodynamic criteria so as to obtain an optimal structure. These methods have limitations, and cannot generate an accurate structure prediction using sequence information alone [36, 37]. A combination of chemical reactivity, nuclease susceptibility and phylogenetic sequence comparison data can be used to predict the structure of an RNA, as demonstrated by the impressive results of Noller & Woese [38] on *E. coli* 16 S rRNA. Comparing Zuker & Stiegler's computer-generated prediction of the 5′-end of 16 S rRNA [36] to the Noller & Woese [38] consensus structure shows considerable differences, even though Zuker & Stiegler take into account some experimental data. It is clear that there is no general method available for unambiguously generating a RNA structure from sequence information alone. A new approach that combines thermodynamics and computer graphics based on interactive methods has recently been described [37]. This method facilitates the input of experimentally derived structural data, and phylogenetic information. It has the potential for providing accurate secondary structures, but is not yet fully developed.

Methods

Removal of m^7G from 5′-terminus of rabbit globin mRNAs

m^7G 'cap' removal was accomplished as previously described [24, 39]. A sample (10–20 μg) of mixed α- and β-rabbit globin mRNAs in 10 μl of buffer containing 50 mM-sodium acetate, pH 6.0, 10 mM-2-mercaptoethanol and 5 units of tobacco alkaline pyrophosphatase was incubated for 2 h at 37°C. After the incubation, enough 0.5 M-Tris/HCl, pH 8.5, was added to give a final concentration of 75 mM. Calf alkaline phosphatase was added to a final concentration of 0.1 unit/ml. After incubation for 1 h at 37°C the reaction mixture was immediately deproteinized with a phenol/chloroform (1 : 1, v/v) mixture and the mRNA was precipitated with ethanol.

5′-End-group labelling of mRNA with ^{32}P

Decapped and dephosphorylated mRNA was labelled with ^{32}P using [γ-^{32}P]ATP (2000–9000 Ci/mmol) and T_4-induced polynucleotide kinase [24, 28, 39]. Up to 0.4 A_{260} units of mRNA was labelled in a 10 μl reaction mixture containing 25 mM-Tris/HCl, pH 8.5, 10 mM-$MgCl_2$, 10 mM-dithiothreitol, 2–4 nmol of [^{32}P]ATP and 2–4 units of T_4 polynucleotide kinase. After incubation for 30–60 min at 37°C, the reaction mixture was brought to a final concentration of 25 mM-EDTA, 8 M-urea, 0.025% Bromphenol Blue and 0.05% Xylene Cyanol and was heated to 50°C for 2 min prior to loading on a preparative polyacrylamide gel.

Preparative two-dimensional polyacrylamide-gel electrophoresis of 5′-^{32}P labelled rabbit α- and β-globin mRNA

5′-Labelled rabbit α- and β-globin mRNAs were electrophoresed in the first dimension on a 3.5% polyacrylamide gel containing 7 M-urea [24, 28], at room temperature and 600–800 V until the Xylene Cyanol marker dye had migrated approx. 37–40 cm. The partially resolved α- and β-globin mRNA bands were located by autoradiography and a 1 cm x 4 cm strip containing the bands was cut out and transferred to a second set of gel plates oriented at 90° to the direction of electrophoresis. A 7%-polyacrylamide/7 M-urea gel in the same buffer system was poured around the first-dimension gel slice and allowed to polymerize for 10–15 h. Following cooling of the gel to 4–5°C electrophoresis was performed for 24 h at 5 W. Radiolabelled spots corresponding to α- and β-globin mRNA were located by autoradiography and excised from the gel. Elution of the labelled globin mRNA was performed in a volume of 1.5 ml containing 0.3 M ammonium acetate, pH 5.6, and 1 mM-EDTA for 12 h at 37°C. The elution solution was isolated by centrifugation of the gel buffer suspension through silanized glass wool in a 3 ml plastic syringe. To this solution was added 40 μg of carrier mixed tRNAs and the total RNA was isolated by addition of 3 vol. of 95% (v/v) ethanol followed by chilling to −80°C and centrifugation at 22000 *g* for 30 min. The pellet was redissolved in 100 μl of 0.3 M-ammonium acetate/1 mM-EDTA, placed in a 1.5 ml polypropylene tube with 300 μl of 95% ethanol and stored at −20°C for use.

Deadenylylation of rabbit α- and β-globin mRNAs

Ribonuclease H directed by one of several oligodeoxyribonucleotides was used to deadenylylate selectively mixed rabbit α- and β-globin mRNA [22, 41]. Globin mRNA (100–300 pmol) was dissolved in 20 μl of distilled deionized sterile water. To this solution was added 3000 pmol of oligodeoxyribonucleotide in 10 μl of water. It was heated to 75°C for 2 min. Sufficient 1 M-KCl was then added to give a final potassium concentration of 80 mM. Following an additional 1 min of incubation at 75°C the solution was cooled in a thermal reservoir of sufficient capacity to bring the sample to 37°C in approx. 30 min. To this mixture was added one-tenth vol. of RNAase H-buffer 2 [22] resulting in a final concentration of 50 mM-Tris/HCl, pH 8.3, 25 mM-$MgCl_2$, 0.08 M-KCl, 0.5 mg of bovine serum albumin/ml, and 0.001 M-dithiothreitol. Following further incubation of the mixture at 25°C for 10 min, 1–1.5 units of calf thymus RNAase H was added and the reaction was allowed to proceed for 90 min at 33°C for the $(dT)_8$-dG-dC and $(dT)_{11-17}$ oligodeoxynucleotides. Reactions directed by $(dT)_4$-dG were conducted at 15°C for 20 h with 8–10 units of RNAase H enzyme.

Isolation of deadenylylated mRNA

Purification of mRNA and removal of oligomer was accomplished by passage of the reaction mixture over a 1.5 ml Sephadex G-75 column equilibrated in 0.3 M-sodium acetate, pH 5.6. Fractions corresponding to the void volume were pooled and the RNA was precipitated with 3 vol. of ethanol.

3′-End labelling of deadenylylated globin mRNA with 5′-[^{32}P]pCp

Deadenylylated globin mRNA was labelled with 5′-[^{32}P]pCp using T_4-induced RNA ligase. Deadenylylated globin mRNA (100 pmol) was dissolved in 20 μl of buffer containing 10% dimethyl sulphoxide, 50 mM-Hepes, pH 7.5, 10 mM-$MgCl_2$, 5 mM-dithiothreitol, 5 μM-ATP, approx. 100–200 pmol of 5′-[^{32}P]pCp and 400 units of T_4 RNA ligase/ml. The reaction was allowed to proceed for 24–48 h at 4°C and incorporation was monitored by polyethyleneimine–cellulose chromatography in 0.75 M-phosphate buffer, pH 6.0, 3′-End-labelled α- and β-globin mRNA were purified as described for the 5′-labelled mRNA above.

Determination of 3′- and 5′-end heterogeneity

The terminal heterogeneity of 5′- and 3′-end-labelled RNA was determined by total digestion of labelled RNA with T_1 ribonuclease followed by polyacrylamide-gel electrophoresis and of the products and detection of labelled bands by autoradiography. Reactions containing 10 mM-Tris/HCl, pH 8.0, 1 mM-EDTA, 3000–5000 c.p.m. (Cerenkov) of 5′- or 3′-end-labelled mRNA and 0.01 units of T_1 ribonuclease in a volume of 10 μl were incubated for 60 min at 55°C. To the reaction was added an equal volume of sample buffer giving a final concentration of 8 M-urea, 0.025% Bromphenol Blue and 0.025% Xylene Cyanol. Electrophoresis was performed on a 20% polyacrylamide gel (0.15 cm × 20 cm × 40 cm) in 1 × Tris/borate/EDTA buffer at 600–800 V until the Bromphenol Blue had migrated 50% of the length of the gel, followed by detection of digestion products by autoradiography.

Partial digestion of end-labelled α- and β-globin mRNA

Structural analysis of 5′-end-labelled α- and β-globin mRNA using the structure-specific enzymes S_1 nuclease and cobra venom ribonuclease was performed essentially as described previously [27, 28], except that RNA samples were occasionally heated to 60°C in the absence of salts and slowly cooled in the presence of salts to 37°C.

Computer aided analysis of mRNA structural data

Data was recorded and hypothetical RNA models were produced by several computer assisted methods as described in detail elsewhere [37].

Results and Discussion

Preparation and purification of 5′-labelled α- and β-globin mRNA

The methods described provide highly labelled α- and β-globin mRNA with little degradation. Due to the heterogeneity in the lengths of α- and β-globin mRNA molecules, polyacrylamide-gel electrophoresis in one dimension does not achieve adequate separation of the mRNA species present. By taking advantage of a change in the relative mobility of the α- and β-globin mRNA molecules (α-globin mRNA migrates faster in the first dimension, β-globin mRNA faster in the second), an adequate separation of labelled mRNA species is achieved, as seen in Fig. 1. T_1 RNAase digestion of α- and β-globin mRNA shows that the prominent digestion products correspond to the nine-nucleotide 5′-α-globin mRNA T_1 fragment and the seven-nucleotide-long 5′-β-globin T_1 fragment, respectively (data not shown).

Deadenylylation of rabbit α- and β-globin mRNAs

The presence of deadenylylated globin mRNAs is immediately evident in the homogeneity of labelled bands in the first and second dimensions of preparative gel electrophoresis, as seen in Fig. 2. Direct analysis of the deadenylylation reaction

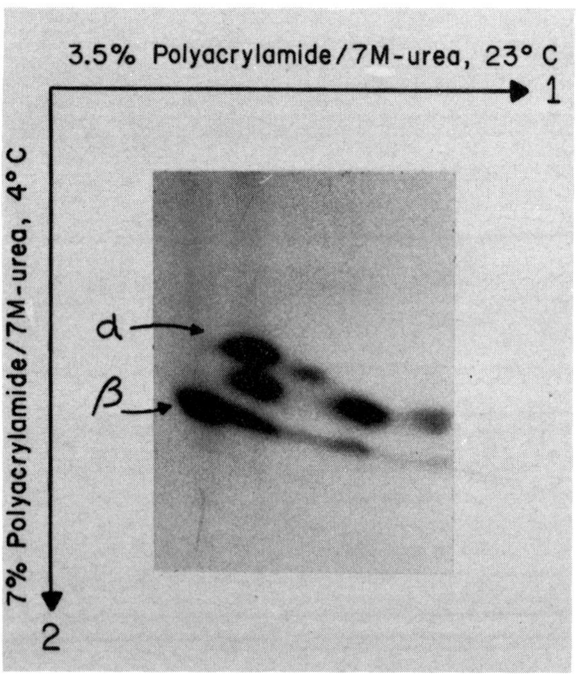

Fig. 1. *Two-dimensional gel electrophoresis of 5′-labelled rabbit globin mRNAs*

Electrophoretic separation of 5′-end labelled rabbit α- and β-globin mRNA was accomplished using a 7% polyacrylamide gel 7 M in urea and run at 5 W constant power at 4°C in the second dimension. (1) Direction of first dimension polyacrylamide-gel electrophoresis in a 3.5% polyacrylamide gel 7 M in urea. (2) Direction of electrophoresis in the second dimension. All gels were run in buffer consisting of 50 mM-Tris/borate, pH 8.3, 1 mM-EDTA.

is accomplished by a total T_1 digestion pattern of 3′-labelled mRNA molecules. Fig. 2 shows the oligo(A) distribution at the 3′-termini of α- and β-globin mRNA molecules digested with RNAase H directed by $(dT)_{11-17}$ and $(dT)_8$-dG-dC. It is evident that the specificity of the cleavage is enhanced when the smaller oligodeoxynucleotide is used. It is also evident that a limiting cleavage pattern is obtained in which several sites are present and serve as acceptors for RNA ligase labelling with pCp. Since the size of the T_1 released oligonucleotides is known the limiting site sizes obtained with $(dT)_{11-17}$ and $(dT)_8$-dG-dC can be determined, as follows:

$(dT)_8$-dG-dC:

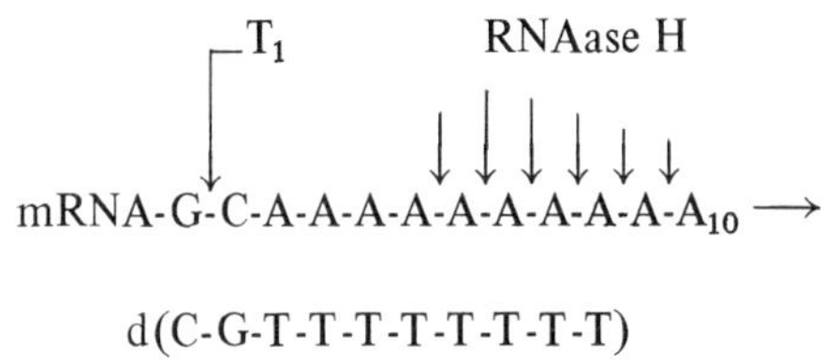

$(dT)_{11-17}$:

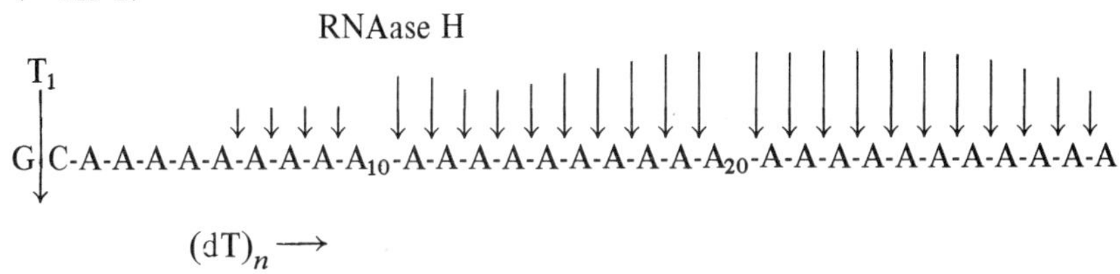

These data indicate that heterogeneity in RNAase H catalysed cleavage of the poly(A) sequence is due to the size of the oligonucleotide used to direct the cleavage. The reason for this is not entirely clear, since the oligonucleotides may be expected to bind anywhere along the length of the adenylate tract and may participate in site–site exclusion caused by adjacent molecules of oligodeoxyribonucleotides overlapping to some extent. It is interesting to note that in neither case do cleavages occur closer than five nucleotides from the T_1 nuclease cleavage point. This may be due to failure of RNAase H to cleave near misaligned DNA–RNA junctions or may reflect the minimum site size requirements of the enzyme, or both. It should be noted that no detectable random fragmentation of α- or β-globin mRNA molecules occurred under these conditions.

RNAase H cleavage of globin mRNA molecules directed by $(dT)_4$-dG

In order to reduce further the 3′-end heterogeneity a smaller oligonucleotide primer, $(dT)_4$-dG, was used. Using the same conditions as for the $(dT)_8$-dG-dC or $(dT)_{11-17}$ primed reactions, only slight hydrolysis of the mRNA molecules occurred with $(dT)_4$-dG (as primer, as seen in Fig. 3A). This situation was not improved with increased salt or decreased temperature. Longer incubation at increased enzyme concentration and at lowered temperature resulted in the digestion pattern observed in Fig. 3B. T_1 digestion of the bands in Fig. 3B following end labelling with pCp

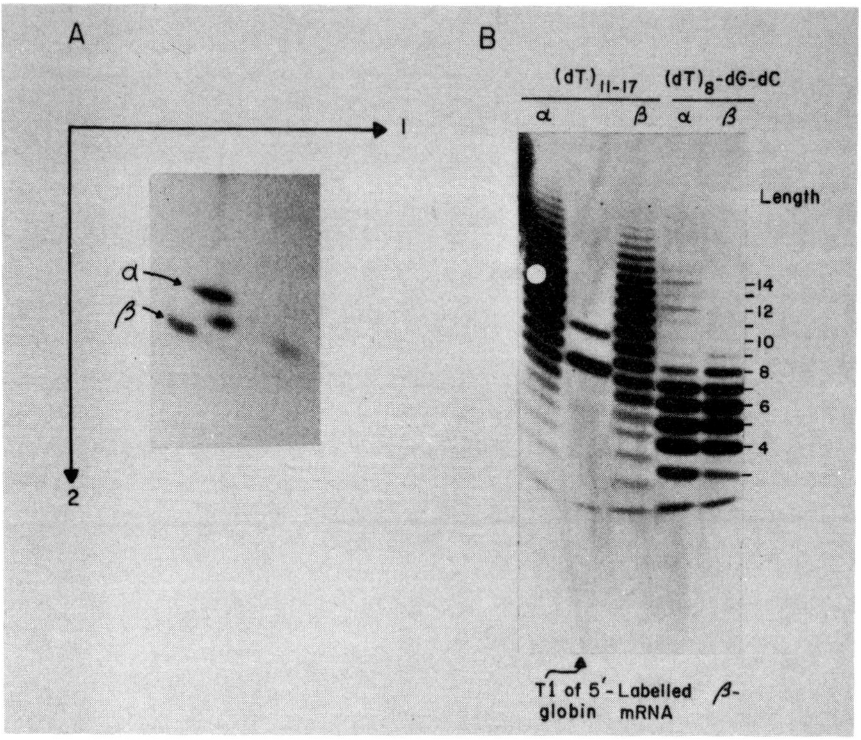

Fig. 2. *Deadenylylation of rabbit globin mRNAs*

Rabbit globin mRNAs were deadenylylated by the action of calf thymus ribonuclease H directed to the mRNA–poly(A) junction using the primers $(dT)_{11-17}$ and $(dT)_8$-dG-dC. Following hybridization and deadenylylation the purified RNA was labelled at its 3′-terminus with T_4RNA ligase and 5′-[^{32}P]pCp as described above. A. Two-dimensional gel electrophoretic purification of 3′-end labelled α- and β-globin mRNAs was performed as described in Methods and in Fig. 1. B. End-analysis of 3′-end-labelled products was achieved by a complete T_1 RNAase digestion (see Methods) followed by electrophoresis of the digestion products on a 20% polyacrylamide/ 8.3 M-urea gel. The size distribution of labelled T_1 oligonucleotides reveals the distance and frequency of cleavages made by RNAase H relative to the terminal G residue in either the α- or β-globin mRNA.

indicated a considerable decrease in heterogeneity such that most bands from the fragmented mRNA profile yield predominantly one or two bands on total T_1 digestion (results not shown). This method provides, therefore, starting material for sequencing and structure mapping polyadenylated mRNA molecules from the 3′-end inward. Work is now in progress to determine the identity of the sites of RNAase H hydrolysis directed by $(dT)_4$dG.

Structural analysis of 5′-end labelled α- and β-globin mRNAs

Figs. 4 and 5 show the results of partial hydrolysis of the 5′-end labelled α- and β-globin mRNA 5′-proximal regions by the single-strand specific S_1 nuclease and the base-pair specific cobra venom ribonuclease (CVR) [28]. Of particular interest is the region surrounding the α- and β-globin mRNA initiator AUG sequences. It can be seen

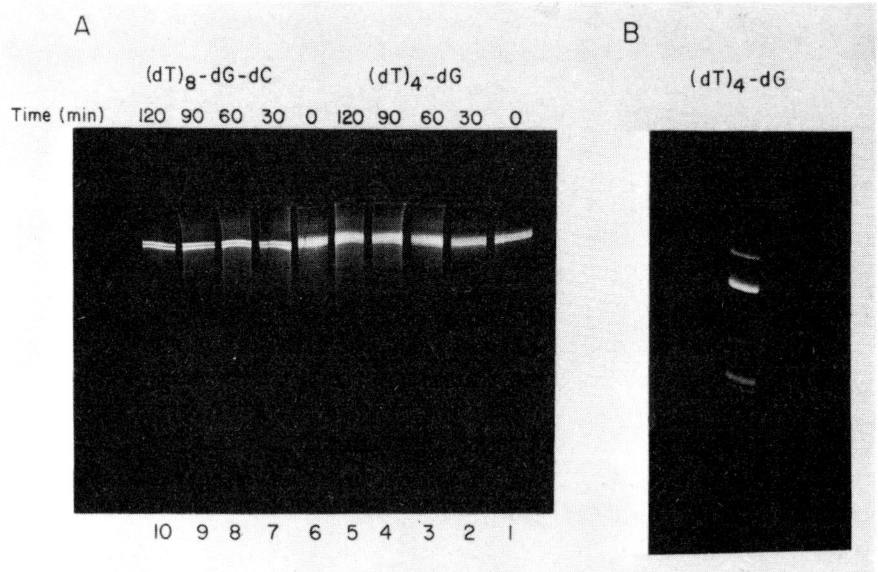

Fig. 3. *Time course of deadenylylation of rabbit globin mRNAs using RNAase H directed by $(dT)_4$-dG*

A. Rabbit globin mRNA (0.5–1 μg) was hybridized to an excess of $(dT)_8$-dG-dC or to $(dT)_4$-dG and digested with RNAase H for 17 h at 15°C as described in Methods. The production of fragments was followed as a function of time by electrophoresis of aliquots of the reaction on a 6% polyacrylamide/8 M-urea gel for 6 h at 600 V. The progress of the reaction, in the case of $(dT)_8$-dG-dC, is seen by the loss in heterogeneity due to removal of poly(A) sequences. B. Discreet α- and β-globin mRNA fragments were generated by $(dT)_4$-dG-directed cleavage of rabbit α- and β-globin mRNAs with RNAase H. These fragments are resolved by electrophoresis on a 3.5% polyacrylamide/8 M-urea gel as described in Methods. The first dimension of the usual two-dimensional electrophoretic technique used for polyadenylated mRNAs is sufficient to prepare homogeneous RNAase H fragments for further characterization.

that the β initiator AUG lies in an S_1-susceptible region of the molecule and is symmetrically flanked by CVR-sensitive areas, indicating that this AUG may be exposed in a hairpin or bulge structure in the β-globin mRNA molecule. The corresponding region of the α-globin mRNA molecule shows that the AUG is also exposed but not to such an extent as in the case of the β-globin mRNA, since S_1 hits here only lightly. Again this AUG is flanked by CVR-sensitive regions of the β-globin mRNA molecule. Gels such as this have been used to map >90% of the β-globin mRNA and approx. 60% of the α-globin mRNA and the following trends were observed:

1. Both molecules have regions of strong preference for S_1 nuclease and other regions which are susceptible to CVR.
2. Some regions of both α- and β-globin molecules appear to be masked from both enzymes, possibly indicating a globular or buried conformation.
3. So-called control sites seem to lie in or very near single-stranded exposed regions as determined by S_1 sensitivity and relative CVR nuclease insensitivity. These regions include the α- and β-globin mRNA splice points for the first intervening sequence, the β-globin mRNA splice junction for the second inter-

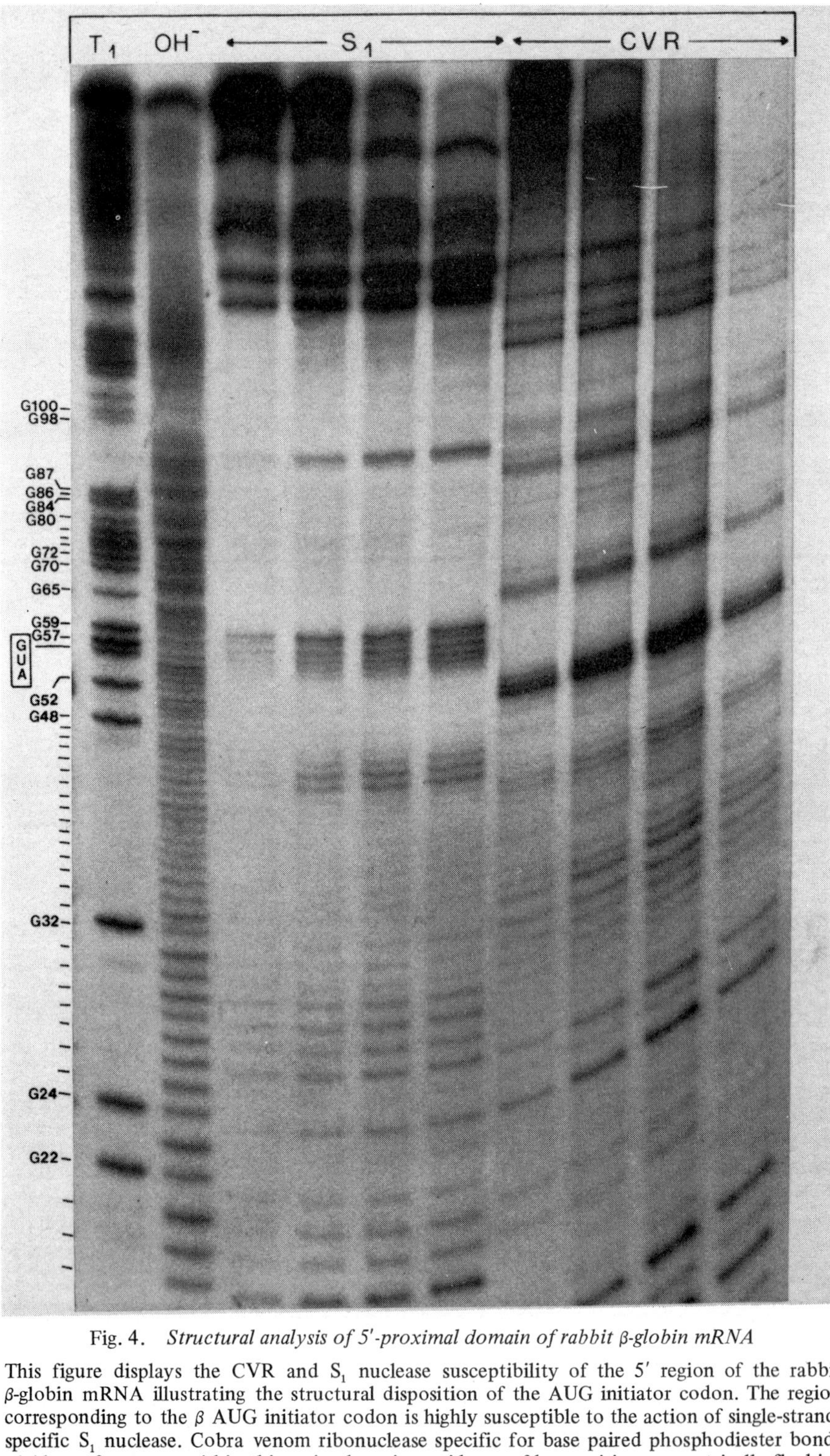

Fig. 4. *Structural analysis of 5′-proximal domain of rabbit β-globin mRNA*

This figure displays the CVR and S_1 nuclease susceptibility of the 5′ region of the rabbit β-globin mRNA illustrating the structural disposition of the AUG initiator codon. The region corresponding to the β AUG initiator codon is highly susceptible to the action of single-strand-specific S_1 nuclease. Cobra venom ribonuclease specific for base paired phosphodiester bonds yields no fragments within this region but gives evidence of base pairing symmetrically flanking the AUG initiation codon. The apparent increased accessibility of the AUG codon may be due to its location in the loop of a hairspin or bulge structure in the mRNA.

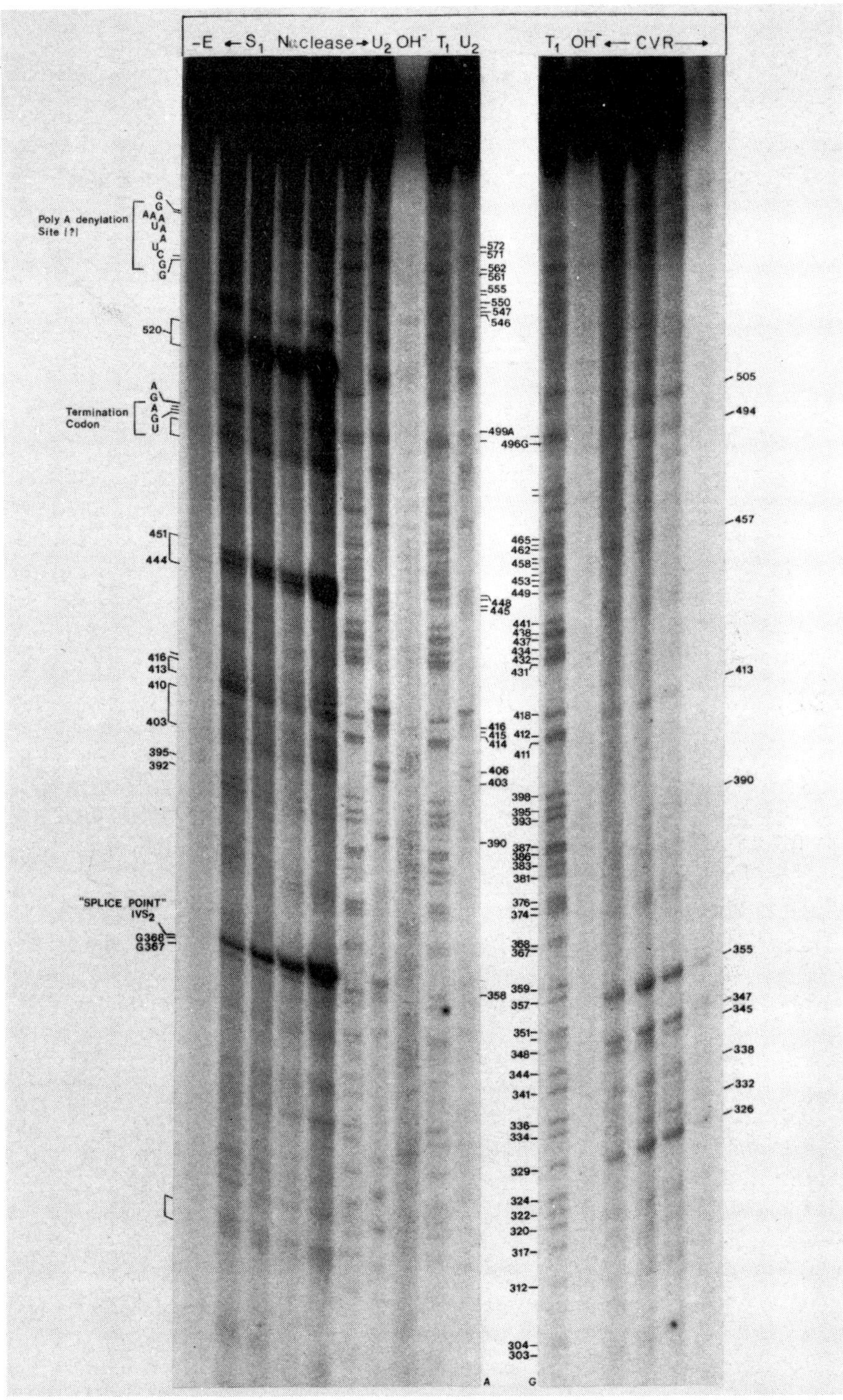

Fig. 5. *Structural analysis of the 3′-proximal domain of rabbit β-globin mRNA*

This figure displays the CVR and S_1 nuclease susceptibility of the 3′ region of 5′-labelled rabbit β-globin mRNA. The data indicates that, in addition to the AUG initiation codon (Fig. 4), the UGA termination codon, intervening sequence splice point 2 and especially the polyadenylation recognition sequence AAUAA are located in regions of high S_1 nuclease susceptibility near regions of helical structure.

vening sequence, the α and β initiator AUG codons, the β terminator codon, and the presumptive poly(A) addition signal A-A-U-A-A-A in the β-globin mRNA.

It is tempting to suggest that presumptive regulatory sequences need to maintain exposed single-stranded configurations for their various functions in mRNA expression and processing. Indeed, recent models for the interaction of small nuclear RNAs with splice junctions in mRNA precursors [42] would suggest that the regions near the splice junctions are single stranded before and most likely after splicing has taken place.

Secondary structure models

Several computer assisted techniques, mentioned earlier [34–37], exist for predicting secondary structure of RNA from primary sequence, enzymic cleavage and thermodynamic data. Figs. 6, 8 and 9 show models of rabbit β-globin mRNA obtained by three different computer methods. The structure in Fig. 6 was generated by using a computer graphics approach on the PROPHET computer system [37]. No thermodynamic energy calculations were used, and the model was constructed to optimize agreement with nuclease susceptibility data. The structure prediction in Fig. 8 was obtained by using the method of Pavlakis *et al.* [24], which is a modification of previously described thermodynamic calculations [34, 35]. The prediction was made by generating a minimum free-energy structure for 200 base sequences, starting at the 5′-end and moving in 50-base steps to the 3′-end of the molecule. No nuclease digestion data was used as input in generating this structure, the S1 data having been added later. The structure in Fig. 9 was obtained by using the RNA2 program of Zuker [36, 37] which is also based on thermodynamic calculations. Nuclease data were added to the model after the minimum free-energy structure was obtained. The Zuker RNA2 approach is designed to feature short-range hairpin structures [37]. In general, the nuclease susceptibility data are in good agreement with the computer prediction, but not as well as in Fig. 6. It is important to note that surprisingly dissimilar structures were generated by the three different computer methods.

Ribosome transit and secondary structure

Experiments to determine the size distribution of nascent chain oligopeptides in a reticulocyte translation system *in vitro* were performed by C.P.H.V. & A.J. Morris (unpublished observations and [11]). An example of the results of such a study is shown in Fig. 7 for rabbit β-globin-mRNA-programmed protein synthesis. These data demonstrate that the oligopeptide β-globin nascent chain size profile is non-uniform. Such a result is likely to be generated by ribosome pausing, which may reflect mRNA structure. Figs. 8 and 9 show correlations of secondary structure models and the nascent chain data. It is indeed surprising that the presumptive ribosome pause sites in Figs. 8 and 9 occur at the 5′-side of stable local hairpin structures. This correlation suggests that ribosome pausing due to secondary structure within mRNA coding regions may be a subtle protein synthesis regulatory feature.

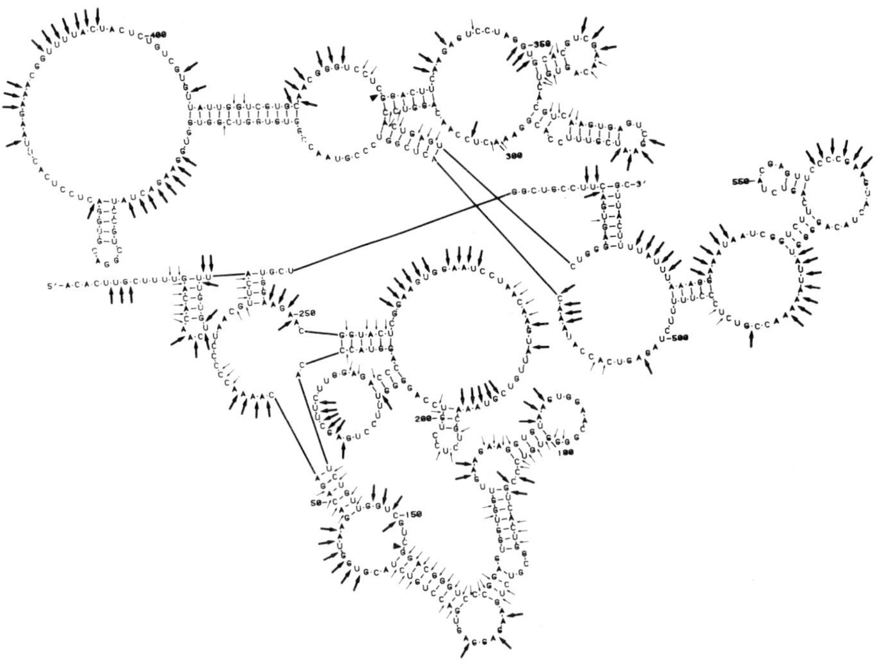

Fig. 6. *Proposed secondary structure for rabbit β-globin mRNA generated during an interactive PROPHET session*

The six selectively displayed components of the molecule as shown by PROPHET are connected by hand-drawn lines. Arrows that indicate nuclease digestion sites were also added. Thick arrows indicate S_1 nuclease cleavage sites, while thin arrows indicate cobra venom ribonuclease digestion sites. The two triangles indicate the positions where two intervening sequences are excised in processing the mRNA.

Conclusions and Future Directions

Enzymic digestions *in vitro* of rabbit α- and β-globin mRNAs have revealed the presence of extensive secondary structure in these molecules. These studies are points of departure in the study of mRNA structure and its influence on functions such as translational initiation, elongation, and termination.

The overall pattern of secondary structures, the exposure of control sequences and functional regions such as the initiator codons, intervening splice point junctions and polyadenylation sequences have mechanistic relevance and provide a basis for comparison of the dispositions of these regions in different mRNA molecules. Certain problems must be addressed in the study of mRNA structure by these methods. These include the assumption that structure is not lost during isolation procedures which subject the RNA to denaturation. Though evidence exists to support this contention the only satisfactory answer must await the development of methods which probe the structure of mRNA molecules in their native state. Methods have been developed to probe RNA in ribonucleoprotein complexes. These involve the use of small chemical probes [32, 33] which can penetrate ribonucleoprotein particles such as ribosomes and ribosomal subunits and react with the

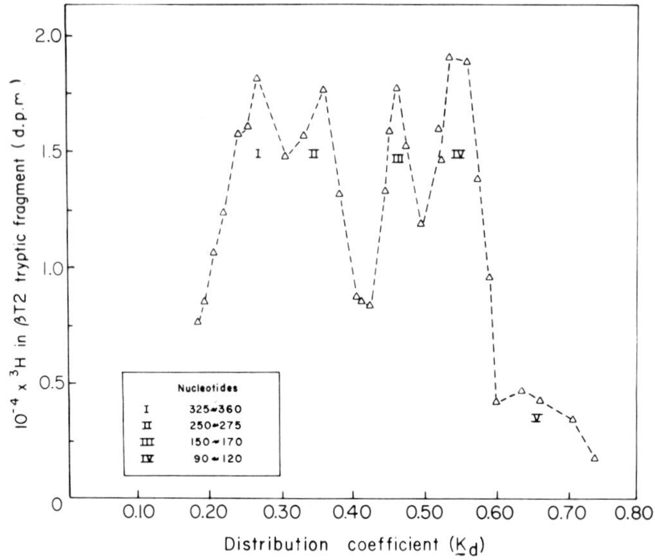

Fig. 7. *Size distribution profile of rabbit β-globin nascent polypeptides*

L-[^{3}H]Tryptophan (Amersham-Searle) labelled nascent globin polypeptides were chromatographed on a Bio-Gel A-0.5 m agarose column (1.5 cm × 90 cm) developed with 6 M-guanidinium chloride/100 mM-2-mercaptoethanol, pH 6.5. The distribution coefficient, K_d, is defined as:

$$K_d = \frac{(V_e - V_0)}{(V_i - V_0)}$$

where V_e is the mass of solvent which corresponds to the volume in which the solute elutes from the column, and V_0 and V_i are the void volume and included volume expressed as mass of solvent, respectively. The column used in this study was found to separate α- and β-globin cyangen bromide and tryptic peptides yielding the following relationship between the distribution coefficient (K_d) and weight-average molecular weight (M_w):

$$K_d^{0.333} = (-2.01 \times 10^{-3})\, M_w^{0.555} + 1.03$$

which is in agreement with the derivation of K_d in [11]. Nascent α- and β-globin oligopeptides were labelled with L-[^{3}H]tryptophan in a reticulocyte lysate programmed only with endogenous α- and β-globin mRNA for 10 min at 37°C. This is two to three times longer than needed to effect steady state labelling of the polysomes [9, 11]. Translating polysomes were quenched by rapid mixing with an iso-osmotic solution containing 0.21 mM-sparsomycin and 0.06 mM-cyclohexamide. The nascent polypeptide size distribution obtained in this way is independent of the type of elongation inhibitor used, and is identical to that obtained in the whole reticulocyte or in a wheat germ cell free protein synthesis system programmed with poly(A)$^+$-reticulocyte mRNA. The β-globin nascent polypeptide size distribution was determined by quantification of the abundance of the [^{3}H]βT_2 tryptic peptide by high pressure liquid chromatography on Pellionex strong cation exchange resin equilibrated in 50 mM-pyridinium acetate (pH 4.0). Regions I–V indicate the nascent oligopeptide peaks in the nascent β-globin peptide population. The insert indicates the positions on the β-globin mRNA that correspond to peaks I–IV, where the A of the initiator AUG is position 1.

RNA in a structure-specific manner. Unfortunately these methods, derived from chemical sequencing technology [32], rely on the availability of 3′-labelled RNA. This possibility is precluded in the case of most eukaryotic mRNAs due to the presence of a polyadenylic acid tract of heterogenous length. We illustrate, above,

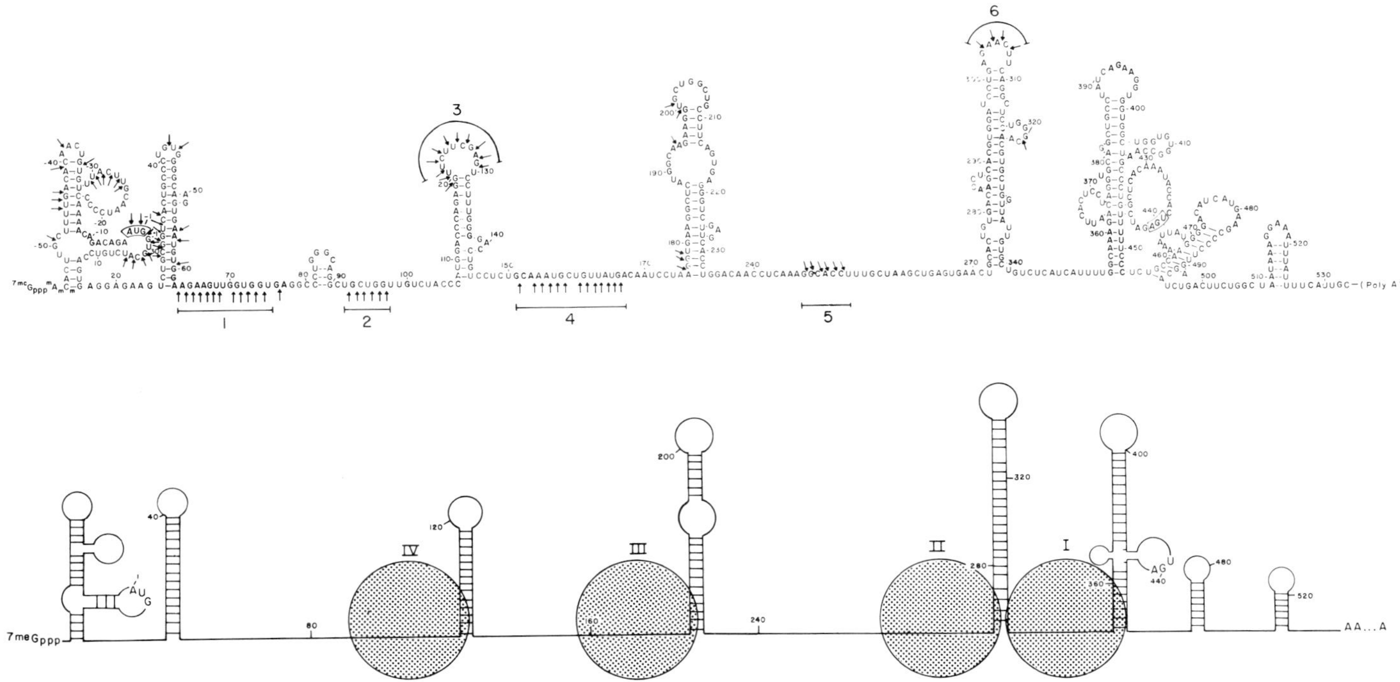

Fig. 8. *Secondary structure model of rabbit β-globin mRNA (upper) and a schematic diagram (lower) of that structure indicating the apparent positions of ribosome accumulation as determined from the data in Fig. 7*

Arrows indicate S_1 and/or T_1 nuclease susceptible regions of 5′-^{32}P-end labelled β-globin mRNA. Six such regions (labelled 1–6) are indicated in the upper figure. Shaded circular areas in the lower diagram are numbered I–IV and correspond to persistent ribosome positions that generate the peaks I–IV in Fig. 7. The positions of the ribosomes were calculated from the apparent molecular weights of the predominant nascent oligopeptide size classes shown in Fig. 7.

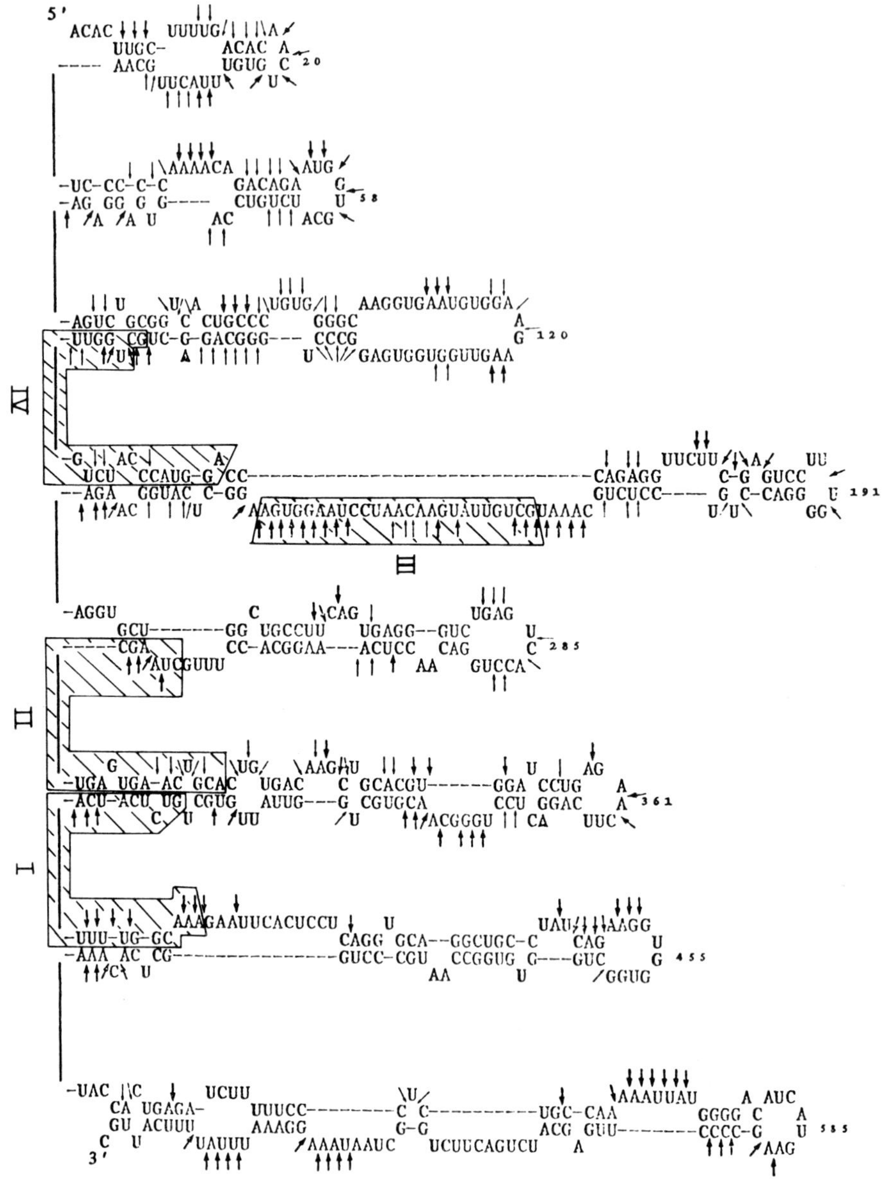

Fig. 9. *Secondary structure for rabbit β-globin mRNA proposed by the RNA2 program*

Bonds susceptible to cleavage by cobra venom ribonuclease are indicated by thin arrows, while those that are susceptible to cleavage by S_1 nuclease are indicated by thick arrows. The two triangles indicate the positions where the two intervening sequences are excised in processing the mRNA. Cross-hatched areas I–IV correspond to the same persistent ribosome positions as shown in Fig. 8.

the use of RNAase H directed to the mRNA poly(A)-mRNA junction with the primer $(dT)_4$-dG to produce mRNA which is specifically deadenylylated. This approach provides a general method for production of mRNA which can be 3′-labelled and sequenced enzymically or chemically, thus providing a doubling of the length range for mRNA sequence analysis. In addition, this methodology allows sequencing and structure mapping methodologies with chemical probes to be brought to bear on problems of mRNA structure in ribonucleoprotein complexes *in situ*. This is possible since the base- or structure-specific modification reaction of the RNA is separate from the cleavage reaction. Therefore, RNA may be modified *in vivo* with structure-specific probes such as dimethyl sulphate or diethyl pyrocarbonate, extracted, end-labelled following deadenylylation, and then cleaved at the sites of modification by treatment with aniline and sodium borohydride. An alternative and complementary scheme would employ 3′-labelled mRNA obtained following deadenylylation to probe the structure of mRNA involved in 40 S preinitiation complexes, 80 S initiation complexes and polysomes. Since 3′-labelled mRNA possesses an intact cap structure, unlike 5′-labelled mRNA, it also provides an opportunity to measure the rate at which labelled mRNA progresses through the stages of preinitiation and initiation complex formation and into polysomes, relative to a standard competing mRNA under mRNA limiting conditions. This will provide a functional correlation which will be useful in relating the structural disposition of the initiator AUG codon and surrounding sequences to a functional parameter such as the relative efficiency of incorporation of labelled mRNA into preinitiation and initiation complexes.

The proposal by Kozak [43] that initiation of protein synthesis in eukaryotic cells proceeds by a scanning mechanism, whereby the 40 S ribosomal subunit attaches at the 5′-end and migrates to the correct AUG initiator codon, postulates that a specific, universal, initiator sequence exists that includes the AUG. According to the scanning model, secondary structure in the 5′-noncoding region would be inhibitory to initiation, since it could retard the movement of the 40 S subunit. Recently, Sonenberg [44, 45] has suggested that cap binding proteins required for initiation may be acting to unwind the 5′-end structure of capped eukaryotic mRNAs, thus making it possible for 40 S subunits to 'scan' to the AUG. Thus, 5′-end structure could act to modulate protein synthetic rates at the initiation step such that mRNAs having less stable 5′-end structure would initiate with a faster rate. Structural studies with cap binding protein–mRNA complexes of the sort mentioned earlier may yield some insight on these issues.

This work was supported by grants from N.I.H. (GM 22280) and N.S.F. (PCM 8004620).

References

1 Lodish, H.F. (1970) *J. Mol. Biol.* **50**, 689–702
2 Steitz, J.A. (1969) *Nature (London)* **224**, 957–967
3 Fiers, W., Contreras, R., Duerinck, F., Haegeman, G., Isertant, D., Merregaert, J., Min Jou, W., Molemans, F., Raeymaekers, A., Van den Berghe, A., Volckaert, G. & Ysebaert, M. (1976) *Nature (London)* **260**, 500–507
4 Johnston, H.M., Barnes, W.M., Chumley, F.G., Bossi, L. & Roth, J.R. (1980) *Proc. Natl. Acad. Sci. U.S.A.* **77**, 508–512

5 Oxender, D.L., Zurawski, G. & Yanofsky, C. (1979) *Proc. Natl. Acad. Sci. U.S.A.* **76,** 5524–5528
6 Keller, E.B. & Calvo, J.B. (1979) *Proc. Natl. Acad. Sci. U.S.A.* **76,** 6186–6190
7 Isertant, D. & Fiers, W. (1980) *Gene* **9,** 1–12
8 Shine, J. & Dalgarno, L. (1974) *Proc. Natl. Acad. Sci. U.S.A.* **71,** 1342–1346
9 Protzel, A. & Morris, A.J. (1974) *J. Biol. Chem.* **249,** 4594–4600
10 Lizardi, P.M., Mahdavi, V., Shields, D. & Candelas, G. (1979) *Proc. Natl. Acad. Sci. U.S.A.* **76,** 6211–6215
11 Vary, C.P.H. (1981) Ph.D. Dissertation, Michigan State University
12 Alquist, P., Dasgupta, R., Shih, D.S., Zimmern, D. & Kaesburg, P. (1979) *Nature (London)* **281,** 277–282
13 Schroeder, H.W., Liarkos, C.D., Gupta, R.C., Randerath, K. & O'Malley, B.W. (1979) *Biochemistry* **18,** 5798–5808
14 Pinck, M., Fritsch, C., Ravelonandro, M., Thivent, C. & Pinck, L. (1981) *Nucleic Acids Res.* **9,** 1087–1100
15 Legon, S., Model, P. & Robertson, H.D. (1977) *Proc. Natl. Acad. Sci. U.S.A.* **74,** 2692–2696
16 Kozak, M. (1980) *Cell* **19,** 79–90
17 von Heijne, G., Nilsson, L. & Blomberg, C. (1978) *Eur. J. Biochem.* **92,** 397–402
18 Bergmann, J.E. & Lodish, H.F. (1979) *J. Biol. Chem.* **254,** 459–468
19 Holder, J.W. & Lingrel, J.B. (1975) *Biochemistry* **14,** 4209–4215
20 Van, N.T., Monahan, J.J., Woo, S.L.C., Means, A.R. & O'Malley, B.W. (1977) *Biochemistry* **16,** 4090–4100
21 Kallenbach, N.R., Brentani, M.M. & Brentani, R.R. (1979) *Biopolymers* **18,** 1515–1531
22 Vournakis, J.N., Flashner, M.S., Katopas, M., Kitos, G.A., Vamuakopoulos, N.C., Sell, M.S. & Wurst, R.A. (1976) *Prog. Nucleic Acid Res. Mol. Biol.* **19,** 233–252
23 Flashner, M. & Vournakis, J.N. (1977) *Nucleic Acids Res.* **4,** 2307–2319
24 Lockard, R. E., Vamvakopoulos, N., Rieser, K., RajBhandary, U.L. & Vournakis, J.N. (1980) *Cell* **19,** 91–102
25 Lodish, H.F. & Jacobsen, M. (1972) *J. Biol. Chem.* **247,** 3622–3629
26 Rosa, M.D. (1981) *J. Mol. Biol.* **147,** 55–71
27 Wurst, R., Vournakis, J. & Maxam, A. (1978) *Biochemistry* **17,** 4493–4499
28 Vournakis, J. H., Celentano, J., Finn, M., Lockard, R.E., Mitra, T., Paulakis, G., Troutt, A., van den Berg, M. & Wurst, R. (1981) in *Gene Amplification and Analysis. Vol. 2: Analysis of Nucleic Acids by Enzymatic Methods* (Chirikjian, J. & Papas, T., eds.), pp. 267–298, Elsevier/North-Holland, Amsterdam
29 Dunn, J.J., Buzash-Pollert, E. & Studier, F.W. (1978) *Proc. Natl. Acad. Sci. U.S.A.* **75,** 2741–2745
30 Roberts, T.M., Kacich, R. & Ptashne, M. (1979) *Proc. Natl. Acad. Sci. U.S.A.* **76,** 760–764
31 Maxam, A. & Gilbert, W.A. (1977) *Proc. Natl. Acad. Sci. U.S.A.* **74,** 560–564
32 Peattie, D.A. & Gilbert, W.A. (1980) *Proc. Natl. Acad. Sci. U.S.A.* **77,** 4679–4682
33 Peattie, D.A. & Herr, W. (1981) *Proc. Natl. Acad. Sci. U.S.A.* **78,** 2273–2277
34 Pipas, J.M. & McMahon, J.E. (1975) *Proc. Natl. Acad. Sci. U.S.A.* **72,** 2017–2021
35 Studnicka, G.M., Rahn, G.M., Cummings, I.W. & Salser, W.A. (1978) *Nucleic Acids Res.* **5,** 3365–3387
36 Zuker, M. & Stiegler, P. (1981) *Nucleic Acids Res.* **9,** 133–148
37 Auron, P.E., Rindone, W.F., Vary, C.P.H., Celantano, J. & Vournakis, J. (1982) *Nucleic Acids Res.* **10,** 403–419
38 Noller, H.F. & Woese, C.R. (1981) *Science* **212,** 403–411
39 Lockard, R., Rieser, L. & Vournakis, J.N. (1981) in *Gene Amplification and Analysis. Vol. 2: Analysis of Nucleic Acids by Enzymatic Methods* (Chirikjian, J. & Papas, T., eds.), pp. 229–251, Elsevier/North-Holland, Amsterdam
40 Lockard, R.E., Alzner-Deweerd, B., Heckman, J.E., MacGee, J., Tabor, M.W. & Raj-Bhandary, U.L. (1978) *Nucleic Acids Res.* **5,** 37–52
41 Stavrianopoulos, J.G. & Chargaff, E. (1978) *Proc. Natl. Acad. Sci. U.S.A.* **75,** 4140–4144
42 Lerner, M.R., Boyle, J.A., Mount, S.M., Wolin, S.L. & Steitz, J.A. (1980) *Nature (London)* **283,** 220–224
43 Kozak, M. (1981) *Nucleic Acids Res.* **9,** 5233–5252
44 Sonenberg, N. (1981) *Nucleic Acids Res.* **9,** 1644–1657
45 Sonenberg, N., Guertin, D., Cleveland, D. & Trachsel, H. (1981) *Cell* **27,** 563–572

Biochem. Soc. Symp. **47**, 79–93
Printed in Great Britain

Studies on the Structure and Biogenesis of Yeast Ribosomes

MICHAEL CANNON

Department of Biochemistry, King's College, University of London, Strand, London WC2R 2LS, U.K.

The ultimate aim of research work on ribosomes is to unravel the complex relationship between structure and function in this organelle. Bacterial 70S ribosomes have been studied in great detail and rapid advances have been made towards characterizing the structures of the individual RNA and protein components (for reviews see Nomura *et al.*, 1974; Stöffler & Wittmann, 1977; Chambliss *et al.*, 1980). In contrast, far less is known about the more complex eukaryotic 80S particle although it is clear that there are significant structural differences between it and its prokaryotic counterpart (for review see Wool, 1979). In the typical eukaryote *Saccharomyces* the ribosome has four mature rRNA species and approx. 80 individual ribosomal proteins. Furthermore, the nucleic acids and proteins are generally larger than the corresponding components in 70S particles. The reasons for this increase in complexity are obscure since all ribosomes carry out the same function of synthesizing cellular proteins. As pointed out by Wool (1979), however, the presence of extra components in eukaryotic ribosomes argues for the future discovery of eukaryotic ribosomal functions not as yet described. Thus some of the extra ribosomal proteins may be needed for interaction with receptors in the endoplastic reticulum and others may be involved in regulating mRNA translation. A particularly attractive hypothesis is that there may be a need for extra proteins to control the complicated process of eukaryotic ribosome biogenesis with its associated need for two-way traffic between the cytoplasm and the nucleus (Wool, 1979).

In ribosome research the ideal achievement would be to identify the precise location of each molecule in the particle and to define its exact function(s). Studies on structure-function relationships in ribosomes necessitates the choice of a suitable organism which can be grown easily in large amounts, provides a rich source for preparation of the particles and can be subjected to convenient biochemical and genetic analysis. Yeast fulfils these requirements well. Furthermore, yeast mutants can easily be isolated – an important point since advances in research on prokaryotic ribosomes have relied heavily upon analysis of mutant bacterial strains isolated as either resistant to antibiotics or temperature-sensitive (*ts*) or having defects in ribosome assembly (Stöffler & Wittmann, 1977; Butler *et al.*, 1980; Isono, 1980). Accordingly, information has been gained in bacteria on the possible functions of individual ribosomal proteins within the complex ribosome particle, the genetic organization and control of ribosome assembly and the biosynthesis of its component parts. Detailed knowledge of ribosome topography, function and biosynthesis is vital to allow a better understanding of the regulation of protein synthesis. Furthermore, there are clear implications for the identification of antibiotic receptor sites and the subsequent development of new antibacterial and antieukaryotic agents that have the ribosome as their target site and may have important uses in chemotherapy.

The first section of this brief article will concentrate on describing the isolation and analysis of mutant strains of the yeast *Saccharomyces cerevisiae*. This work has involved studies on both antibiotic-resistant and *ts* mutants. The second section deals with the analysis of ribosomes isolated from a number of yeast strains of the two genera *Saccharomyces* and *Kluyveromyces*. Many of these strains have the intriguing property of being naturally resistant to a large number of antibiotics *in vivo* (Adoutte-Panvier *et al.*, 1980).

Many antibiotic-resistant strains of *Escherichia coli* possess ribosomes with altered proteins (Ozaki *et al.*, 1969) and the genetic loci controlling synthesis of such proteins have, in many cases, been identified (Isono, 1980). With the knowledge of the inhibitory action of a given drug and an identification of its ribosomal target site a possible direct, or at least associated, functional role can be assigned to a particular ribosomal protein. The situation in *S. cerevisiae* is, however, more difficult. Although a wide range of antibiotics are known to be specific inhibitors of 80S ribosome function (Vázquez & Jiménez, 1980), many of these drugs fail to enter intact yeast cells and isolation of mutants resistant to these drugs is thus impossible. Nevertheless, mutants of *S. cerevisiae* have been isolated as resistant to the alkaloid cryptopleurine (Skogerson *et al.*, 1973; Grant *et al.*, 1974; Vázquez, 1979), the antibiotics cycloheximide (Cooper *et al.*, 1967; Vázquez, 1979; Stocklein & Piepersberg, 1980) and anisomycin (Schindler *et al.*, 1974; Jiménez *et al.*, 1975) and the fungal toxin trichodermin (Schindler *et al.*, 1974; Jiménez *et al.*, 1975). All these mutants are resistant to the relevant drug at the ribosome level and, as for *E. coli*, some of the genetic loci involved have been identified (Jiménez *et al.*, 1975; Grant *et al.*, 1976). Resistance to trichodermin is clearly associated with 60S ribosomal subunits and cryptopleurine resistance with the 40S subparticle. However, neither for these two inhibitors nor for anisomycin has a mutant of *S. cerevisiae* been isolated with a ribosomal protein shown to migrate differently on two-dimensional gel electrophoresis relative to wild-type preparations. A breakthrough has, however, been achieved in yeast in the case of cycloheximide resistance. This antibiotic inhibits elongation specifically on 80S ribosomes, probably by blocking the translocation step in protein synthesis, and it exerts its action primarily on the 60S ribosomal subunit. Stocklein & Piepersberg (1980) have recently analysed a spontaneous high-level cycloheximide-resistant mutant of *S. cerevisiae* and have claimed that large subparticles derived from cytoplasmic ribosomes have an alteration in protein L29 that can be detected electrophoretically. This mutant is apparently the first *S. cerevisiae* strain to be described where a defined drug-resistant phenotype is associated with an altered ribosomal protein.

Over the last few years we have studied the modes of action of the 12,13-epoxytrichothecenes, a group of sesquiterpenoid fungal antibiotics that are potent and selective inhibitors of protein synthesis on eukaryotic ribosomes (for references see Carter & Cannon, 1977). The group includes the compounds T-2 toxin, fusarenon-X and trichodermin and we have concluded from our work that all the inhibitors have the ribosomal peptidyl transferase centre as their target site. More recently we have focused our attention on selected spontaneous mutants of *S. cerevisiae* isolated as resistant to trichodermin. These mutants are designated TR-1 (α,*trp-5'*,*his-4*$^-$,*tcm-1*^{r}) and CLP-8 (*a*,*leu-2*$^-$,*tcm-1*^{r}) and are derived respectively from wild-type strains designated Y166 (α,*trp-5*$^-$,*his-4*$^-$) and A224A (*a*,*leu-2*$^-$). As implied above, however, we have found no ribosomal protein in either of these two mutants strains that has an altered electrophoretic mobility relative to the corresponding wild-type protein. Nevertheless, trichodermin resistance does appear to be controlled by a ribosomal protein gene. Thus Fried & Warner (1981) have recently isolated a plasmid which carries a fragment of yeast DNA specifying resistance to trichodermin. They conclude that the relevant gene codes for the 60S ribosomal protein L3. Warner (1982) and his co-workers have now isolated the genes for 16 of the yeast ribosomal proteins including that for the protein conferring resistance to cycloheximide. Clearly their techniques should prove to be powerful tools in structure-function studies on yeast ribosomes.

Upon routine analysis of ribosomes from CLP-8 on sucrose density gradients we observed that cytoplasmic extracts from this strain had an unusual distribution of native ribosomal subunits (Carter *et al.*, 1980). Thus, a large relative excess of 60S subparticles was apparent with little or no material sedimenting at 40S. In contrast, the yeast strain TR-1 had a normal distribution of native ribosomal subunits, although both TR-1 and CLP-8 were equally resistant to inhibition by trichothecene antibiotics as assayed *in vivo* (Carter *et al.*, 1980) and both mutants appear to have the same genetic locus controlling the antibiotic-resistant lesion (Jiménez *et al.*, 1975). CLP-8 was found also to have a further unusual property in that both *in vivo* and *in vitro* it has a reduced ability to generate 40S ribosomal subunits from 80S particles (Carter *et al.*, 1980). We deduce that there is some abnormality in the protein factor(s) normally required for the dissociation process.

In view of their observations on the abnormal distribution of ribosomal subunits in the cytoplasm of CLP-8, Carter & Cannon (1980) considered that the mutant might be defective in some aspect of ribosome biogenesis. This process, in both bacteria and eukaryotes, involves many complex interactions normally designed to produce the various rRNAs and associated proteins in roughly equimolar amounts (Huynh-Van-Tan *et al.*, 1971). These multiple components are assembled to produce pre-ribonucleoprotein (pre-rRNP) particles which then undergo a series of modification steps resulting in the final appearance of mature ribosomes (Warner & Soeiro, 1967; Trapman *et al.*, 1975; Kruiswijk *et al.*, 1978). The process of ribosome biogenesis in eukaryotes necessitates a high level of communication between nucleus and cytoplasm (Udem & Warner, 1972) since ribosomal proteins are synthesized in the latter and ribosome assembly takes place in the former (Hadjiolov & Nikolaev, 1976). The assembled ribosomal subunits have themselves finally to be transferred from nucleus to cytoplasm. In *S. cerevisiae* the genes for rRNA are highly repetitive and tightly linked (Cramer *et al.*, 1976), with approx. 120 of them forming a single tandem array on chromosome XII (Petes, 1979). The repeating unit contains approx. 9000 base pairs and encodes two separate transcription units, one for 5S rRNA and one for 5.8S, 18S and 25S rRNAs (Planta *et al.*, 1980). The last three rRNA types derive from a common precursor RNA (pre-rRNA) molecule that is less than 7000 nucleotides long and has a sedimentation coefficient of 35S (Udem & Warner, 1972; Trapman & Planta, 1975). The *s* values cited in this article for the various pre- and mature rRNA types are from Warner and his co-workers (see Udem & Warner, 1972) although to prevent confusion the reader should note that Planta and his co-workers (see Veldman *et al.*, 1981), using *S. carlsbergensis*, employ a slightly different nomenclature for these components (see later).

The primary pre-rRNA transcript (35S) is processed through a number of steps into the mature rRNA types (Sillevis Smith *et al.*, 1972; Udem & Warner, 1973; Trapman & Planta, 1975). Processing is carried out not on naked RNA but on the pre-rRNP particles. The correct assembly of these particles and the terminal processing of their pre-rRNA components apparently requires the latter to be extensively methylated at highly specific sites (Wejksnora & Haber, 1974). Thus, after transcription is complete approx. 60 methyl groups are added to the 35S pre-rRNA with methylation being mainly confined to ribose units (Retèl *et al.*, 1969). A few later base methylations may occur during the final cytoplasmic processing of the smaller pre-rRNP particle (Retèl *et al.*, 1969; Udem & Warner, 1973). The 35S pre-rRNA never leaves the nucleus and is cleaved rapidly, possibly via a short-lived 32S pre-rRNA intermediate, to form two pre-rRNA species with sedimentation coefficients of 27S and 20S respectively (Udem & Warner, 1972). The 27S pre-rRNA is a nuclear species and is rapidly processed to form both the mature 25S and 5.8S rRNA types characteristic of mature 60S ribosomal subparticles (Udem & Warner, 1972). The production of 5.8S rRNA requires the transient formation of a pre-rRNA species that sediments at 7S (Helser & McLaughlin, 1975; Trapman & Planta, 1975). Maturation of the 20S pre-rRNA species to the mature 18S form characteristic of 40S ribosomal units is exclusively a cytoplasmic event (Udem & Warner, 1973; Trapman & Planta, 1976) with the 20S pre-rRNA being trimmed from its 3′-end (DeJonge *et al.*, 1977). The overall process in *S. cerevisiae* can be summarized as follows:

```
              20S → 18S
             ↗
35S → 32S
             ↘
              27S → 25S
                    ↘
                     7S → 5.8S
```

The remaining mature rRNA species sediments at 5S. It arises from an independent transcription unit (Udem & Warner, 1972) and is transcribed in the leftward direction (Maxam *et al.*, 1977). It is finally incorporated into the mature 60S subparticle Approx. 2000 nucleotides are lost during the processing of 35S pre-rRNA although the methyl groups appear to be conserved (Retèl *et al.*, 1969). It appears that no splicing events take place during the processing (Philippsen *et al.*, 1978).

Veldman *et al.* (1981) have recently and elegantly determined the nucleotide sequences of part of a cloned yeast ribosomal RNA operon extending from the 5.8S rRNA gene downstream into the 5′-terminal region of the 26S (their nomenclature) rRNA gene. They have mapped the pertinent processing sites and have established common sequence features possibly involved in the recognition, by the cleavage enzyme(s), of nucleolytic processing sites in yeast pre-rRNA. From these data they have constructed a model for the processing of yeast pre-rRNA, the essential features of which add some degree of complexity to the Warner processing scheme described above. The main details of the Veldman *et al.* (1981) model are summarized as follows, starting with a primary transcript of 37S pre-rRNA:

$$
\begin{array}{l}
37S \nearrow 18S \rightarrow 17S \\
37S \searrow 29S(a) \searrow 29S(b) \rightarrow 26S \\
29S(b) \searrow 7S \rightarrow 5.8S
\end{array}
$$

The Warner scheme and the Veldman scheme can be compared as follows with the Warner nomenclature first in each case:

35S = 37S   27S = 29S   25S = 26S   20S = 18S   18S = 17S

For the sake of clarity, however, I shall continue to use the Warner nomenclature in interpreting our own results.

It has now been shown conclusively by Carter & Cannon (1980) that CLP-8 is defective in a particular aspect(s) of ribosome biogenesis. The mutant accumulates 20S pre-rRNA in the nucleus (presumably in the form of a pre-rRNP particle) and there is a resultant slowing of the transport of the particle to the cytoplasm. Furthermore, the cytoplasmic maturation of 20S pre-rRNA to mature 18S rRNA is also slowed, although no significant instability in this precursor was detected. CLP-8 is thus a most unusual mutant. Although it is derived from an apparently normal wild-type strain of *S. cerevisiae* (A224A) by what we assume to be a single step spontaneous mutation, it has acquired a multiplicity of lesions. In summary, these include trichodermin resistance, an abnormal distribution of cytoplasmic ribosomal subunits, a defective anti-association factor activity, a slowed transport of 20S pre-rRNA from nucleus to cytoplasm and a slowed cytoplasmic cleavage of this pre-rRNA component into the mature 18S form.

In attempting to consider how these various lesions may be related, we have carried out a series of genetic crosses between selected strains (Y166, A224A, TR-1, CLP-8) of *S. cerevisiae*. Diploids were formed between the appropriate mating types and after sporulation the resulting tetrads were dissected. All the various strains were analysed and the results obtained are summarized in Table 1. As expected, of the four diploids constructed, only that between TR-1 and CLP-8 was trichodermin resistant. Furthermore, the level of resistance observed was indistinguishable, in experiments both *in vivo* and *in vitro*, from that displayed by the two trichodermin-resistant haploid strains. This confirms, not surprisingly, that trichodermin sensitivity is expressed in dominant fashion over trichodermin resistance. Tetrad analysis

Table 1. *Formation and tetrad analysis of diploids from selected genetic crosses*

Diploids were formed between selected strains as indicated and after sporulation four haploid strains were isolated from each tetrad. All strains were checked both *in vivo* and *in vitro* for trichodermin resistance (T^r) and trichodermin sensitivity (T^s). Cell extracts were analysed for possession of an abnormal distribution of cytoplasmic ribosomes (AP). Under tetrad analysis are the numbers of haploid cells per tetrad with the property indicated.

	Diploid analysis			Tetrad analysis		
Genetic cross	T^r	T^s	AP	T^r	T^s	AP
CLP-8/Y166	—	+	—	2	2	1(T^r)
CLP-8/TR-1	+	—	+/—	4	-	2(T^r)
A224A/TR-1	—	+	—	2	2	?
A224A/Y166	—	+	—	-	4	-

of the various diploids was then undertaken. Predictably, two out of the four strains isolated from tetrads formed from the diploid CLP-8/Y166 were trichodermin resistant. Surprisingly, however, only one strain from each tetrad analysed had the abnormal distribution of cytoplasmic ribosomal subunits that is characteristic of CLP-8, and this strain was always trichodermin resistant (Carter *et al.*, 1980). In a corresponding analysis of tetrads from the diploid TR-1/CLP-8 each strain was, as predicted, trichodermin resistant and two out of the four strains had the abnormal distribution of cytoplasmic ribosomal subunits characteristic of CLP-8. As suggested by Carter *et al.* (1980), these results indicate that the parent yeast strain (A224A) from which CLP-8 is derived contains a lesions(s) whose phenotypic expression is allowed only in the presence of a second mutation that confers trichodermin resistance. Thus the two lesions in combination cause in addition to trichodermin resistance the cytoplasmic imbalance in ribosomal subunits. The situation is complicated further by the results of Carter & Cannon (1980) since three additional (related?) lesions have also been described in CLP-8 all apparently arising solely from the acquisition of trichodermin resistance by A224A. It is stressed (Carter *et al.*, 1980) that the postulated lesion(s) in A224A is not present in Y166, from which the trichodermin-resistant strain TR-1 is derived, since the latter has a normal ribosome profile.

We have now shown (A. Ferro & M. Cannon, unpublished work) that the TR-1/CLP-8 diploid has, in addition to being trichodermin resistant, the characteristic abnormality in the distribution of its cytoplasmic ribosomal subunits (see Carter *et al.*, 1980, for ribosome profiles for TR-1 and CLP-8). Since this diploid will contain genes derived from both TR-1 and CLP-8 it is clear that the normal ribosome profile characteristic of TR-1 is not expressed in dominant fashion in this diploid. We have not, however, rigorously excluded the possibility that the ribosome profile in the TR-1/CLP-8 diploid may be intermediate between the two extreme cases. For example, it is possible that anti-association factors are in some way concerned with the transport of pre-rRNP particles through the nuclear membrane into the cytoplasm and may even control their final maturation. We have shown that CLP-8 has a defective anti-association factor activity and this lesion might well cause the observed defect(s) in ribosome biogenesis in this strain. The effective result is to slow the rate at which mature 40S subparticles are released into the cytoplasm prior to entering polyribosomes in association with 60S subparticles.

Accordingly, the latter accumulate in the cytoplasm where they await the (slowed) arrival of their 40S counterpart. Carter & Cannon (1980) and Carter *et al.* (1980) have indeed observed that even in A224A the amount of anti-association factor(s) present is limiting with all available protein already associated with the small pool of native 40S ribosomal subunits, and the same situation applies in TR-1. In combination with a defective anti-association factor activity in CLP-8 the TR-1/CLP-8 diploid may well be unable to produce mature 40S subparticles at a rate sufficient to prevent a cytoplasmic build up of 60S subparticles, although the cytoplasmic accumulation of the latter may occur to a lesser extent than in the case of CLP-8 alone. To consolidate this hypothesis we have now studied ribosome processing in the TR-1/CLP-8 diploid. Preliminary results (A. Ferro & M. Cannon, unpublished work) indicate that the rate of processing of the pre-rRNP particle (containing 20S pre-rRNA) into the mature form is intermediate between the rates observed for TR-1 and CLP-8 respectively. This observation clearly supports the above interpretation. Perhaps the definitive experiments will come from studies on the TR-1/A224A diploid and its haploid products. Tetrads from this diploid contain two trichodermin-resistant spores. We predict that one of these will have an abnormal distribution of cytoplasmic ribosomal subunits and will have all the associated lesions characteristic of CLP-8. Ultimately we shall insert into A224A the plasmid (Fried & Warner, 1981) that carries the gene controlling trichodermin resistance and encoding ribosomal protein L3. It would indeed be remarkable and exciting if such a simple genetic transfer involving one particular ribosomal protein were to produce all five lesions characteristic of our CLP-8 strain. We are confident that our studies will aid our understanding of the complex processes involved in ribosome biogenesis in *S. cerevisiae* and possibly elucidate the role(s) played by specific ribosomal proteins. We hope too to be able to define the exact functions of the intriguing anti-association factors present in yeast cytoplasmic extracts (Pêtre, 1970).

I should now like to consider, albeit briefly, some current experimental approaches involving the isolation and analysis of *ts* mutants of *S. cerevisiae*. The possibility of genetically analysing the processes of transcription and translation is a major consideration in yeast molecular biology. Indeed earlier studies using *ts* mutants from *S. cerevisiae* revealed alterations in aminoacyl-tRNA synthetases (Hartwell & McLaughlin, 1968; McLaughlin & Hartwell, 1969) and deficiencies in elongation (Hartwell & McLaughlin, 1968) and initiation (Hartwell & McLaughlin, 1969) of protein synthesis. However, to date no *ts* mutants have been isolated from yeast that specifically affect ribosome function although many *ts* mutants have been described that are defective in the regulation of ribosome biogenesis (Hartwell, 1967; Hartwell *et al.*, 1970; Gorenstein & Warner, 1976). The primary effect of many of these mutations is to block the synthesis of nearly all the ribosomal proteins at the restrictive temperature, usually by preventing the synthesis of their mRNAs (Warner & Gorenstein, 1977; Hereford & Rosbash, 1977). Although the relevant genes are dispersed throughout the genome (for review see McLaughlin, 1974) it seems clear that yeast have developed co-ordinate control for synthesis of their ribosomal proteins and some common factor is thought to regulate the synthesis of their mRNAs. Analysis of further *ts* mutants is likely to be of use in elucidating some of these problems. Another approach is, however, of interest to us and was inspired by the work of Isono *et al.* (1977) on *E. coli*. As indicated earlier, identification of changed ribosomal proteins in mutants of *S. cerevisiae* has proved to be extremely difficult. We decided, therefore, to apply to yeast the method of Isono *et al.* (1977) whereby *E. coli* was heavily mutagenized with nitrosoguanidine to produce more than 5000 *ts* mutants. These mutants carry multiple lesions and random analysis has allowed the identification of strains carrying mutations in at least 35 different ribosomal proteins (Isono, 1980). We have now subjected *S. cerevisiae* to the same treatment and have selected more than 300 *ts* mutants. Approx. 100 of these have now been screened by two-dimensional gel electrophoresis for the presence of altered ribosomal proteins. To date we have identified three mutants designated *ts*205, *ts*212 (both derived from Y166) and *ts*417 (derived from A224A) all with altered electrophoretic ribosomal protein patterns as detected on urea/acrylamide gels (Kaltschmidt & Wittmann, 1970). It is curious that the same ribosomal proteins seem to be affected in each case (M. Cannon & L. Gritz, unpublished work). An example of an electrophoretic analysis of ribosomal proteins isolated from 80S particles of *S. cerevisiae* strains A224A and *ts*417 is illustrated in Fig. 1 (courtesy of L. Gritz). Several differences between the two preparations are apparent, as indicated by the arrows on the gel photographs.

The three *ts* strains are currently being investigated by both biochemical and genetic analysis and we have already observed (J. Mitlin, L. Gritz & M. Cannon, unpublished work) that they have abnormal distributions of their cytoplasmic ribosomal subunits (cf. CLP-8). Thus, at the permissive temperature (23°C) *ts*205 and *ts*417 have an apparent excess of material sedimenting at 60S and a relative decrease in the 40S population. The profile for *ts*212 is more extreme with a large amount of material sedimenting at 60S and relatively little material sedimenting at 40S. At the restrictive temperature (37°C), the ribosome profiles for *ts*212 and *ts*417 show even fewer 40S particles than at 23°C although the profile for *ts*205 is similar at both temperatures. At least one of these strains (*ts*212) has a defect in ribosome biogenesis (J. Mitlin, unpublished work). Thus, the conversion of 20S pre-rRNA to mature 18S rRNA is unusually rapid and we have evidence that both these RNA components may be undermethylated. It must be emphasized, however, that the three *ts* strains carry multiple mutations and as yet we have no evidence that there is a direct relationship between the presence of a changed ribosomal protein and the defect in ribosome biogenesis observed for *ts*212. We have, however, isolated a single step revertant of *ts*212 that has a normal ribosomal protein pattern as shown by two-dimensional gel electrophoresis. This strain will be analysed to determine if the lesion in biogenesis is still present and hence not directly related to the presence of a changed ribosomal protein(s). We hope that this and similar studies will help to elucidate the role(s) of individual ribosomal proteins in the function and/or assembly of the yeast ribosome *in vivo*. Furthermore, isolation and genetic analysis of more *ts* mutants of *S. cerevisiae* might allow the various genetic loci for individual ribosomal proteins in yeast to be established. We stress, however, the importance of studying ribosome profiles in the initial screening of yeast mutants, since as indicated for both CLP-8 and *ts*212 such studies can provide good indications for possible defects in ribosome biogenesis.

Finally I shall consider the phenomenon of natural resistance to antibiotics in different species of the yeast genera *Saccharomyces* and *Kluyveromyces*. Cundliffe

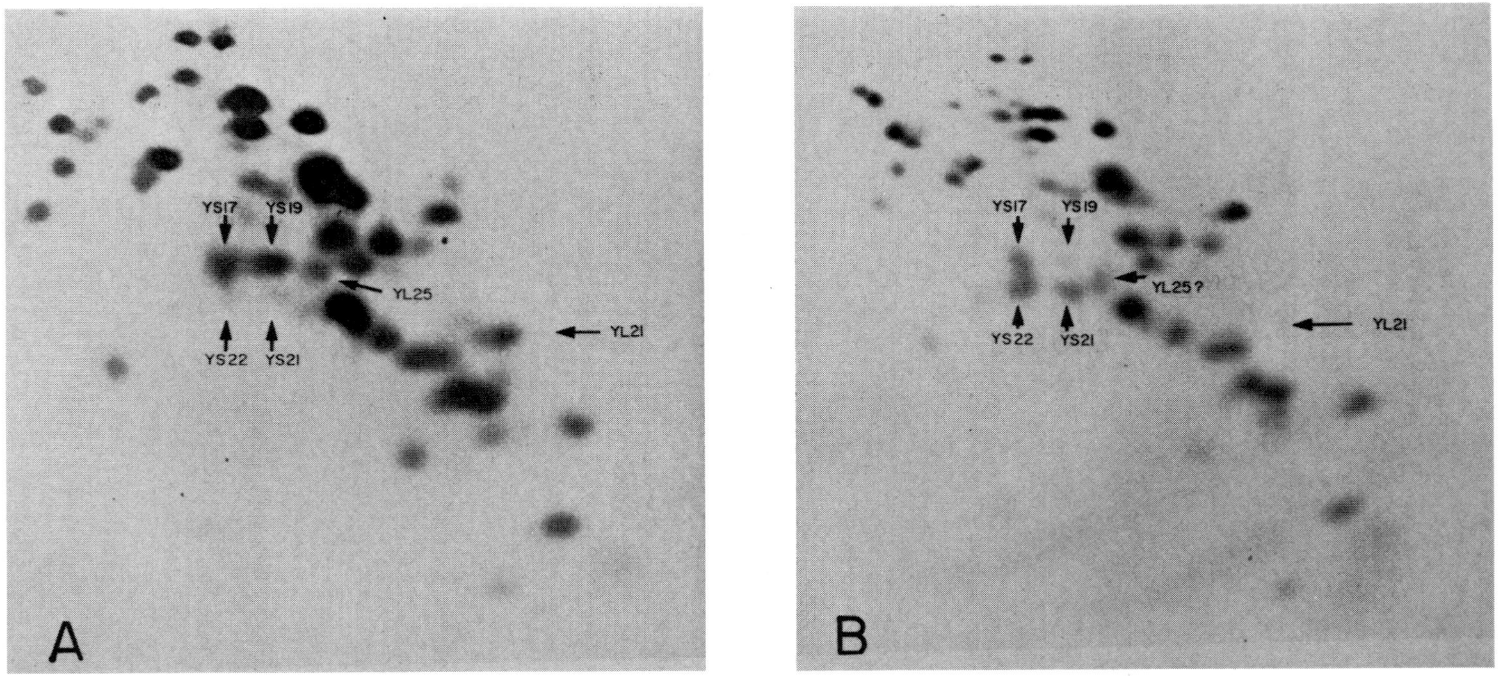

Fig. 1. *Separation of ribosomal proteins from 80S particles of S. cerevisiae strains A224A and ts417 by two-dimensional gel electrophoresis*

Experimental conditions are described in detail in Cannon *et al.* (1977). Migration in the first dimension was from left to right and in the second from top to bottom. (A) Ribosomal proteins from A224A; (B) ribosomal proteins from *ts*417. The arrows on the photographs indicate the main differences between the two preparations. YS refers to proteins that are associated with the 40S ribosomal subunits and YL to proteins that are associated with 60S ribosomal subunits. This Figure illustrates unpublished work by M. Cannon & L. Gritz. Photograph by courtesy of L. Gritz (Biogen, S.A., Geneva, Switzerland).

(1978) has demonstrated elegantly that certain antibiotic-producing organisms are resistant to the inhibitors they elaborate. Many such antibiotics are secreted into the culture medium by the producing organisms and resistance may result from a failure of the inhibitor to enter the cell. However, resistance is not always caused by a permeability block and in certain cases the ribosome itself is insensitive to inhibition by the relevant drug. The antibiotic thiostrepton provides an excellent example of this phenomenon. The compound is produced by *Streptomyces azureus* and the organism protects itself against its own toxic product by modifying its ribosomes. *S. azureus* actually methylates its 23S rRNA in such a way that thiostrepton can not associate with its normal receptor site (Cundliffe & Thompson, 1979). Natural resistance towards antibiotics in the relevant drug-producing organism has also been described for *Streptomyces erythreus* which produces erythromycin. Again methylation of 23S rRNA is implicated (Graham & Weisblum, 1979).

Yeast do not elaborate antibiotics but natural resistance to certain inhibitors is widespread. Thus, all members of the genus *Kluyveromyces* are apparently insensitive to cycloheximide (Van der Walt, 1970) and, at least in the case of *Kluyveromyces fragilis*, resistance is expressed at the ribosome level with an unidentified alteration(s) in the 60S subparticle being responsible (Rao & Grollman, 1967). Work on natural resistance to antibiotics in yeast has now been thoughtfully extended by Adoutte-Panvier *et al.* (1980). Selected drugs, added to culture plates at concentrations that totally inhibit the growth of wild-type *S. cerevisiae*, sometimes failed to inhibit the growth of many of the yeast strains tested. Some typical results (see also Adoutte-Panvier *et al.*, 1980) are shown in Table 2, which also includes unpublished work from B. Littlewood and from our own laboratory. These results *per se* do not distinguish between resistance that results from an inability of a given drug to enter the cells and resistance caused by an alteration(s) in some intracellular component(s). Accordingly, cell-free protein-synthesizing systems from selected strains were prepared and checked for sensitivity towards relevant antibiotics. Extracts from *S. cerevisiae* (strain SR17) are resistant to cycloheximide at the ribosome level with some component(s) of the 60S subparticle being responsible (see also Rao & Grollman, 1967). Similar studies have now been extended to *Saccharomyces kluyveri*; protein-synthesizing systems from this strain are again clearly resistant to inhibition this time by blasticidin S and there are strong indications that resistance is a property of the 60S ribosomal subparticle (A. Adoutte-Panvier & M. Cannon, unpublished work) supporting the notion that blasticidin S inhibits the peptidyl transferase centre (Vázquez, 1979) that is itself located there. These results are most promising but disappointingly it has been claimed (A. Adoutte-Panvier, unpublished work) that for anisomycin there is marked inhibition of protein synthesis in cell-free extracts from either *Saccharomyces bisporus* or *Saccharomyces globosus*. This is not, however, the case for *S. cerevisiae* strains CLP-8 and TR-1 which, although isolated in the laboratory as trichodermin-resistant, are also resistant to anisomycin both *in vivo* and *in vitro* (Grant *et al.*, 1976). Nevertheless, both alphasarcin and paromomycin inhibit cell-free protein synthesis in *S. kluyveri* extracts (M. Cannon, unpublished work) and the resistance shown by this strain towards these antibiotics added *in vivo* is apparently again caused by a permeability block. Furthermore, since of the strains listed in Table 2 all are

Table 2. *Natural resistance to antibiotics in selected strains from the genera Saccharomyces and Kluyveromyces*

All the yeast strains were tested for resistance (R) or sensitivity (S) *in vivo*. Antibiotic key: A, alpha sarcin (100 μg/ml); B, Anisomycin (100 μg/ml); C, blasticidin S (20 μg/ml); D, cryptopleurine (5 μg/ml); E, cycloheximide (20 μg/ml); F, paromomycin (200 μg/ml); G, trichodermin (20 μg/ml); H, trichodermol (30 μg/ml).

	Resistance or sensitivity							
Species Antibiotic tested...	A	B	C	D	E	F	G	H
S. cerevisiae Y166	R	S	S	S	S	R	S	S
S. cerevisiae A224A	R	S	S	S	S	R	S	S
S. cerevisiae CLP-8	R	R	R	S	S	R	R	R
S. cerevisiae TR-1	R	R	R	S	S	R	R	R
S. cerevisiae SR17	R	S	S	S	R	R	S	S
S. cerevisiae CRY6	R	S	S	R	S	R	S	S
S. bailii	R	S	R	S	S	S	S	S
S. bayanus	R	S	S	S	S	S	S	S
S. bisporus	R	R	R	S	S	R	R	R
S. carlsbergensis	R	S	S	S	S	R	S	S
S. delbrueckii	R	S	R	S	S	S	S	S
S. diastaticus	R	S	R	S	S	R	S	S
S. globosus	R	R	R	S	S	R	R	R
S. italicus	R	S	S	S	S	R	S	S
S. kluyveri	R	S	R	S	S	R	S	S
S. rosei	R	S	S	S	S	S	S	S
S. rouxii	R	S	S	S	S	S	S	S
S. uvarum	R	S	S	S	S	R	S	S
K. dobzhanskii	R	S	R	S	R	S	S	S
K. lactis	R	S	R	S	R	S	S	S
K. marxianus	R	S	R	S	R	R	S	S
K. wickerhami	R	S	R	S	R	S	S	S

resistant to alphasarcin and most are resistant to paromomycin, it appears that both antibiotics may be unable to enter intact cells of many different yeasts. It is clear that much work is still required to identify possible antibiotic resistance *in vitro* for more of the strains listed in Table 2.

Adoutte-Panvier *et al.* (1980) have also analysed by one-dimensional gradient polyacrylamide gel electrophoresis in the presence of SDS, cytoplasmic ribosomal proteins from a selection of yeast strains. Extensive differences were observed in the various ribosomal protein patterns for both 40S and 60S subparticle preparations. Adoutte-Panvier *et al.* (1980) consider that under the analytical conditions employed minor variations in molecular weight can be detected as can substitution, in ribosomal proteins, of amino acid residues. Their results are certainly dramatic and are similar in many ways to the extensive variation of ribosomal proteins found in both the *Bacillaceae* (Geisser *et al.*, 1973*a*, *b*) and in the *Enterobacteriaceae* (Hori & Osawa, 1978). As a general rule, however, ribosomal proteins in most eukaryotes appear to be strongly conserved, and accordingly there is little difference between those from lizard, rat, bird and man (DeLaunay *et al.*, 1973). Variation in lower eukaryotes (i.e. yeast) is presumably indicative of the short generation times involved in cell growth. Adoutte-Panvier *et al.* (1980) have also attempted to estimate strain relatedness in yeast by comparing the various ribosomal protein patterns in their preparations. Their results indicate that ribosomal protein relatedness can correlate well with classification of yeast based on nucleic acid hybridization studies (Bicknell & Douglas, 1970) although inevitably there are a few surprises. Clearly, however, this is a valid and useful approach. It should be stressed, though, that most of the ribosomal protein variations observed by Adoutte-Panvier *et al.* (1980) are likely to be functionally neutral, and considerable fluctuation is clearly not always detrimental to ribosome function.

We have now analysed ribosomal proteins from *S. cerevisiae*, *S. bisporus*, *S. globosus* and *S. kluyveri* by studying their separation by two-dimensional gel electrophoresis. Although such analyses may provide no further information on strain relatedness it is vital to obtain a reliable correlation between the one-dimensional gel electrophoretic patterns of Adoutte-Panvier *et al.* (1980) and a standard two-dimensional electrophoretic pattern for which a ribosomal protein numbering system has been described (Kruiswijk & Planta, 1974). Only in this way will it be possible to define which particular ribosomal proteins are common to selected yeast strains and which are changed in electrophoretic mobility or even missing. Ideally we should like also to obtain a direct relationship, for a particular yeast, between a drug-resistant phenotype *in vitro* and an associated change in a ribosomal protein. Accordingly, preparations of proteins from 80S ribosomes were subjected to two-dimensional gel electrophoresis in urea/acrylamide (Kaltschmidt & Wittmann, 1970). *S. cerevisiae* and *S. bisporus* have ribosomal protein patterns that are almost identical under these analytical conditions (D. Savin & M. Cannon, unpublished work), a result that parallels the similarities observed using one-dimensional SDS gels (Adoutte-Panvier *et al.*, 1980). The ribosomal proteins from the two species do, however, show only partial immunochemical cross-reactivity and some differences are, therefore, indicated (Adoutte-Panvier *et al.*, 1980). At first sight the electrophoretic similarity shown by the two strains is somewhat surprising, since the patterns of antibiotic resistance are very different (Table 2). Thus, *S. bisporus* is resistant *in vivo* to six out of the eight antibiotics tested, although as stressed earlier anisomycin resistance *in vitro* is not observed for *S. bisporus*. In contrast to these results the ribosomal protein patterns for both *S. kluyveri* (K. Fairburn & M. Cannon, unpublished work) and *S. globosus* (D. Savin & M. Cannon, unpublished work) are markedly different in the two-dimensional analysis relative to those for *S. cerevisiae* and *S. bisporus* again reflecting, at least for *S. globosus*, the evidence from one-dimensional gel electrophoresis. The differences between *S. bisporus* and *S. globosus* are particularly interesting since the antibiotic resistance patterns for the two strains are identical (see Table 2). Although it is, as yet, too early to draw detailed conclusions from these studies there are indications that certain ribosomal proteins migrate in very similar positions in preparations from the various strains. This might imply a high degree of conservation of certain ribosomal protein amino acid sequences between species and might indicate a necessity for ribosomal proteins of a given structure to allow a fulfilment of function. Electrophoretic analyses of ribosomal proteins from 80S particles of *S. cerevisiae* strain Y166 and *S. kluyveri* are illustrated in Fig. 2. (This Figure illustrates unpublished work by M. Cannon & K. Fairburn).

Although our electrophoretic analyses are intriguing they are also in one way disappointing. The detection of so many changes in electrophoretic mobility makes it very difficult to identify particular proteins. In turn no hard facts as yet emerge

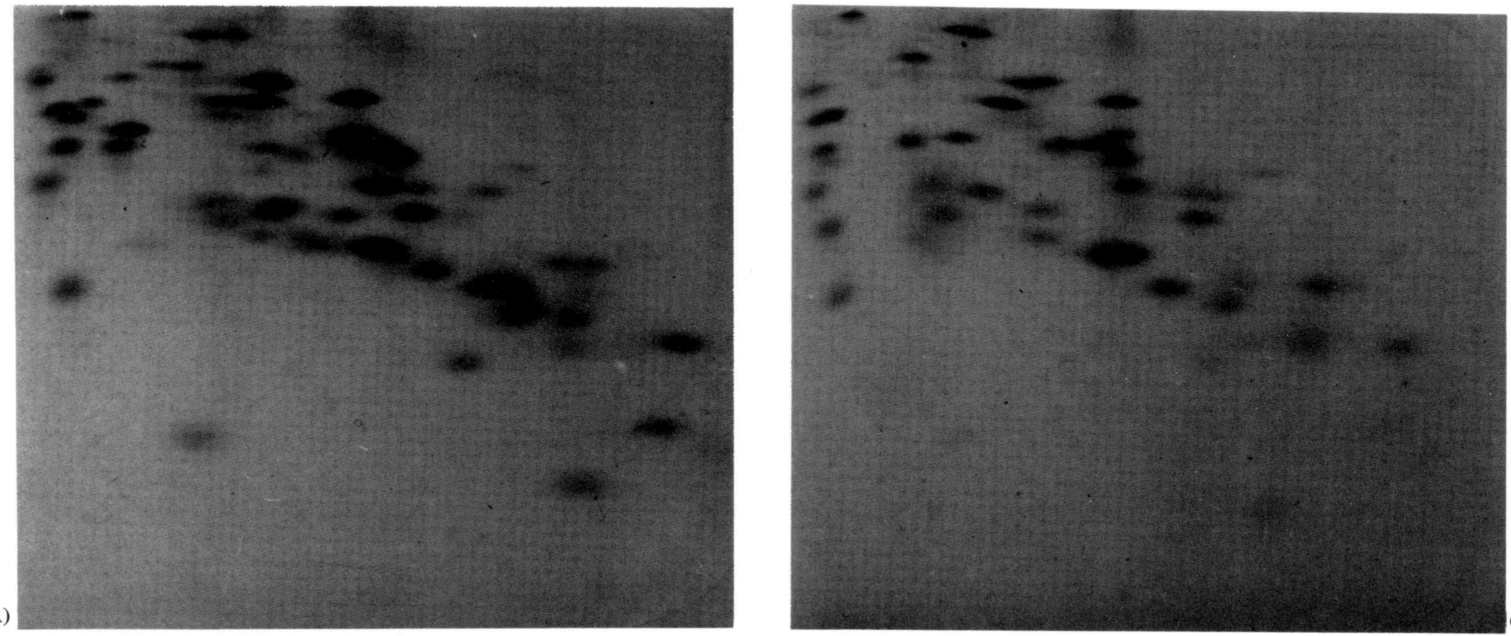

(A) (B)

Fig. 2. *Separation of ribosomal proteins from 80S particles of S. cerevisiae strain Y166 (A) and S. kluyveri (B) by two-dimensional gel electrophoresis*

Experimental details are as indicated for Fig. 1. This Figure illustrates unpublished work by M. Cannon & K. Fairburn. Photograph by courtesy of K. Fairburn (Department of Biochemistry, King's College, University of London).

concerning relationships between these ribosomal protein mobility changes and antibiotic resistance in these strains. We have, nevertheless, carried out one more important experimental check on three of the preparations. Studies on *S. cerevisiae* have previously shown that one particular ribosomal protein (L15) from this strain is very heavily methylated (Cannon *et al.*, 1977) and we have now also screened *S. bisporus* and *S. globosus* for a similar presence (for methodology see Cannon *et al.*, 1977). In all three strains we have identified (D. Savin & M. Cannon, unpublished work) a heavily methylated ribosomal protein and significantly the electrophoretic mobility observed appears to be identical in the three cases (refer to Cannon *et al.*, 1977). It will be interesting to identify this particular protein on the one-dimensional SDS gels of Adoutte-Panvier *et al.* (1980) and ascertain if its mobility and methylation level are constant in the various preparations analysed.

In summary it seems clear that yeast mutants are invaluable probes for studying structure–function relationships in 80S ribosomes and in elucidating the complex mechanisms involved in ribosome biogenesis. Antibiotic-resistant yeast strains are also potentially exploitable in attempting to define the molecular bases of the phenomena and in identification of the particular component(s) involved. Studies on yeast may allow the ultimate goal in ribosome research to be achieved whereby the precise location and function of each molecule in the particle can be defined. As stressed in the introduction to this article, yeast can be subjected to convenient genetic analysis. Yeast can now be used to combine classical genetic methods with the modern technology of gene cloning and transformation to facilitate the types of analysis required to further our understanding of eukaryotic protein synthesis.

I am grateful to those of my colleagues who have allowed me to refer in this article to some of their unpublished data. In this respect I should particularly like to thank Arlette Adoutte-Panvier, Christopher Carter, Kevin Fairburn, Albert Ferro, Linda Gritz, Barbara Littlewood, Judith Mitlin and David Savin.

References

Adoutte-Panvier, A., Davies, J.E., Gritz, L.R. & Littlewood, B. (1980) *Mol. Gen. Genet.* **179**, 273–282
Bicknell, J.N. & Douglas, H.C. (1970) *J. Bacteriol.* **101**, 505–512
Butler, P.D., Sims, P.F.G. & Wild, D.G. (1980) *Biochem. J.* **190**, 157–170
Cannon, M., Schindler, D. & Davies, J. (1977) *FEBS Lett.* **75**, 187–191
Carter, C.J. & Cannon, M. (1977) *Biochem. J.* **166**, 399–409
Carter, C.J. & Cannon, M. (1980) *J. Mol. Biol.* **143**, 179–199
Carter, C.J., Cannon, M. & Jiménez, A. (1980) *Eur. J. Biochem.* **107**, 173–183
Chambliss, G., Craven, G.R., Davies, J., Davis, K., Kahan, L. & Nomura, M. (eds.) (1980) *Ribosomes: Structure, Function and Genetics*, University Park Press, Baltimore
Cooper, D., Banthorpe, D.V. & Wilkie, D. (1967) *J. Mol. Biol.* **26**, 347–350
Cramer, J.H., Farally, F.W. & Rownd, R.H. (1976) *Mol. Gen. Genet.* **148**, 233–241
Cundliffe, E. (1978) *Nature (London)* **272**, 792–795
Cundliffe, E. & Thompson, J. (1979) *Nature (London)* **278**, 859–861
DeJonge, P., Klootwijk, J. & Planta, R.J. (1977) *Nucleic Acids Res.* **4**, 3655–3663
DeLaunay, J., Creusot, F. & Shapira, G. (1973) *Eur. J. Biochem.* **39**, 305–312
Fried, H.M. & Warner, J.R. (1981) *Proc. Natl. Acad. Sci. U.S.A.* **78**, 238–242
Geisser, M., Tischendorf, G.W., Stöffler, G. & Wittmann, H.G. (1973*a*) *Mol. Gen. Genet.* **127**, 111–128
Geisser, M., Tischendorf, G.W. & Stöffler, G. (1973*b*) *Mol. Gen. Genet.* **127**, 129–145
Gorenstein, C.G. & Warner, J.R. (1976) *Proc. Natl. Acad. Sci. U.S.A.* **73**, 1547–1551
Graham, M.Y. & Weisblum, B. (1979) *J. Bacteriol.* **137**, 1464–1467
Grant, P., Sánchez, L. & Jiménez, A. (1974) *J. Bacteriol.* **120**, 1308–1314

Grant, P., Schindler, D. & Davies, J.E. (1976) *Genetics* **83**, 667–673
Hadjiolov, A.A. & Nikolaev, N. (1976) *Prog. Biophys. Mol. Biol.* **31**, 95–144
Hartwell, L.H. (1967) *J. Bacteriol.* **93**, 1662–1670
Hartwell, L.H. & McLaughlin, C.S. (1968) *Proc. Natl. Acad. Sci. U.S.A.* **59**, 422–428
Hartwell, L.H. & McLaughlin, C.S. (1969) *Proc. Natl. Acad. Sci. U.S.A.* **62**, 468–474
Hartwell, L.H., McLaughlin, C.S. & Warner, J.R. (1970) *Mol. Gen. Genet.* **109**, 42–56
Helser, T.L. & McLaughlin, C.S. (1975) *J. Biol. Chem.* **250**, 2003–2007
Hereford, L. & Rosbash, M. (1977) *Cell* **10**, 463–467
Hori, H. & Osawa, S. (1978) *J. Bacteriol.* **133**, 1089–1095
Huynh-Van-Tan, DeLaunay, J. & Schapira, G. (1971) *FEBS Lett.* **17**, 163–167
Isono, K. (1980) in *Ribosomes: Structure, Function and Genetics* (Chambliss, G., Craven, G.R., Davies, J., Davis, K., Kahan, L. & Nomura, M., eds.), pp. 641–669, University Park Press, Baltimore
Isono, K., Cumberlidge, A.G., Isono, S. & Hirota, Y. (1977) *Mol. Gen. Genet.* **152**, 239–243
Jiménez, A., Sánchez, L. & Vázquez, D. (1975) *Biochim. Biophys. Acta* **383**, 427–434
Kaltschmidt, E. & Wittmann, H. G. (1970) *Anal. Biochem.* **36**, 401–412
Kruiswijk, T. & Planta, R. J. (1974) *Mol. Biol. Rep.* **1**, 409–415
Kruiswijk, T., Planta, R.J. & Krop, J.M. (1978) *Biochim. Biophys. Acta* **517**, 378–389
Maxam, A., Tizard, R., Skryabin, K.G. & Gilbert, W. (1977) *Nature (London)* **267**, 643–645
McLaughlin, C.S. (1974) in *Ribosomes* (Nomura, M., Tissières, A. & Lengyel, P., eds.), pp. 815–828, Cold Spring Harbor Laboratory, Cold Spring Harbor, New York
McLaughlin, C.S. & Hartwell, L.H. (1969) *Genetics* **61**, 557–566
Nomura, M., Tissières, A. & Lengyel, P. (eds.) (1974) *Ribosomes*, Cold Spring Harbor Laboratory, Cold Spring Harbor, New York
Ozaki, M., Mizushima, S. & Nomura, M. (1969) *Nature (London)* **222**, 333–339
Petes, T.D. (1979) *Proc. Natl. Acad. Sci. U.S.A.* **76**, 410–414
Pêtre, J. (1970) *Eur. J. Biochem.* **14**, 399–405
Philippsen, P., Thomas, M., Kramer, R.A. & Davis, R.W. (1978) *J. Mol. Biol.* **123**, 387–404
Planta, R.J., Klootwijk, J., Raué, H.A., Brand, R.C., Veldman, G.M., Stiekema, W.J. & DeJonge, P. (1980) in *Genetics and Evolution of RNA Polymerase, tRNA and Ribosomes* (Osawa, S., Ozeki, H., Uchida, H. & Yura, T., eds.), pp. 451–470, University of Tokyo Press
Rao, S.S. & Grollman, A.P. (1967) *Biochem. Biophys. Res. Commun.* **29**, 696–704
Retèl, J., Van Den Bos, R.C. & Planta, R.J. (1969) *Biochim. Biophys. Acta* **195**, 370–380
Schindler, D., Grant, P. & Davies, J. (1974) *Nature (London)* **248**, 535–536
Sillevis Smith, W.W., Vlak, J.M., Schiphof, R. & Rozijn, T.H. (1972) *Exp. Cell Res.* **71**, 33–40
Skogerson, L., McLaughlin, C.S. & Wakatama, E. (1973) *J. Bacteriol.* **116**, 818–822
Stocklein, W. & Piepersberg, W. (1980) *Current Genet.* **1**, 177–186
Stöffler, G. & Wittmann, H.G. (1977) in *Molecular Mechanisms of Protein Biosynthesis* (Weissbach, H. & Pestka, S., eds.), pp. 117–202, Academic Press, New York
Trapman, J. & Planta, R.J. (1975) *Biochim. Biophys. Acta* **414**, 115–125
Trapman, J. & Planta, R.J. (1976) *Biochim. Biophys. Acta* **442**, 265–274
Trapman, J., Retèl, J. & Planta, R.J. (1975) *Exp. Cell Res.* **90**, 95–104
Udem, S.A. & Warner, J.R. (1972) *J. Mol. Biol.* **65**, 227–242
Udem, S.A. & Warner, J.R. (1973) *J. Biol. Chem.* **248**, 1412–1416
Van der Walt, J.P. (1970) in *The Yeasts – A Taxonomic Study* (Lodder, J., ed.), pp. 1–33, North-Holland, Amsterdam
Vázquez, D. (1979) *Inhibitors of Protein Biosynthesis*, Springer-Verlag, Berlin
Vázquez, D. & Jiménez, A. (1980) in *Ribosomes: Structure, Function and Genetics* (Chambliss, G., Craven, G.R., Davies, J., Davis, K., Kahan, L. & Nomura, M., eds.), pp. 847–869, University Park Press, Baltimore
Veldman, G.M., Klootwijk, J., Van Herrikhuizen, H. & Planta, R.J. (1981) *Nucleic Acids Res.* **9**, 4847–4862
Warner, J.R. (1982) in *The Molecular Biology of the Yeast Saccharomyces cerevisiae* (Strathern, J.N., Jones, E.W. & Broach, J.R. (eds.)), Cold Spring Harbor Laboratory, New York, in the press
Warner, J.R. & Soeiro, R. (1967) *Proc. Natl. Acad. Sci. U.S.A.* **58**, 1984–1990
Warner, J.R. & Gorenstein, C.G. (1977) *Cell* **11**, 201–212
Wejksnora, P.J. & Haber, J.E. (1974) *J. Bacteriol.* **120**, 1344–1355
Wool, I.G. (1979) *Annu. Rev. Biochem.* **48**, 719–754

Biochem. Soc. Symp. 47, 95–112
Printed in Great Britain

Translation Mechanism in Prokaryotes: Structure and Expression of *Escherichia coli* Initiation Factor IF3 Gene

M. GRUNBERG-MANAGO, M. SPRINGER and J.A. PLUMBRIDGE

Institut de Biologie Physico-Chimique, 13 rue Pierre et Marie Curie, 75005 Paris, France

and S. BLANQUET, G. FAYAT and C. SACERDOT

Laboratoire de Biochimie, Ecole Polytechnique, 91128 Palaiseau-Cedex, France

Introduction

The mechanism of initiation of protein synthesis in bacteria has been studied intensively for the past 15 years and important advances have been achieved which were recently reviewed in detail (Grunberg-Manago *et al.*, 1978; Grunberg-Manago, 1980; Hershey, 1980). The macromolecular components of protein synthesis have been identified, purified and we have now a more or less coherent picture of the localization of the different components on the ribosome.

Several schemes have been proposed to explain how the various components interact during protein synthesis. First, the 70S ribosome dissociates into 30S and 50S subunits. The 30S subunit forms a preinitiation complex with fMet-tRNA and mRNA. This is joined by the 50S subunit to complete formation of the 70S initiation complex. During these steps two critical events occur: the ribosomes select a specific mRNA from among numerous species of mRNAs and the fMet-tRNA interacts with a specific initiation site on the mRNA to assure proper translation in the correct phase. These reactions are promoted or catalysed by three protein initiation factors called IF1, IF2 and IF3. Their interactions involve the hydrolysis of GTP (Fig. 1).

Under physiological ionic conditions, step 1 requires two initiation factors IF1 and IF3. IF1 (M_r 9500) serves to increase the rate constants for both association

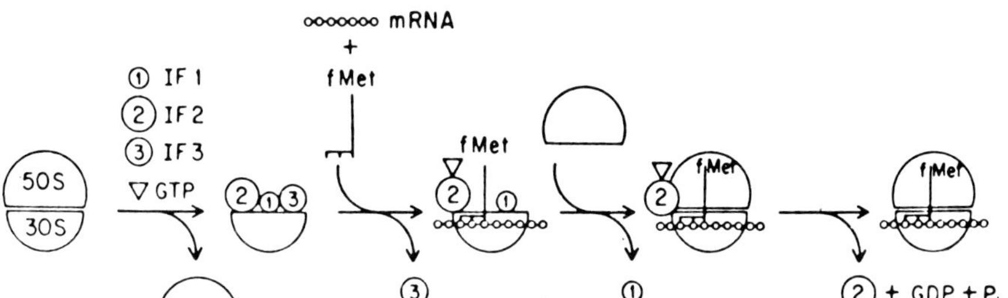

Fig. 1. *Pathways for initiation of protein synthesis in prokaryotes (Grunberg-Manago, 1980)*

and dissociation of the two ribosomal subunits, and IF3 (M_r 20700) shifts the equilibrium toward dissociation by strongly binding to the 30S particle. Step 2 requires the third factor IF2 (M_r 115000) and GTP to direct fMet-tRNA binding. Step 3 is accompanied by the release of all the three factors. The release of IF2 is dependent upon GTP hydrolysis and results in the formation of the initiation complex.

While we have an understanding in broad terms of the general pathway, there are many gaps within our knowledge at the molecular level. Is there a unique pathway for the assembly of the 30S preinitiation complex or do several parallel pathways operate simultaneously, i.e. is there an order in the binding of fMet-tRNA or mRNA to ribosomes? Does IF2 bind first to ribosome or does it form a binary (IF2 · fMet-tRNA) or ternary (IF2 · fMet-tRNA · GTP) complex as happens in eukaryotes? Factor IF2 can be seen as an aminoacyl-tRNA carrier protein which ensures the correct binding of the fMet-tRNA within the ribosomal peptidyl transferase centre. In accordance with this model, it has been found that IF2 specifically protects the formylated form of the initiator tRNA against the spontaneous hydrolysis of the aminoacyl ester (Petersen *et al.*, 1979) (Fig. 2); it also protects specific internal regions of initiator tRNA against hydrolysis by a double-strand specific ribonuclease (Petersen *et al.*, 1981). However, IF2 binds also very strongly in the absence of the initiator tRNA to the 30S ribosomal subunit and it can be alternatively postulated that it activates a binding site for the initiator tRNA on the ribosome.

The role of IF3 is even more mysterious with apparently two distinct functions. It does not act only as an anti-association factor, but it also stimulates the formation of the preinitiation complex involving the small subunit when a long polymer is used as mRNA but not with the messenger initiation triplet codon (Grunberg-Manago *et al.*, 1978) (Table 1). It is therefore believed to be an essential component for the specific recognition of initiation regions of the mRNA which are presumably longer than the initiation triplet.

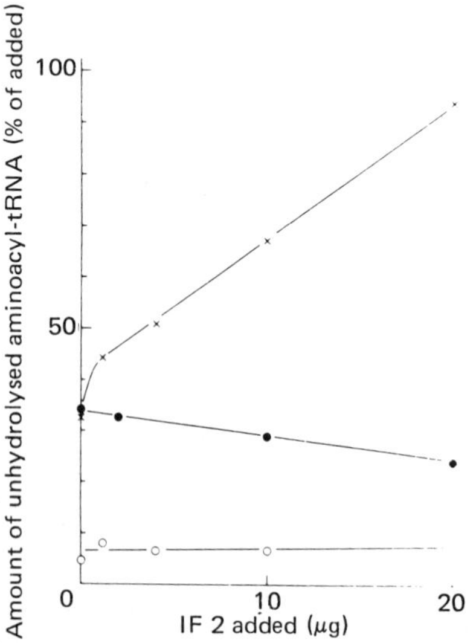

Fig. 2. *Effect of IF2 on the non-enzymic hydrolysis of the aminoacyl ester bonds of fMet-tRNA$_f^{Met}$ (x), Met-tRNA$_f^{Met}$ (○) and Met-tRNA$_m^{Met}$ (●)*

Table 1. *Stimulation of fMet-tRNA binding to 30S subunits by initiation factors (Grunberg-Manago, 1980)*

Mixtures contained in 100 μl: 50 mM-Tris/HCl (pH 7.4), 100 mM-NH_4Cl, 5 mM-magnesium acetate, 7 mM-2-mercaptoethanol, 1 mM-GTP, 0.1 A_{260} units of poly (A,G,U) or 0.5 A_{260} units of AUG, 1.0 A_{260} units of [^{3}H]fMet-tRNA (sp. radioactivity 3000 c.p.m. pmol^{-1}), 0.35 A_{260} units of 30S subunits, and initiation factors as indicated: 15 pmol of IF1, 10 pmol of IF2 and 10 pmol of IF3. Mixtures were incubated for 10 min at 37°C, diluted with cold buffer and filtered through Millipore HA nitrocellulose filters; the filters were dried and immersed in scintillation fluid for counting. The results are reported as c.p.m., following the subtraction of blanks (minus all factors) of 855 c.p.m. for AUG and 2677 c.p.m. for poly (A,U,G).

	Binding (c.p.m.)	
Additions	Poly (A,G,U)	AUG
IF2	511	553
IF2 + IF3	3167	956
IF2 + IF1	2597	2923
IF1 + IF2 + IF3	5517	2102

The mRNA primary structure participates in at least two RNA–RNA interactions during the positioning of the ribosome in the initiation complex: specific portions of the message can form base pairs with the initiator tRNA on one hand and with the 3′-end of 16S RNA on the other hand (Shine and Dalgarno base-pairs) (Shine & Dalgarno, 1974). IF3 could be involved in the recognition of either one of these message regions. There are some indications *in vitro* that a mRNA with a strong Shine and Dalgarno sequence requires less initiation factor (Grunberg-Manago, 1980). In agreement with this view is the observation that in Gram-positive bacteria mRNA with a very strong Shine and Dalgarno sequence preceding the initiation codon have a reduced translational dependence on initiation factor IF3 in a cell-free system (McLaughlin *et al.*, 1981).

In contrast to eukaryotic mRNAs, prokaryotic mRNAs are polycistronic and contain internal ribosome sites. As a result, during the ribosome binding process, mRNA secondary structure can function as a controlling element covering unwanted internal initiators and affecting the availability of authentic initiation sites. IF3 could control the secondary structure of mRNA, and several regulatory mechanisms have been ascribed to IF3, but the evidence is weak (Grunberg-Manago *et al.*, 1978).

One advantage of a prokaryotic system is the possibility of selecting mutants and important advances have been made in our knowledge concerning RNA–RNA interactions by analysis of mutants. mRNA mutants with a change either in the initiation codon or in the Shine and Dalgarno sequence have been isolated and confirm the importance of these two interactions in the selection of the correct initiation signals (Grunberg-Manago *et al.*, 1978; Grunberg-Manago, 1980; Steitz, 1980). Furthermore, the effects of some mutations are most probably explained by an alteration in the mRNA secondary structure such that the accessibility of the ribosome interaction site is reduced (Steitz, 1980; Gold *et al.*, 1981). Unfortunately, we are lacking mutations in initiation factors, and while we have solid evidence for a strict requirement *in vitro* for the three initiation factors (Table 2), such evidence is not available *in vivo*.

Furthermore, while a wealth of information exists about levels and synthesis rates of ribosomal components (Yates *et al.*, 1980; Maaløe, 1979; Miura *et al.*,

Table 2. *Factor dependencies for β-galactosidase synthesis using a purified E. coli cell-free system (Kung et al., 1977)*

The cell free system contains principally washed ribosomes, tRNA, salts, nucleotides, amino acids, an energy generating system and the protein components listed in the Table. It is programmed by DNA from a λ lac phage. The ascites extract is used to provide aminoacyl-tRNA synthetases. L and L_{α} are partially purified fractions from *E. coli* extracts which stimulate β-galactosidase synthesis in the *E. coli* cell-free system. For further details, see Kung *et al.* (1977).

Protein components	Activity (%)
Complete	100
Ascites S-100 alone	<10
Other synthetases alone	<10
Complete – RNA polymerase	<10
– L	22
– L_{α}	30
– IF-1	15
– IF-2	<10
– IF-3	<10
–EF-Tu	<10
– EF-Ts	70
– EF-G	15
– release factors 1 and 2	>50
– cyclic AMP receptor	<10
– Transformylase	83

1981), with respect to initiation factors very little is known. Some information is available about changes at the cellular level: the expression of the initiation factors is co-ordinated with the level of ribosomes (Howe *et al.*, 1978; J.G. Howe & J.W.B. Hershey, unpublished work), but nothing is known about the control mechanisms affecting rates of initiation factor synthesis. In order to address these questions, as well as to try to understand the molecular mechanisms of action of initiation factors, we turned our efforts towards the isolation of mutants. In this review we would like to report studies on the structure, expression and sequence of the gene for *E. coli* translational initiation factor IF3.

Studies of the Region of the *E. coli* Chromosome Containing the *infC* Gene

A thermosensitive mutant with a thermolabile initiation factor IF3 was isolated several years ago from a collection of mutants obtained after nitrosoguanidine mutagenesis and enrichment for thermosensitivity by using the 'tritiated amino acid suicide method' (Springer *et al.*, 1977*a*).

The mutation is located near 38 min on the *E. coli* genetic map. This mutant was not only modified in the IF3 gene but also had a defective small subunit of phenylalanyl-tRNA synthetase (EC 6.1.1.20) (Springer *et al.*, 1977*a*; Hennecke *et al.*, 1977). This mutant enabled us to isolate a λ transducing phage (λp2) complementing its thermosensitivity (Springer *et al.*, 1977*b*). Such a phage integrated in the mutant chromosome reversed both the IF3 and phenylalanyl-tRNA synthetase defects.

The genetic content of the *E. coli* insert in λDNA was studied by SDS/polyacrylamide-gel analysis of the proteins synthesized after infection of u.v.-irradiated lysogenic cells. As the u.v.-irradiated cells are lysogenic (where host protein synthesis is stopped) the only genes expressed from λ are those under the control of an *E. coli* promoter. The *E. coli* insert was shown to carry the following structural genes: *infC* for initiation factor IF3 (M_r 20700), *pheS* and *pheT* for the small and large subunit of phenylalanyl-tRNA synthetase (M_r 38000 and 94000 respectively), *thrS* for threonyl-tRNA synthetase (EC 6.1.1.3; M_r 78000) and P12 for an M_r 12000 protein of unidentified function (Fig. 3) (Springer *et al.*, 1977*b*). These genes, all coding for components of protein synthesis, are clustered on the *E. coli* chromosome and recall the situation of genes for elongation factors clustered with some ribosomal proteins.

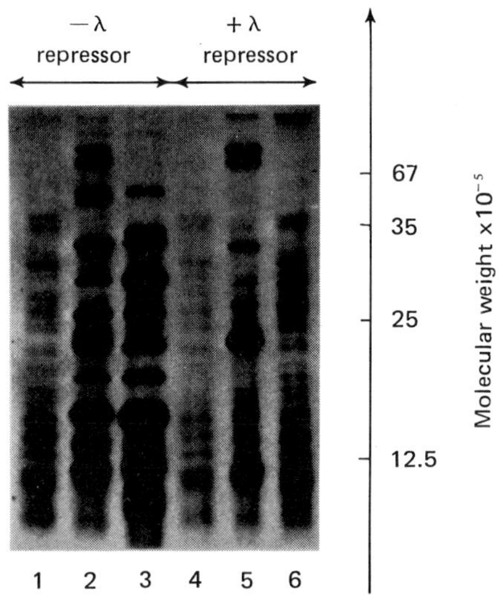

Fig. 3. *SDS/polyacrylamide-gel analysis of sonicated extracts of u.v.-irradiated and then infected cells labelled with [¹⁴C]leucine in the presence and absence of λ repressor*

Sample 1, uninfected 159 *E. coli* strain; sample 2, 159 infected with λp2; sample 3, 159 infected with Y199 (phage carrying several deletions in the λb2 region where *E. coli* DNA was inserted); sample 4, uninfected 159 (λind⁻); sample 5, 159 (λind⁻) infected with λp2; sample 6, 159 (λind⁻) infected with Y199 (from Springer *et al.*, 1977*b*).

Starting from the original phage, λp2, carrying the whole cluster of genes, a set of phages deleted for some of the genes in the cluster (Springer *et al.*, 1979) was isolated. This set of phages was studied genetically by complementation of *pheS*, *pheT* and *thrS* mutants. The coding content of these phages was also characterized by using the u.v.-irradiated and then infected cell system (Fig. 4). This type of deletion mapping gave the order of the *E. coli* genes on the original transducing phage, this order is: *thrS infC P12 pheS pheT*.

Direction of Transcription of the Cluster of Genes in the *infC* Region

An *Eco*RI/*Hind*III fragment (called B1) (Plumbridge *et al.*, 1980) internal to the *E. coli* insert of λp2 was cloned between the *Eco*RI and *Hind*III sites of pBR322 (Fig. 5). The resulting plasmid (called pB1) was shown by genetic complementation to carry *pheS*, *pheT* and *thrS*. pB1 was also shown to carry *infC* because strains

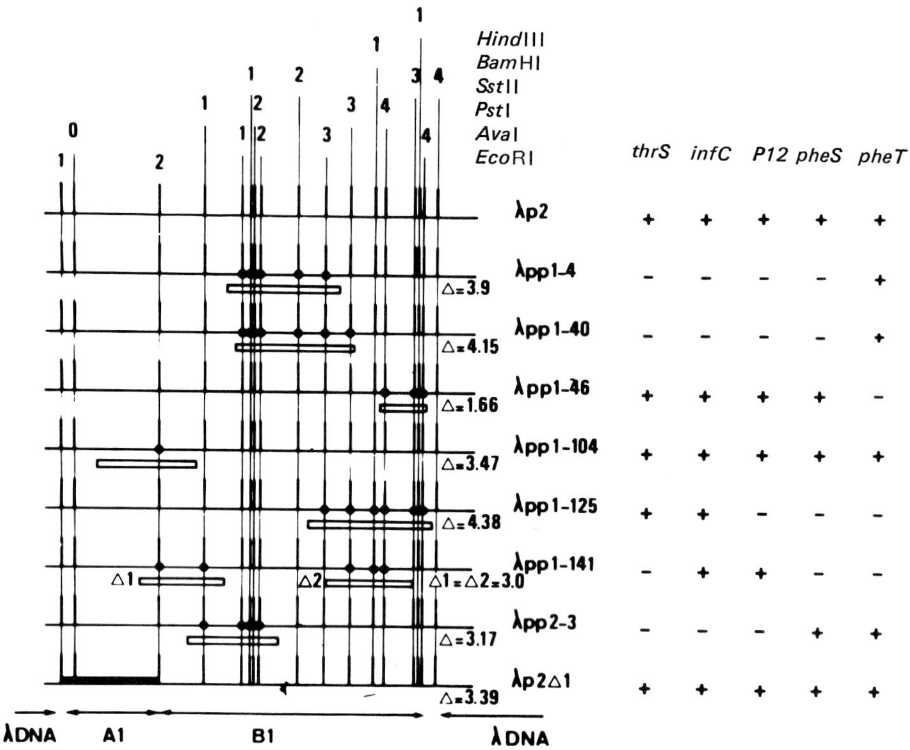

Fig. 4. *Correlation of the restriction endonuclease maps of the DNA of λp2 and the derived deleted phages and the pattern of expression of the genes carried by these phages*

The left part of the figure shows the restriction maps of the central part of the phages corresponding to the region of *E. coli* DNA insert. The restriction sites are represented by vertical lines and individual sites are numbered from left to right. The restriction sites of λp2 eliminated by deletions in the derived phages are indicated by black dots. The lengths of the deletions (Δ) are indicated in the figure in units of 1000 bases and are drawn to scale. However, the precise endpoints of these deletions are not known unless the deletion is closely bracketed by restriction sites. In λp2Δ1, A1 is deleted and so both the length and the end points of the deletion are accurately defined. A1 designates the DNA fragment between *Eco*RI$_1$ and *Eco*RI$_2$ and B1 the fragment between *Eco*RI$_2$ and *Hind*III. The right part of the figure shows the pattern of expression of the different phages. Symbol '+' indicates that the gene is expressed from the corresponding phage '–' means that the gene is not expressed (Springer *et al.*, 1979).

carrying this plasmid overproduce a protein identified as IF3. The fragment B1 was also cloned within an integration proficient λ vector and shown to express *thrS*, *pheS* and *pheT* under lysogenic conditions. Thus, the structural genes and their promoters are located between *Eco*RI and *Hind*III sites. In order to find out the direction of transcription [^{3}H]uridine-labelled mRNA was isolated from a strain carrying pB1. This labelled mRNA was found to hybridize exclusively to the heavy DNA strand of λp2 (Fig. 6). In λ^+ the heavy DNA strand gives rise to the rightward transcripts starting from the promoters p_R and p_Q, allowing expression of the late genes in the direction A to J. In λp2 the *E. coli* DNA is inserted so that the gene order is A → J *thrS-infC-pheS-pheT*-N. Assuming that the heavy strand of λp2 corresponds to that of λ^+, the *E. coli* genes are thus expressed in the direction *thrS* to *pheT* (Plumbridge & Springer, 1980).

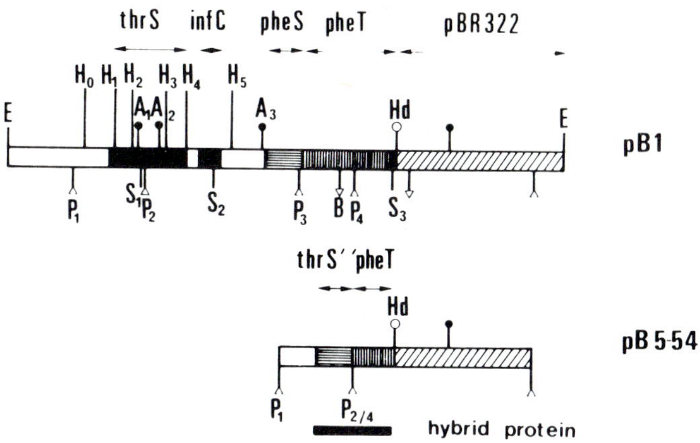

Fig. 5. *Structure of plasmids derived from different fragments of B1 cloned in pBR322*

The symbols used for the different restriction endonuclease sites are indicated in the Figure. The sites are also named in the following way: *Eco*RI, E; *Hpa*I, H; *Ava*I, A; *Hind*III, Hd; *Pst*I, P; *Sac*II, S; *Bam*HI, B; *Sal*I, Sa (Springer *et al.*, 1982).

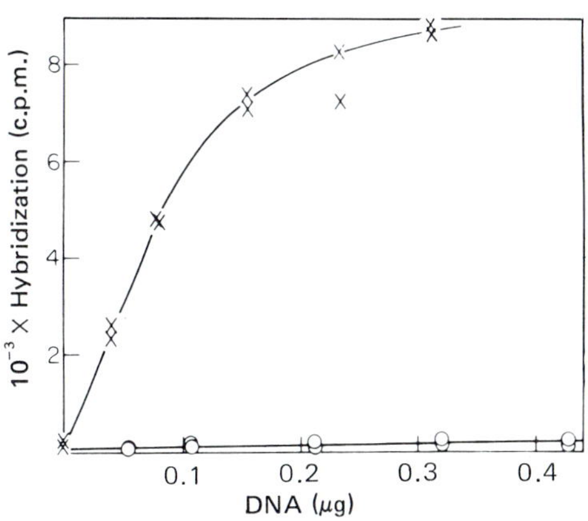

Fig. 6. *Hybridization of [^{3}H]uridine-labelled mRNA from IBPC 1365 pB1 to the separated strands of λp2 DNA*

Liquid hydridization was carried out using a fixed amount of [^{3}H]uridine-labelled mRNA and increasing amounts of heavy (x) and light (○) strand λp2 DNA (Plumbridge & Springer, 1980).

Localization of the Genes on the Restriction Map

The different genes have been located on the restriction map by subcloning the *Pst*I fragments of pB1 into the *Pst*I site of pBR322 (Plumbridge *et al.*, 1980) (Fig. 7). Since *infC* is expressed by pB5-57, the gene is localized between the sites *Pst*I_2-*Pst*I_3. Removal of the *Ava*I_2-*Ava*I_3 fragment eliminates *infC* expression (pB8-7 and pB9-1) but its reinsertion in the opposite orientation (pB8-9) allows *infC* expression. The gene *infC* is therefore located between *Ava*I_2-*Ava*I_3.

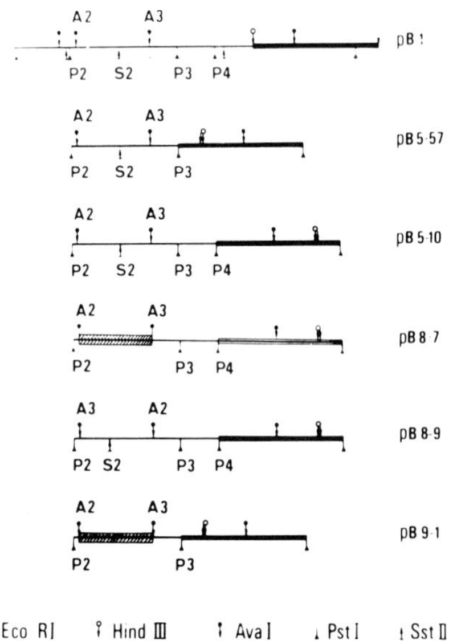

Fig. 7. *Structure of several plasmids derived from different fragments of B1 cloned in pBR322*
Restriction site designations are indicated, and the sites are numbered as in Fig. 5. DNA derived from pBR322 is indicated by a heavy bar. DNA which has been deleted by *in vitro* recombination between *Ava*I sites is indicated by a hatched bar (Plumbridge *et al.*, 1980).

It has been shown that the genes *pheS* and *pheT*, which account for a minimum 3470 bases of DNA, are accommodated between the *Ava*I_3 and *Hind*III restriction sites (Plumbridge *et al.*, 1980).

The fragment *Ava*I_2-*Ava*I_3 (2300 bases) is considerably longer than the expected size of the *infC* gene (550 bases). As the *Sac*II_2 site subdivides the *Ava*I_2-*Ava*I_3 fragment, the *Sac*II_1-*Sac*II_2 and *Sac*II_2-*Sac*II_3 fragments were cloned in the unique *Sac*II site of λY199 (Fig. 8). The genetic content of these recombinants was studied after infection of u.v.-irradiated lysogenic cells (Springer *et al.*, 1982). Neither the recombinant carrying the *Sac*II_1-*Sac*II_2 fragment (λYS) nor the *Sac*II_2-*Sac*II_3 fragment (λYSP) promotes IF3 synthesis (Fig. 9). This shows that the whole *infC* gene is not expressed by the *Sac*II_1-*Sac*II_2 nor by the *Sac*II_2-*Sac*II_3 fragments. The only possible solution is that the *Sac*II_2 site cuts the *infC* gene. As *infC* is only 550 bases long, this result locates the gene within 500 ± 100 bases.

The *thrS* gene has also been precisely located on the restriction map of pB1. This has been possible due to the existence of an hybrid protein formed between

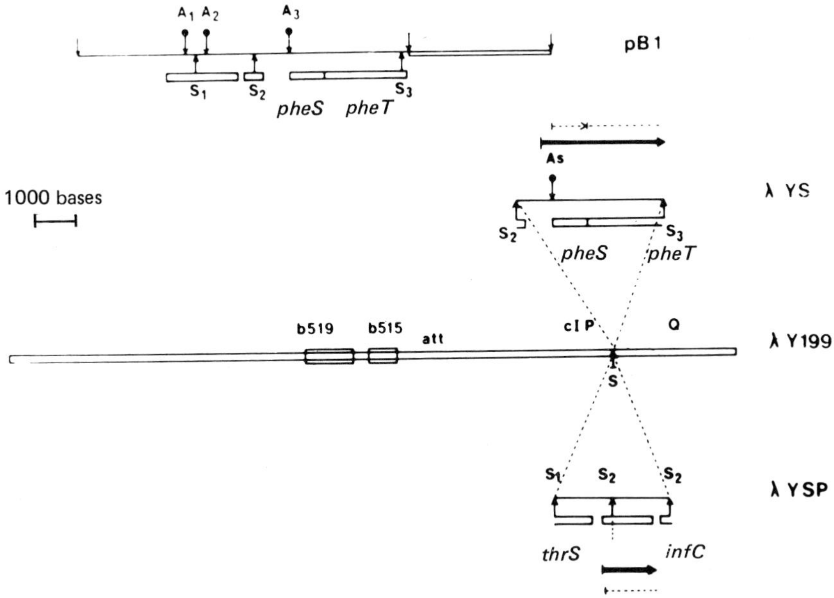

Fig. 8. *Structure of the two phages containing SacII$_1$–SacII$_2$ and SacII$_2$–SacII$_3$ fragments*
Restriction site designations are indicated, and the sites are numbered as in Fig. 5 (Springer *et al.*, 1982).

threonyl-tRNA synthetase and the large subunit of phenylalanyl-tRNA synthetase expressed by the recombinant plasmid pB5-54 (Springer *et al.*, 1982) (Fig. 5). This plasmid has the two *Pst*I fragments of pB1, *Pst*I$_1$–*Pst*I$_2$ and *Pst*I$_4$–*Pst*I$_5$ joined together and is responsible for the overproduction of an M_r 78000 protein. This M_r 78000 protein was proved to be an hybrid between threonyl-tRNA synthetase and the large subunit of phenylalanyl-tRNA synthetase by immunoprecipitation from extracts of ^{35}S-labelled cells using anti-(threonyl-tRNA synthetase) and anti-(phenylalanyl-tRNA synthetase large subunit). As both *thrS* and *pheT* are transcribed in the same direction, this protein has to be hybrid made from the *N*-terminal part of threonyl-tRNA synthetase and the *C*-terminal end of the large subunit of phenylalanyl-tRNA synthetase. We know that *pheT* extends to the *Hind*III site at the *E. coli*-pBR322 boundary of pB1; hence the *pheT* part of the hybrid gene corresponding to the hybrid protein is of the size of the 1300 base *Pst*I$_4$–*Hind*III fragment. An M_r 78000 protein requires 2130 bases of DNA, hence $2130 - 1300 = 830$ bases must be supplied by the *Pst*I$_1$–*Pst*I$_2$ fragment. Thus, this locates the start of the structural gene of *thrS* at 830 bases to the left of *Pst*I$_2$.

The *infC* Gene is Expressed Independently of *thrS*

Genes whose products are involved in translation are often expressed as long transcription units. We investigated whether all the genes (*thrS*, *infC*, *pheS* and *pheT*) were transcribed from a single promoter or not. The recombinant λYSP carries two *Sac*II$_1$–*Sac*II$_2$ fragments in tandem (Fig. 8) and promotes the synthesis

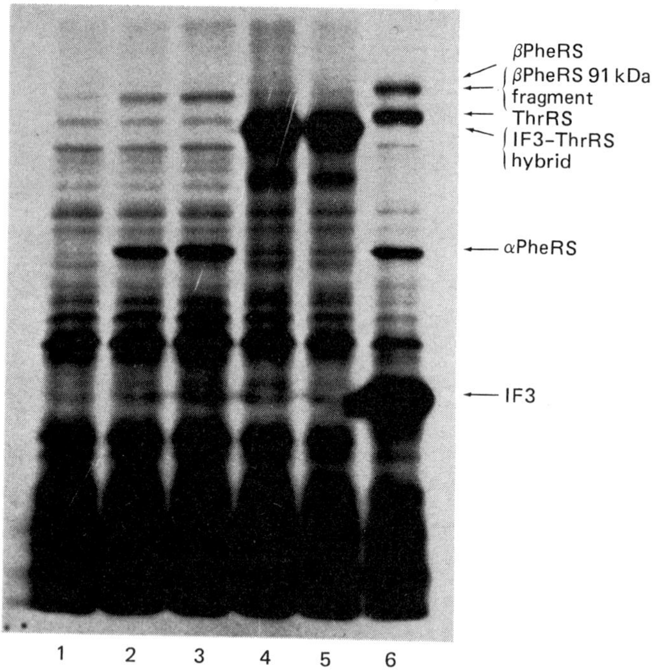

Fig. 9. *Proteins synthesized in u.v.-irradiated cells after infection with λYS and λYSP*

The strain 159 (λind⁻) was grown, irradiated, infected and labelled as described in Springer *et al.* (1977*b*) except that [^{35}S]methionine was used. The infected cells were lysed, the extracts analysed by electrophoresis on SDS/polyacrylamide slab gels, the gels stained and processed for fluorography. Infecting phages: lane 1, λY199; lane 2, λYS; lane 3, λYS; lane 4, λYSP; lane 5, λYSP; lane 6, λp2, λYSP lysates contain λY199. The small and large subunits of phenylalanyl-tRNA synthetase and respectively indicated as αPheRS and βPheRS (Springer *et al.*, 1982). ThrRS, threonyl-tRNA synthetase.

of an M_r 70000 protein in u.v.-irradiated lysogenic cells (Fig. 9). This protein is precipitated by both anti-IF3 and anti-(threonyl-tRNA synthetase) sera (Springer *et al.*, 1982). As it is expressed in lysogenic u.v.-irradiated cells, it has to be expressed from an *E. coli* promoter. The translational start of *thrS* is located at 830 bases to the left of *Pst*I$_2$, not on the *Sac*II$_1$–*Sac*II$_2$ fragment (see above). This fragment, however, carries both a promoter and a translational start site for the IF3-*thrS* hybrid to be expressed. The only gene which can provide these signals is *infC*. The hybrid gene coding for the M_r 70000 protein must start with *infC* in the left *Sac*II$_1$–*Sac*II$_2$ fragment and finish in the right *Sac*II$_1$–*Sac*II$_2$ fragment with *thrS*. The promoter of *infC* is thus active in the cloned *Sac*II$_1$–*Sac*II$_2$ fragment which does not contain the *thrS* promoter. As the promoter of *infC* is active in the absence of the promoter of *thrS* it is likely that *infC* is not expressed from the promoter of *thrS*. The 1900 base *Sac*II$_1$–*Sac*II$_2$ fragment can, at the maximum, code for an M_r 70000 protein; as the IF3-threonyl-tRNA synthetase hybrid has about this size, it must come from an hybrid *infC–thrS* gene which covers a length almost equivalent to that of the *Sac*II$_1$–*Sac*II$_2$ fragment. This shows that *thrS* has to end very close to the place where *infC* starts. The existence of such an hybrid *infC–thrS* gene also indicates that the two genes are transcribed in the same direction.

The *pheS-pheT* Multicistronic Operon is Expressed Independently of *infC*

It has been previously shown that *pheS* and *pheT* constitute an operon and that there is no promoter between the two structural genes *pheS* and *pheT* (Plumbridge & Springer, 1980). They are transcribed from a unique promoter which is carried on the $PstI_2$-$PstI_3$ fragment. In order to find out if the operon is expressed independently of the *infC* promoter, the $SacII_2$-$SacII_3$ fragment was cloned in λY199. It has been shown that *pheS* is expressed by the resulting phage (λYS) after infection of u.v.-irradiated cells. In addition to the synthesis of the small subunit of phenylalanyl-tRNA synthetase, an M_r 91000 protein just smaller than the large subunit of phenylalanyl-tRNA synthetase is synthesized (Fig. 9). This protein is the *pheT* gene product lacking the *C*-terminal end, which has been deleted because $SacII_3$ cuts at the end of the gene, confirming that the gene extends to the *Hind*III site. The synthesis of the *pheS* gene product from λYS gene product and the truncated form of the *pheT* gene product in the absence of the promoter of *infC* shows that *pheS* and *pheT* are expressed independently of *infC.*

Nucleotide Sequence of *infC*

The nucleotide sequence of a 1260 base restriction fragment, $HpaI_4$-$BglI_3$ (the $BglI_3$ site is close to the $HpaI_5$ site) (Fig. 5), of the plasmid pB5-10, which includes the gene for IF3, has been determined (Sacerdot *et al.*, 1982). A sequence with an open reading frame of the size of IF3 is found, which closely corresponds to the localization predicted by the genetic studies (Fig. 10). The amino acid sequence of the protein determined by Brauer & Wittmann-Liebold (1977) is in very good agreement with the coding properties of the nucleotide sequence, with very few exceptions which may reflect the different *E. coli* strains used for protein preparation and cloning. The utilization of codons within the coding sequence closely parallels the frequencies of codon usage in those *E. coli* proteins which are highly expressed (Post & Nomura, 1980; Grosjean & Fiers, 1982).

The important feature of the coding DNA sequence, however, is the presence of an AUU as the translational initiation codon. This is the first instance in which AUU, normally coding for isoleucine, acts as the initiation codon of a normal protein of prokaryotic origin. Note, however, that AUU has been recognized as the possible initiation codon of an unidentified protein coded by an open reading frame in the human mitochondrial DNA sequence (Anderson *et al.*, 1981). In the case of native *E. coli* IF3 (MRE 600), the *N*-terminal amino acid was determined as *N*-monomethylmethionine (Brauer & Wittmann-Liebold, 1977). From experiments *in vivo* and *in vitro*, it can be inferred that methionine is well incorporated into the nascent protein IF3 in response to the ambiguous AUU codon. In fact, all other examples known of ambiguous initiation codons, GUG, UUG or AUA, incorporate formylmethionine efficiently (Grunberg-Manago *et al.*, 1978; Poulis *et al.*, 1981; Steege, 1977; Mackie, 1981).

As we shall discuss below, the primary sequence in the region of a start codon may also play a more important role in selection of the site for protein chain initiation.

```
AACGAATTGACGCAAAAACTATCAAATGCGGGCATTCGTGTTAAAGCAGACTTGAGAAATGAGAAGATTGGCTTTAAAATCCGCGAGCAC
TTGCTTAACTGCGTTTTTGATAGTTTACGCCCGTAAGCACAATTTCGTCTGAACTCTTTACTCTTCTAACCGAAATTTTAGGCGCTCGTG
                       0030                                   0060                                   0090
ACTTTGCGTCGCGTCCCATATATGCTGGTCTGTGGTGATAAAGAGGTGGAATCAGGCAAAGTTGCCGTTCGCACCCGCCGTGGTAAAGAC
TGAAACGCAGCGCAGGGTATATACGACCAGACACCACTATTTCTCCACCTTAGTCCGTTTCAACGGCAAGCGTGGGCGGCACCATTTCTG
                       0120                                   0150                                   0180

CTGGGAAGCATGGACGTAAATGAAGTGATCGAGAAGCTGCAACAAGAGATTCGCAGCCGCAGTCTTAAACAATTGGAGGAATAAGGTATT
GACCCTTCGTACCTGCATTTACTTCACTAGCTCTTCCACGTTGTTCTCTAAGCGTCGGCGTCAGAATTTGTTAACCTCCTTATTCCATAA
                       0210                                   0240                                   0270
LysGlyGlyLysArgValGlnThrAlaArgProAsnArgIleAsnGlyGluIleArgAlaGlnGluValArgLeuThrGlyLeuGluGly
AAAGGCGGAAAACGAGTTCAAACGGCGCGCCCTAACCGTATCAATGGCGAAATTCGCGCCCAGGAAGTTCGCTTAACAGGTCTGGAAGGC
TTTCCGCCTTTTGCTCAAGTTTGCCGCGCGGGATTGGCATAGTTACCGCTTTAAGCGCGGGTCCTTCAAGCGAATTGTCCAGACCTTCCG
                       0300                                   0330                                   0360
GluGlnLeuGlyIleValSerLeuArgGluAlaLeuGluLysAlaGluGluAlaGlyValAspLeuValGluIleSerProAsnAlaGlu
GAGCAGCTTGGTATTGTGAGTCTGAGAGAAGCTCTGGAGAAAGCAGAAGAAGCCGGAGTAGACTTAGTCGAGATCAGCCCTAACGCCGAG
CTCGTCGAACCATAACACTCAGACTCTCTTCGAGACCTCTTTCGTCTTCTTCGGCCTCATCTGAATCAGCTCTAGTCGGGATTGCGGCTC
                       0390                                   0420                                   0450
ProProValCysArgIleMetAspTyrGlyLysPheLeuTyrGluLysSerLysSerSerLysGluGlnLysLysLysGlnLysValIle
CCGCCGGTTTGTCGTATAATGGATTACGGCAAATTCCTCTATGAAAAGAGCAAGTCTTCTAAGGAACAGAAGAAAAAGCAAAAAGTTATC
GGCGGCCAAACAGCATATTACCTAATGCCGTTTAAGGAGATACTTTTCTCGTTCAGAAGATTCCTTGTCTTCTTTTTCGTTTTTCAATAG
                       0480                                   0510                                   0540
GlnValLysGluIleLysPheArgProGlyThrAspGluGlyAspTyrGlnValLysLeuArgSerLeuIleArgPheLeuGluGluGly
CAGGTTAAGGAAATTAAATTCCGTCCTGGTACAGATGAAGGCGACTATCAGGTAAAACTCCGCAGCCTGATTCGCTTTCTCGAAGAGGGT
GTCCAATTCCTTTAATTTAAGGCAGGACCATGTCTACTTCCGCTGATAGTCCATTTTGAGGCGTCGGACTAAGCGAAAGAGCTTCTCCCA
                       0570                                   0600                                   0630
AspLysAlaLysIleThrLeuArgPheArgGlyArgGluMetAlaHisGlnGlnIleGlyMetGluValLeuAsnArgValLysAspAsp
GATAAAGCCAAAATCACGCTGCGTTTCCGCGGTCGTGAGATGGCGCACCAGCAAATCGGTATGGAAGTGCTTAATCGCGTGAAAGACGAT
CTATTTCGGTTTTAGTGCGACGCAAAGGCGCCAGCACTCTACCGCGTGGTCGTTTAGCCATACCTTCACGAATTAGCGCACTTTCTGCTA
                       0660                                   0690                                   0720
LeuGlnGluLeuAlaValValGluSerPheProThrLysIleGluGlyArgGlnMetIleMetValLeuAlaProLysLysLysGln
TTGCAAGAACTGGCAGTGGTCGAATCCTTCCCAACGAAGATCGAAGGCCGCCAGATGATCATGGTGCTCGCTCCTAAGAAGAAACAGTAA
AACGTTCTTGACCGTCACCAGCTTAGGAAGGGTTGCTTCTAGCTTCCGGCGGTCTACTAGTACCACGAGCGAGGATTCTTCTTTGTCATT
                       0750                                   0780                                   0810
GGCCTTCAAGTAGCACTGTCTGTGGAGCCTTCGGGCTCACAGCTTTGTTCGCCTTTGTTTCGTTTATTAACAATGCGAAGTGGAAGTTAT
CCGGAAGTTCATCGTGACAGACACCTCGGAAGCCCGAGTGTCGAAACAAGCGGAAACAAAGCAAATAATTGTTACGCTTCACCTTCAATA
                       0840                                   0870                                   0900
TAAAATGCCAAAAATTAAAGACCGTACGCGGTGCGTCTAAGCGCTTCAAAAAAACCGGTAAAGGTGGTTTTAAGCACAAGCACGCTAACC
ATTTTACGGTTTTTAATTTCTGGCATGCGCCACGCAGATTCGCGAAGTTTTTTTGGCCATTTCCACCAAAATTCGTGTTCGTGCGATTGG
                       0930                                   0960                                   0990
TGCGTCACATTCTGACCAAAAAAGCGACCAAACGTAAACGTCACCTGCGTCCGAAAGCCATGGTTTCCAAAGGCGATCTGGGCCTGGTAA
ACGCAGTGTAAGACTGGTTTTTTCGCTGGTTTGCATTTGCAGTGGACGCAGGCTTTCGGTACCAAAGGTTTCCGCTAGACCCGGACCATT
                       1020                                   1050                                   1080
TCGCGTGCCTGCCGTACGCATAAGCCGTTAACGTTTTTAACTTTTTAATTAGAATATAGATACAGGAGAGCACATATGGCTCGCGTAAAA
AGCGCACGGACGGCATGCGTATTCGGCAATTGCAAAAATTGAAAAATTAATCTTATATCTATGTCCTCTCGTGTATACCGAGCGCATTTT
                       1110                                   1140                                   1170
CGTGGTGTTATTGCACGTGCACGTCACAAGAAAATTTTGAAACAAGCTAAAGGCTACTACGGTCGGCGTTCCGCGTATACGCCGTTGCC
GCACCACAATAACGTGCACGTGCAGTGTTCTTTTAAAACTTTGTTCGATTTCCGATGATGCCAGCCGCAAGGCGCATATGCGGCAACGG
                       1200                                   1230                                   1260
```

Fig. 10. *Nucleotide sequence of the gene for IF3 and its flanking regions*

The sequence is numbered from the *Hpa*I$_4$ to the *Bgl*I$_3$ cleavage sites. The amino acid sequence deduced from the DNA sequence is superimposed on the non-transcribed DNA strand (Sacerdot *et al.*, 1982).

Plasmid-encoded IF3 is strongly expressed both *in vivo* and *in vitro* (Lestienne *et al.*, 1982). The high level of expression of IF3 can be correlated with the existence of a GGAGG sequence (bases 255 to 259 in the DNA sequence of Fig. 1) upstream of the AUU initiation codon. This sequence can pair with the sequence CCUCC near the 3′-terminus of 16S rRNA, as initially proposed by Shine & Dalgarno (1974). Considering mRNA–fMet-tRNA interactions, we note that the sequence indicates an A residue on the 3′-side of the starter codon. An A residue following the AUU codon may play a role in ribosome binding by participating in a fourth base-pair with the sequence UACU found in the anticodon loop of fMet-tRNA. Experimental evidence that an A residue in this position aids ribosome binding has come from the work of Taniguchi & Weissmann (1978) with the Qβ coat protein ribosome binding site. The presence of an A residue on the 3′-side of the initiation codon has also been invoked by Belin *et al.* (1979) to explain the use of an altered initiation codon (AUA instead of AUG).

The sequence also predicts a U on the 5′-side of the IF3 initiation codon which may help in the use of the AUU codon. It has been shown by Ganoza *et al.* (1978) that bases adjacent to the 5′-end of an AUG have considerable influence on the formation of the initiation complex: tetramers of composition UAUG or CAUG initiate well, AAUG and GAUG do not. The preference for pyrimidines over purines at the 5′-side of an initiator AUG in native mRNA was already noted (Steitz *et al.*, 1977). U on the 5′-side of the initiator codon might base-pair with the A on the 3′-side of the fMet-tRNA anticodon.

Finally, another interesting feature is the occurrence of a UAA terminator between the Shine and Dalgarno region and the initiation codon of IF3, two codons upstream from the AUU initiation codon. The frequent occurrence of in-phase or out-of-phase nonsense codons 5′ to initiator AUG was noted first by Ganoza (1977). Atkins (1979) has proposed that UAA or UGA close to the initiation codon may be part of the ribosome recognition signal.

All these features, a strong Shine and Dalgarno interaction, a possible engagement in base pairing with initiator tRNA of the nucleotides adjacent to the starter codon and the occurrence of an UAA terminator 5′ to the initiator, could offset a reduced codon–anticodon interaction in initiation complex formation and be the origin of the high efficiency of translation of IF3.

Gene Dosage Effect on IF3 Synthesis

It has been shown that the cellular concentration of initiation factor IF3 increases with increasing growth rate and is co-ordinated with the level of ribosomes (J.G. Howe & J.W.B. Hershey, unpublished work). In view of the autogenous regulation of synthesis of ribosomal proteins, it was important to determine whether synthesis of the initiation factors is subject to the same type of control. As the synthesis of ribosomal proteins in the cell is not subject to dependent gene dosage, the gene dosage response of IF3 synthesis (Lestienne *et al.*, 1982) was determined. For that purpose a strain haploïd for *infC* was modified to carry additional copies of the *infC* gene. Cells diploïd for *infC* were constructed by lysogenization with λB1 or by introduction of an F′ episome KLF48; a triploïd strain was made by introduction of the F′ episome into the λB1 lysogen. Finally, polyploïd strains were achieved by the presence of the multicopy plasmids pB5-57 and pB1 expressing the *infC* gene. The level of initiation factor was determined in crude lysates of these different strains by an immunoblotting technique which enables the quantitative measurement of the amount of IF3 (Howe *et al.*, 1978; Howe & Hershey, 1981; Lestienne *et al.*, 1982). Fig. 11 shows autoradiography of the nitrocellulose blot tested with anti-IF3 sera followed by visualization with ^{125}I-labelled *Staphylococcus aureus* protein A. The amount of radioactivity present was measured by counting the excised bands (Table 3). The average values for IF3 were increased two-fold in

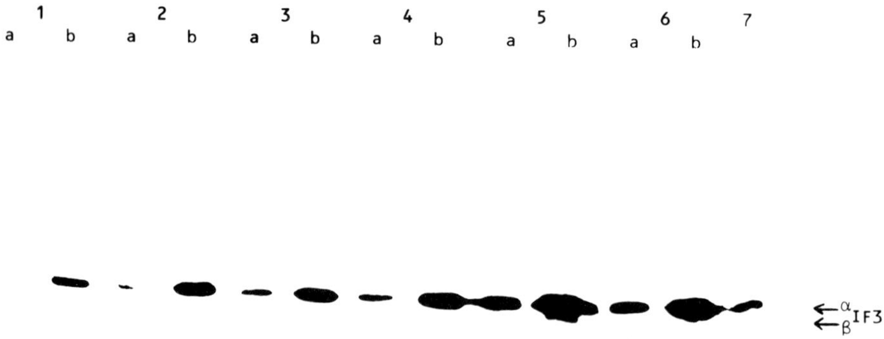

Fig. 11. *Autoradiogram of immunoblots of cells carrying different numbers of copies of the IF3 gene*

The strain used was IBPC 1671 (haploïd *infC*) (lane 1) modified as follows: by the presence of KLF48 (diploïd) (lane 2); λB1 lysogen (diploïd) (lane 3); KLF48, λB1 (triploïd) (lane 4); pB5-57 (polyploïd) (lane 5); pB1 (polyploïd) (lane 6); lane 7 contains 0.068 μg of IF3α and 2 μg of bovine serum albumin. Each lane 'a' contains about 2 μg of total protein and each lane 'b' 10 μg of total protein. The SDS gel was electrophoretically blotted on to nitrocellulose and treated with antibodies to IF3 and subsequently with ^{125}I-labelled protein A. Under the conditions of lysis used here, IF3 appears predominantly as a single band, IF3α (Lestienne *et al.*, 1982).

Table 3. *Level of IF3 in cells carrying different numbers of copies of the infC gene*

The radioactive bands on the nitrocellulose blot located by autoradiography (Fig. 11) were cut out and counted in a γ counter. The assay for each strain was performed in duplicate with two samples of about 2 μg and 10 μg of protein. The radioactivity is expressed as an average of the two measurements except for IBPC 1671 carrying pB5-57 or pB1, where only the measurement with 2 μg of protein is shown because non-linearity was observed at higher protein concentration. Values in parentheses indicate the relative quantity of IF3 in different strains.

Strains	^{125}I (c.p.m./μg of protein)	
IBPC 1671 (haploïd)	850 (1)	
IBPC 1671 KLF 48 (diploïd)	1410 (1.65)	(2.06)
IBPC 1671 λB1 (diploïd)	2115 (2.48)	
IBPC 1671 KLF 48 λB1 (triploïd)	2869 (3.37)	
IBPC 1671 pB5-57 (polyploïd)	8987 (10.5)	
IBPC 1671 pB1 (polyploïd)	8079 (9.5)	

the diploïd strain and three-fold in the triploïd; about a 10-fold amplification was found in the polyploïd strains. This indicates a quantitative dependence of IF3 levels on gene dosage. As a control, IF2 levels were also determined and were not found to vary in a significant way. IF3 synthesis is therefore not subject to the type of autogenous regulation shown by some ribosomal protein operons. This was confirmed by studying IF3 synthesis in a cell-free coupled transcriptional–translation system dependent on exogenous DNA carrying the *infC* gene. Addition of excess IF3 to the system *in vitro* does not inhibit IF3 synthesis (Lestienne *et al.*, 1982; Elhardt *et al.*, 1982).

Purification of IF3 from Overproducing Strains

As was shown above, the presence of a multicopy plasmid causes considerable overproduction of IF3. This enables the purification of large quantities of IF3 from a cell carrying the plasmid pB5-57 (Lestienne *et al.*, 1982). When pB5-57 lysates are prepared from freshly grown cells in SDS buffer, the IF3 protein can be detected directly as a single stained protein band. When such lysates are prepared in non-denaturing buffer to make an 'S30' extract and analysed subsequently, two forms of IF3 corresponding to IF3α and IF3β are readily detected in the strain carrying pB5-57 (Fig. 12) (Lestienne *et al.*, 1982). IF3β is a shorter form lacking a hexa-

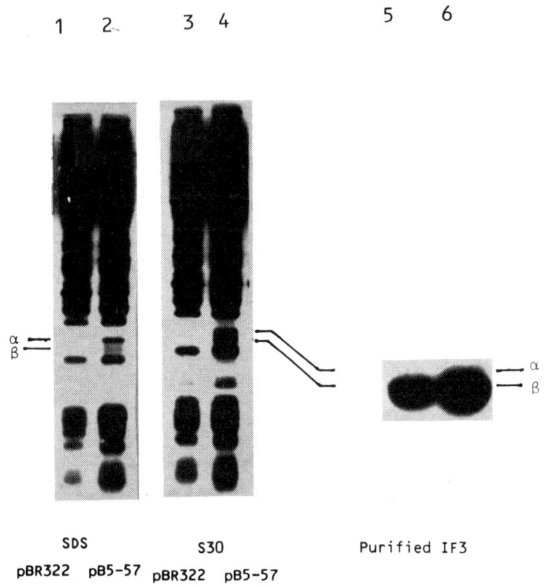

Fig. 12. *Analysis of proteins of E. coli strain IBPC 1671 carrying plasmids by SDS/polyacrylamide gel electrophoresis*

Lane 1, SDS lysate of IBPC 1671 carrying pBR322 (20 μg of protein); lane 2, SDS lysate of IBPC 1671 carrying pB5-57 (20 μg of protein); lane 3, 'S30' lysate of IBPC 1671 carrying pBR322 (27 μg of protein); lane 4, 'S30' lysate of IBPC 1671 carrying pB5-57 (27 μg of protein); lane 5, purified IF3 from IBPC 1671 carrying pB5-57 (8.6 μg of protein); lane 6, purified IF3 from IBPC 1671 carrying pB5-57 (25 μg of protein). Electrophoresis was done in 12.5% (lanes 1–4) or 15% (lanes 5–6) polyacrylamide gel. Proteins were stained by Coomassie Blue (Lestienne *et al.*, 1982).

peptide at the *N*-terminus and has been reported previously in extracts of *E. coli* MRE 600 (Suryanarayana & Subramanian, 1977). The amino acid analysis of the smaller overproduced protein verified that it indeed corresponds to IF3β (C. Gualerzi, personal communication) and it has been shown that it is due to limited proteolysis of IF3α. IF3β was purified by the previously described method (Hershey *et al.*, 1977) from the strain carrying pB5-57 (harvested at the end of exponential growth). It is worthwhile noting that the majority of the overproduced IF3 is still found associated with the ribosome fraction. Starting from 500 g of cells, 100 mg of IF3β were purified to homogeneity; this corresponds to an amplification of yield around 20-fold compared with the wild-type strain.

Discussion

A crucial problem of the dynamics of cell growth is to understand the mechanism of regulation of the enzymes of the translational apparatus. The synthesis of proteins is one of the major metabolic activities, especially in rapidly growing bacteria. The macromolecular components involved in protein synthesis represent up to 40% of the cell mass. Because synthesis of the translational machinery itself requires a large commitment of the cell's resources it is expected, and observed, that precise controls exist which regulate and co-ordinate the levels of the various components. In this review, we focus on the problem of control of initiation factor IF3 expression in the bacterium *E. coli*. We hope that understanding the mode of expression of this gene will also shed light on the role of IF3 in initiation of translation at the molecular level.

We have found that the gene *infC* for initiation factor IF3 is grouped with three other genes whose products are involved in translation, namely the *thrS* gene coding for threonyl-tRNA synthetase and the *pheS* and *pheT* genes coding for the two subunits of phenylalanyl-tRNA synthetase. This recalls the case of the genes for elongation factors Tu and G, which are transcribed with two genes for ribosomal proteins S7 and S12. However, in contrast with this case, the IF3 gene, although transcribed in the same direction as *thrS*, *pheS* and *pheT*, appears to be on a separate transcriptional unit with an independent promoter. The promoter of *infC* appears, however, to be very close to the structural gene of *thrS*. We have noted that the coding sequence of IF3 is preceded by an open in-phase reading frame of 261 bases which ends at the stop codon UAA, only two codons upstream from the initiator triplet AUU of IF3. If this really represents the end of the *thrS* gene, it correlates well with the closeness of the *thrS* and *infC* genes implied by the λYSP hybrid protein. Such a close proximity of two genes would imply that regulatory signals for the expression of the second gene are located within the structural part of the first gene. Thus, although we have shown the two genes are transcribed independently, there is a possibility that their expression is at least partly interdependent.

The occurrence of the ambiguous AUU initiator codon is a striking feature of the *infC* sequence. Clearly this codon, normally coding for isoleucine, does not impair the expression of the initiation factor IF3 gene *infC* from the plasmid DNA *in vivo* or *in vitro*. Other examples of unexpected ambiguous initiation triplets have recently been reported: AUA (Belin *et al.*, 1979) and UUG (McLaughlin *et al.*, 1981; Poulis *et al.*, 1981; Mackie, 1981). In all these cases the central U of the standard AUG triplet is invariant but either the first or the last was changed. The question remains open if an AUU initiation codon reflects some special feature of the expression or regulation of *infC*. Possibly the efficient use of AUU as the initiator codon is provided by the sequence of the surrounding regions. Perhaps AUU and AUG are equally good as initiation codons and it is the context of the codon which determines the efficiency of initiation complex formation and hence translation.

In the wild-type cell the ratio of IF3 to ribosomes remains constant, both increasing in parallel as the growth rate increases. The same metabolic regulation is true for elongation factor G. However, the way that these levels are controlled must differ. Some ribosomal proteins and EF-G have been shown to be subject to autogenous control at the translational level (Dean & Nomura, 1980; Yates *et al.*, 1980). This is not the case for IF3 whose cellular level *in vivo* responds to gene dosage. The mechanism by which the level of IF3 is regulated has still to be discovered. Hopefully this is not too distant. The cloned gene allows the isolation of mutants affected in IF3 function and synthesis by *in vitro* mutagenesis of a precise region of the chromosome. Comparison of the sequence of mutants and wild type should indicate important structural and control regions.

We are grateful to C. Gualerzi for performing the *N*-terminal amino acid sequence of the overproduced IF3β. This work was supported by the following grants: Centre National de la Recherche Scientifique (G.R. no. 18, L.A. no. 240, ATP 'Biologie Moléculaire du Gène' et 'Microbiologie 1979'), Délégation Générale à la Recherche Scientifique et Technique (Convention 80.E.0872), Ligue Nationale contre le Cancer (Comité de la Seine) and Commissariat à la l'Energie Atomique.

References

Anderson, S., Bankier, A.T., Barrell, B.G., de Bruijn, M.H.L., Coulson, A.R., Drouin, J., Eperon, I.C., Nierlich, D.P., Roe, B.A., Sanger, F., Schreier, P.H., Smith, A.J.H., Staden, R. & Young, I.G. (1981) *Nature (London)* **290**, 457–465

Atkins, J.F. (1979) *Nucleic Acids Res.* **7**, 1035–1041

Belin, D., Hedgpeth, J., Selzer, G.B. & Epstein, R.H. (1979) *Proc. Natl. Acad. Sci. U.S.A.* **76**, 700–704

Brauer, D. & Wittmann-Liebold, B. (1977) *FEBS Lett.* **79**, 269–275

Dean, D. & Nomura, M. (1980) *Proc. Natl. Acad. Sci. U.S.A.* **77**, 3590–3594

Elhardt, D., Wirth, R. & Bock, A. (1982) *Eur. J. Biochem.* in press

Ganoza, M.C. (1977) *Can. J. Biochem.* **55**, 267–281

Ganoza, M.C., Frazer, A.R. & Neilson, T. (1978) *Biochemistry* **17**, 2769–2775

Gold, L., Pribnow, D., Schneider, T., Shinedling, S., Singer, B.S. & Stormo, G. (1981) *Annu. Rev. Microbiol.* **35**, 365–403

Grosjean, H. & Fiers, W. (1982) *Gene* **13**, in press

Grunberg-Manago, M., Buckingham, R.H., Cooperman, B.S. & Hershey, J.W.B. (1978) *Relation between Structure and Function in the Prokaryotic Cell* (Stanier, R.V., Rogers, H.J. & Ward, B.J., eds.), vol. 28, pp. 27–110, Cambridge University Press

Grunberg-Manago, M. (1980) in *Ribosomes: Structure, Function and Genetics* (Chambliss, G., Craven, G.R., Davies, J., Davis, K., Kahan, L. & Nomura, M., eds.), pp. 445–447, University Park Press, Baltimore

Hennecke, H., Springer, M. & Bock, A. (1977) *Mol. Gen. Genet.* **152**, 206–210

Hershey, J.W.B. (1980) *Cell Biol.* **4**, 1–68

Hershey, J.W.B., Yanov, J., Johnston, K. & Fakunding, J.L. (1977) *Arch. Biochem. Biophys.* **182**, 626–638

Howe, J.G., Yanov, J., Meyer, L., Johnston, K. & Hershey, J.W.B. (1978) *Arch. Biochem. Biophys.* **191**, 813–820

Howe, J.G. & Hershey, J.W.B. (1981) *J. Biol. Chem.* **256**, 12836–12839

Kung, H.F., Redfield, B., Tradwell, B.V., Eskin, H., Spears, C. & Weissbach, H. (1977) *J. Biol. Chem.* **252**, 6889–6894

Lestienne, P., Dondon, J., Plumbridge, J.A., Howe, J.G., Mayaux, J.F., Springer, M., Blanquet, S., Hershey, J.W.B. & Grunberg-Manago, M. (1982) *Eur. J. Biochem.* in press.

Maaløe, O. (1979) in *Biological Regulations and Development* (Goldberger, R.F., ed.), vol. 1, pp. 487–542, Plenum Press, New York

Mackie, G.A. (1981) *J. Biol. Chem.* **256,** 8177–8182

McLaughlin, J.R., Murray, C.L. & Rabinowitz, J.C. (1981) *Proc. Natl. Acad. Sci. U.S.A.* **78,** 4912–4916

Miura, A., Krueger, J.H., Itah, S., de Boer, H.A. & Nomura, M. (1981) *Cell* **25,** 773–782

Petersen, H.U., Roll, T., Grunberg-Manago, M. & Clark, B.F.C. (1979) *Biochem. Biophys. Res. Commun.* **91,** 1068–1074

Petersen, H.U., Kruse, T.A., Worm-Leonard, H., Siboska, G.E., Clark, B.F.C., Boutorin, A.S., Remy, P., Ebel, J.P., Dondon, J. & Grunberg-Manago, M. (1981) *FEBS Lett.* **128,** 161–165

Plumbridge, J.A. & Springer, M. (1980) *J. Mol. Biol.* **144,** 595–600

Plumbridge, J.A., Springer, M., Graffe, M., Goursot, R. & Grunberg-Manago, M. (1980) *Gene* **11,** 38–42

Post, L. & Nomura, M. (1980) *J. Biol. Chem.* **255,** 4660–4666

Poulis, M.I., Shaw, D.C., Campbell, H.D. & Young, I.G. (1981) *Biochemistry* **20,** 4178–4185

Sacerdot, C., Fayat, G., Dessen, P., Springer, M., Plumbridge, J.A., Grunberg-Manago, M. & Blanquet, S. (1982) *EMBO J.* in press

Shine, J. & Dalgarno, L. (1974) *Proc. Natl. Acad. Sci. U.S.A.* **71,** 1342–1346

Springer, M., Graffe, M. & Grunberg-Manago (1977*a*) *Mol. Gen. Genet.* **151,** 17–26

Springer, M., Graffe, M. & Hennecke, H. (1977*b*) *Proc. Natl. Acad. Sci. U.S.A.* **74,** 3970–3974

Springer, M., Graffe, M. & Grunberg-Manago, M. (1979) *Mol. Gen. Genet.* **169,** 337–343

Springer, M., Plumbridge, J.A., Trudel, M., Graffe, M. & Grunberg-Manago, M. (1982) *Mol. Gen. Genet.* in press

Steege, D.A. (1977) *Proc. Natl. Acad. Sci. U.S.A.* **74,** 4163–4167

Steitz, J.A., Sprague, K.V., Steege, D.A., Yuan, R.C., Laughrea, M., Moore, P.B. & Wabha, A.J. (1977) in *Nucleic Acid Protein Recognition* (Vogel, H.J., ed.), pp. 491–508. Academic Press, New York

Steitz, J.A. (1980) in *Ribosomes: Structure, Function and Genetics* (Chambliss, G., Craven, G.R., Davies, J., Davis, K., Kahan, L. & Nomura, M., eds.), pp. 479–495, University Park Press, Baltimore

Suryanarayana, T. & Subramanian, A.R. (1977) *FEBS Lett.* **79,** 264–268

Taniguchi, T. & Weissmann, C. (1978) *J. Mol. Biol.* **118,** 533–565

Yates, J.L., Arfsten, A.E. & Nomura, M. (1980) *Proc. Natl. Acad. Sci. U.S.A.* **77,** 1837–1841

Biochem. Soc. Symp. 47, 113–128
Printed in Great Britain

How do Eukaryotic Ribosomes Recognize the Unique AUG Initiator Codon in Messenger RNA?

MARILYN KOZAK

Department of Biological Sciences, University of Pittsburgh, Pittsburgh, PA 15260, U.S.A.

Introduction

Although the pivotal role of the AUG codon in defining translational start sites has long been recognized, only a small fraction of the AUG triplets in a given message function as ribosome binding sites. The question of how ribosomes distinguish between initiator and non-initiator AUG triplets has stimulated considerable experimentation and debate. In the case of prokaryotes, an answer began to emerge in 1974 when Shine and Dalgarno noticed that functional initiator codons are usually preceded by a short sequence that is complementary to the 3′-end of *Escherichia coli* 16 S ribosomal RNA. Steitz & Jakes (1975) lost no time in devising an experiment to demonstrate directly base-pairing between initiation regions of phage mRNA and the 3′-terminal fragment of 16 S rRNA. The importance of this interaction was confirmed when J.J. Dunn *et al.* (1978) isolated mutants in which the 'Shine–Dalgarno sequence' was altered. These pioneering experiments were carried out with *E. coli* ribosomes and various bacterial or phage mRNAs. A few years later, when investigations were undertaken to probe the mechanism of initiation in eukaryotes, a surprisingly different answer emerged. A 'scanning mechanism' has been proposed to explain the peculiar characteristics of initiation in eukaryotes. The hypothesis, briefly stated, is that the 40 S ribosomal subunit (with associated initiation factors) binds initially at or near the 5′-end of a message and subsequently migrates along the RNA chain, stopping when it encounters the first AUG triplet. At that point, the 60 S subunit joins and peptide bond formation begins. What follows is a summary of evidence supporting various aspects of the scanning model. Since inclusion of all of the pertinent literature would be beyond the scope of this review, I have focused on more recent papers. The reader is referred to earlier reviews (Kozak, 1978, 1981*a*) for more extensive discussion of the literature prior to 1980.

The pertinent evidence can be evaluated most easily by considering three steps in the initiation process: (i) binding of 40 S ribosomal subunits at or near the 5′-terminus of the message, followed by (ii) migration, and (iii) recognition of a particular AUG codon as a 'stop signal' by the migrating 40 S ribosome.

A nearby 5′-end is necessary (and perhaps sufficient) for ribosome binding

There is a peculiar restriction on the ability of eukaryotic ribosomes to translate polycistronic mRNAs. In contrast with their prokaryotic counterparts, which initiate quite efficiently at internal cistrons, eukaryotic ribosomes translate only the 5′-proximal cistron in a polycistronic transcript. For example, although the 3′-portion of brome mosaic virus (BMV) RNA-3 contains a complete copy of the coat protein gene, RNA-3 does not direct translation of that protein. The functional message for BMV coat protein is a smaller RNA in which the coat protein coding region lies closer to the 5′-terminus (Shih & Kaesberg, 1973). There are many additional examples of 'silent' downstream cistrons among plant and animal virus mRNAs (Smith, 1977); a few are described in Table 1. In some cases, as indicated in the Table, the initiator codon of the 3′-proximal cistron overlaps the coding sequence of the preceding cistron, and failure to translate the downstream cistron might be blamed on interference from ribosomes translating the first cistron. But this explanation cannot be extended to many other polycistronic mRNAs in which the two cistrons do not overlap (Table 1). The inability of eukaryotic ribosomes to initiate at internal AUG codons extends even to prokaryotic mRNAs. For example, Rosenberg & Paterson (1979) showed that wheat germ ribosomes translate only the 5′-proximal cistron of a polycistronic mRNA from bacteriophage lambda. On the other hand, Berg and his colleagues obtained contradictory results with certain SV40 recombinants that carry foreign genes (Mulligan & Berg, 1981; Subramani *et al.*, 1981). They showed that, with some of their recombinants, a

Table 1. *Plant and animal virus mRNAs which contain a 'silent cistron' downstream from the translationally active cistron*

Arrangement of the cistrons	mRNA	Expressed cistron (5′-proximal)	Silent cistron (3′-proximal)	Reference
Partially overlapping	(1) SV40 late 19S mRNA	VP2/3	VP1	Reddy *et al.*, 1978
	(2) Polyoma late 19S mRNA	VP2/3	VP1	Arrand *et al.*, 1980
	(3) Moloney murine leukaemia virus genome	*gag/pol*	*env*	Shinnick *et al.*, 1981
	(4) Adenovirus-2 region e1B (13S mRNA)	20 kDa protein	Protein IX	Perricaudet *et al.*, 1980
Non-overlapping	(5) Tobacco mosaic virus RNA genome	30 kDa non-structural protein	Coat protein	Guilley *et al.*, 1979
	(6) Alfalfa mosaic virus RNA-3	35 kDa non-structural protein	Coat protein	Koper-Zwarthoff *et al.*, 1980
	(7) Semliki Forest virus genome	Nonstructural proteins	Capsid protein	Garoff *et al.*, 1980*a*
	(8) Rous sarcoma virus 24S mRNA	*env*	*src*	Czernilofsky *et al.*, 1980
	(9) Adenovirus-2 late region 3 (largest mRNA)	pVI	Hexon	Akusjärvi & Persson, 1981*a*
	(10) Adenovirus-2 hexon mRNA	Hexon	23 kDa protein	Akusjärvi *et al.*, 1981
	(11) Adenovirus-2 region e1B (22S mRNA)	67 kDa protein	Protein IX	Perricaudet *et al.*, 1980

foreign gene positioned near the 3′-end of a hybrid transcript was expressed. A possible explanation is that the foreign protein was translated, not from the most abundant form of mRNA (in which the foreign gene lies downstream from SV40 genes), but from minor species of mRNA in which the foreign gene lies closer to the 5′-terminus. Failure to translate the 3′-proximal cistron in polycistronic mRNAs is most simply explained by its being too far from the 5′-end of the message; the initiation mechanism in eukaryotes somehow depends on having a nearby 5′-terminus. [More precisely, it is not the number of upstream nucleotides *per se* that determines whether a given AUG triplet can function in initiation, since the sequence preceding the functional initiator codon in eukaryotic mRNAs varies in length from 10 to over 300 nucleotides (Kozak, 1981*a*). The critical requirement seems to be that the upstream sequence, however long it may be, must be free of other potential initiator codons. Adenovirus late mRNAs provide an interesting illustration. The normal tripartite leader consists of 200 nucleotides and is free of AUG triplets (Zain *et al.*, 1979). There are two instances of adenovirus mRNAs which carry 'extra' leader segments. In one case the extra leader segment (181 nucleotides long, designated *y*) lacks AUG triplets (Zain *et al.*, 1979) and it does not prevent translation from occurring at the normal AUG codon (A.R. Dunn *et al.*, 1978). On the other hand, Akusjärvi & Persson (1981*b*) have shown that a different leader segment (designated *i*), which contains AUG triplets, does interfere with translation of downstream sequences.]

The requirement for a free 5′-terminus was underscored by experiments carried out with circular RNA templates. Circularization of AUG-containing synthetic polymers abolished their ability to bind to wheat germ or reticulocyte (but not to *E. coli*) ribosomes (Kozak, 1979*a*). In control experiments, linear uncapped forms of the same ribopolymers were able to form stable 80 S complexes. Thus, the nucleotide sequence of the circular template was adequate for interaction with ribosomes; it seemed to be the form of the circular RNA molecules that precluded binding. These experiments suggested that mRNA must have a free 5′-terminus in order for eukaryotic ribosomes to bind. Evidence that seemed to contradict this generalization was reported by Golini *et al.* (1980). They found that poliovirus genomic RNA, which is covalently linked at its 5′-end to a small viral-encoded peptide (VPg), can form 80 S initiation complexes with reticulocyte ribosomes. Some of the experiments supporting that conclusion have been called into question, however, by more recent data from the same laboratory. Using a more reliable assay, Dorner *et al.* (1981) found that some VPg is removed from poliovirus RNA during incubation in the reticulocyte lysate.

Whereas circularization of mRNA abolishes binding, manipulations that create new ends create new ribosome binding sites. There are many examples of viral messages which, upon undergoing limited fragmentation, display a large spectrum of initiation sites that were inactive in the intact form of the message. Activation of adventitious ribosome binding sites has been observed *in vitro* upon cleavage of cowpea mosaic virus RNA (Pelham, 1979), foot-and-mouth disease virus RNA (Sangar *et al.*, 1980), reovirus mRNA (Kozak, 1980*b*), and the *src* region of Rous sarcoma (Beemon & Hunter, 1978) and Moloney murine sarcoma virus RNA (Cremer *et al.*, 1981; Papkoff *et al.*, 1981). Internal AUG triplets can also be converted to functional initiator codons when aberrant transcripts are produced, thereby bringing internal AUG codons near the 5′-end of the message. This seems to happen *in vivo* during transcription of the SV40 sequences in certain adenovirus–SV40 hybrids (Lawrence *et al.*, 1979), and during transcription *in vitro* of sequences coding for SV40 VP1 and T antigen (Cepko *et al.*, 1981). Failure to initiate translation at these internal sites in normal, full-sized SV40 mRNAs emphasizes that a nearby 5′-end is needed. On the other hand, the ease with which internal AUG codons can be activated (via cleavage or aberrant transcription) suggests that a nearby 5′-end might be the only signal required for ribosome binding. If there are additional sequence requirements, they cannot be very elaborate in view of the large number of spurious sites that become active in fragmented mRNA. The fact that translation of eukaryotic mRNAs is not abolished by extensively perturbing the 5′-noncoding sequence (Bendig & Folk, 1979; Barkan & Mertz, 1981; Ghosh *et al.*, 1981; Solnick, 1981; Woodworth-Gutai, 1981) reinforces the notion that eukaryotic ribosomes are indifferent to the exact sequence at the 5′-end of the message.

After entering at the 5′-end of a message, 40 S subunits migrate toward the interior until an AUG triplet is encountered

The conclusion that 40 S ribosomal subunits can migrate derives from analysis of initiation complexes formed in the presence of edeine (Kozak & Shatkin, 1978). The antibiotic edeine seems to impair recognition of the AUG codon, thereby allowing 40 S subunits to advance beyond the 5′-proximal AUG triplet which would normally be used to initiate translation. Thus we were able to show that, in the presence of edeine, 40 S subunits bind at the 5′-end of the message and advance all the way to the 3′-end (Kozak, 1980*c*). The ability to migrate does not seem to be induced by edeine; exaggerated migration of 40 S subunits (i.e. beyond the 5′-proximal AUG triplet) also occurs when the magnesium concentration is lowered (Kozak, 1979*b*) or when denatured mRNA is used as template (Kozak, 1980*a*). Thus, the ability to migrate seems to be an inherent property of the 40 S subunit · initiation factor complex. Although many of these experiments were carried out using reovirus mRNAs, some of the observations have been confirmed with plant virus mRNAs (Kozak & Shatkin, 1978, Fig. 11; Kozak, 1980*c*).

Migration of 40 S ribosomes in the presence of edeine is dependent on an ATP regenerating system (Kozak, 1980*c*). This observation was exploited to test a key prediction of the scanning model. If 40 S subunits enter at the 5′-end of the mRNA chain and slide down to the AUG codon, one should be able to trap 40 S subunits upstream from the AUG triplet by depleting the system of ATP. This prediction has been verified (Kozak, 1980*c*).

The scanning model predicts that an obstruction introduced upstream from the AUG initiator codon will inhibit translation. In other words, the entire 5′-noncoding region must be passable in order for the 40 S ribosome to reach the AUG triplet. Although this prediction has not yet been rigorously tested, data from certain hybrid-arrested translation experiments bear on the question. Translation of herpes simplex virus thymidine kinase was inhibited by hybridization with DNA fragments complementary to the 5′-terminal 50 nucleotides of the mRNA (Preston & McGeoch, 1981). Since the functional initiator codon in thymidine kinase mRNA lies around position 110, one may conclude that sequences which lie far upstream from the AUG codon participate (somehow) in ribosome binding. Similar experiments have been carried out with Rous sarcoma virus and avian erythroblastosis virus mRNAs. In both cases, a DNA restriction fragment complementary to a portion of the 5′-noncoding region blocked translation of the 5′-proximal *gag* or *gag*-fusion protein (J.M. Bishop, personal communication). Hybrid-arrest of translation has been achieved with DNA fragments complementary to the 5′-leader region of adenovirus mRNAs, although the efficiency of inhibition was somewhat variable (Miller *et al.*, 1980). In the case of Moloney murine sarcoma virus, however, translation of the *gag* polyprotein was not arrested by annealing with DNA complementary to the 5′-end of the message (Papkoff *et al.*, 1981).

A striking characteristic of eukaryotic mRNAs is that translation nearly always begins at the AUG codon that lies closest to the 5′-end of the message. This generalization holds for 115 out of 130 mRNAs for which sequence information is available (Table 2). A possible explanation for the 15 exceptions is suggested below. The point that merits emphasis here is that the exceptions are few. In 90% of the messages that have been examined, including several in which the leader is hundreds of nucleotides long, the 5′-proximal AUG triplet is the functional initiator codon. The scanning model rationalizes this provocative observation. The hypothesis that a 40 S ribosomal subunit enters at the 5′-end of the message and advances linearly until it comes to the first AUG codon can be tested by introducing an AUG triplet upstream from the normal initiation site. Although this type of experiment has been attempted [with the expected result; i.e. ribosomes seemed to initiate at the adventitious 5′-proximal AUG codon (Kozak, 1980*d*; Stiles *et al.*, 1981)], interpretation of those experiments was complicated by the fact that other changes were introduced in the 5′-noncoding region, in the process of creating the new AUG codon. Thus, this critical test of the scanning model requires further experimentation. Stiles *et al.* (1981) emphasize that they were able to find only one cytochrome *c*-deficient mutant in which the defect could be attributed to introduction of an upstream out-of-phase AUG triplet. They point out that, within the stretch of 100 nucleotides preceding the functional initiator codon, there are nine sites where a single base change could create an AUG triplet; thus it seemed puzzling that such mutants were not obtained with higher frequency. But Faye *et al.* (1981) have demonstrated recently that the 5′-termini of yeast cytochrome *c* mRNA molecules map to seven locations, one of which is located just 29 nucleotides upstream from the AUG initiator codon. Thus, the effective target size for ‘leader mutations’ that inactivate translation of the cytochrome *c* gene is much smaller than had been anticipated. The prediction that an AUG triplet introduced upstream will prevent ribosomes from initiating at the normal downstream site must be restated somewhat if sequences flanking the AUG triplet modulate the efficiency of initiation. This idea is expanded in the following section.

Although many questions remain to be answered, the circumstantial evidence outlined above supports the hypothesis that 40 S subunits enter at or near the 5′-end of the message and migrate down to the AUG initiator codon. The next section addresses the question of whether the AUG triplet alone signals the 40 S subunit to stop, or whether ancillary features embodied in the flanking primary or secondary structure help to identify the initiation site.

Table 2. *Position of the functional initiator codon in eukaryotic mRNAs*

Translation begins at the 5′-proximal AUG triplet[a]

(91) Hagfish preproinsulin (Chan *et al.*, 1981)
(92) Adenovirus-7 early region e1A (Van Ormondt *et al.*, 1980; Dijkema *et al.*, 1980)
(93) Mouse immunoglobulin $V_{\kappa 167}$ gene segment (Selsing & Storb, 1981)
(94) Duck β-globin (Hampe *et al.*, 1981)
(95) Chicken histone H2A (D'Andrea *et al.*, 1981)
(96) Mouse metallothionein-I (Glanville *et al.*, 1981)
(97) Human leucocyte interferon B (Goeddel *et al.*, 1981)
(98) Human leucocyte interferon C (Goeddel *et al.*, 1981)
(99) Human leucocyte interferon H (Goeddel *et al.*, 1981)
(100) Human chorionic gonadotropin (β-subunit) (Fiddes & Goodman, 1980)
(101) Human histone H4 (Heintz *et al.*, 1981)
(102) Chicken VLD lipoprotein II (Wieringa *et al.*, 1981)
(103) Bovine corticotropin/β-lipotropin precursor (Nakanishi *et al.*, 1981)
(104) Mouse α-foetoprotein (Law & Dugaiczyk, 1981)
(105) Rat relaxin (Hudson *et al.*, 1981)
(106) *Dictyostelium* actin gene (clone M6) (McKeown & Firtel, 1981)
(107) Herpes simplex IE mRNA-5 (12 kDa protein) (Watson *et al.*, 1981)
(108) Semliki Forest virus 26S mRNA (Garoff *et al.*, 1982)
(109) Adenovirus-5 DNA binding protein (Kruijer *et al.*, 1981; Baker *et al.*, 1979)
(110) Chicken embryonic ρ-globin (Roninson & Ingram, 1981)
(111) Human influenza virus nucleoprotein (Winter & Fields, 1981)
(112) Human influenza virus neuraminidase (Fields *et al.*, 1981)
(113) Human influenza haemagglutinin (H1 subtype) (Winter *et al.*, 1981)
(114) Fowl plague haemagglutinin (H1 subtype)
(115) Yeast *his3* gene (Struhl & Davis, 1981)

Translation begins at an AUG triplet which is not first in line[b]

- SV40 late 16S and 19S mRNAs (Ghosh *et al.*, 1978*a*, *b*)
- SV40 19S mRNA for T-antigen (late) (Ghosh & Lebowitz, 1981)
- Rous sarcoma virus mRNAs for *gag*, *env*, *src* (Cordell *et al.*, 1978; Hackett *et al.*, 1982; Swanstrom *et al.*, 1982)
- Semliki Forest virus genome (Wengler *et al.*, 1979)
- Mouse α-amylase: salivary, liver (Hagenbüchle *et al.*, 1981)
- Chicken preproinsulin (Perler *et al.*, 1980)
- Chicken α_2-collagen, type I (Vogeli *et al.*, 1981)
- Poliovirus (Kitamura *et al.*, 1981; Racaniello & Baltimore, 1981)
- Moloney murine leukaemia virus (Shinnick *et al.*, 1981)
- Moloney murine sarcoma virus (Reddy *et al.*, 1981; Van Beveren *et al.*, 1981)
- Herpes simplex 6000 base late mRNA (Frink *et al.*, 1981)

[a] The mRNA sequences listed in this column were published or confirmed within the past 12 months. An earlier compilation (Kozak, 1981*a*) listed 90 other eukaryotic mRNAs in which the first AUG triplet is the functional initiator codon. I have included only those messages for which the entire 5′-noncoding sequence is known and in which the AUG initiator codon has been unambiguously identified.

[b] In the case of Semliki Forest virus and herpes simplex 6000 base mRNA, although the functional initiator codon has not been identified, the first AUG triplet is ruled out by presence of an in-frame nonsense codon a short distance downstream. The sequence originally published for bovine preproparathyroid hormone mRNA revealed an AUG triplet upstream from the functional initiation site (Kronenberg *et al.*, 1979). More recent experiments indicate, however, that the upstream AUG codon was created by a sequence rearrangement during cloning (Weaver *et al.*, 1981). Thus bovine preproparathyroid mRNA is not included in this list.

What is the 'stop signal' that halts migration of 40 S ribosomal subunits?

An elegant genetic analysis of mutations in the yeast cytochrome *c* gene revealed that eukaryotic ribosomes are remarkably undemanding regarding the sequence requirements for initiation. When the wild-type initiator codon was inactivated by mutation, introduction of an AUG triplet almost anywhere within a 37-nucleotide span restored translation of cytochrome-*c* (Sherman *et al.*, 1980). The sequences flanking the ectopic AUG codons differed extensively from the wild-type ribosome binding site, suggesting that (at least in yeast) ribosomes initiate at the first AUG triplet irrespective of flanking sequences. Those experiments do not necessarily mean that ectopic AUG triplets function as efficiently as the wild-type initiator codon, however; the experiments simply indicate that ectopic AUG triplets function. Thus, the more circumscribed conclusion is that *some* yeast ribosomes initiate at the first AUG triplet irrespective of flanking sequences. This conclusion is not incompatible with the notion that flanking nucleotides modulate the efficiency with which an AUG codon is recognized as a 'stop signal' by the migrating 40 S ribosomal subunit. Circumstantial evidence supporting the latter hypothesis was obtained by surveying the sequences flanking initiator and non-initiator AUG triplets in eukaryotic mRNAs (Kozak, 1981*b*). The sequences bordering functional AUG initiator codons are not random; purines occur with very high frequency in positions −3 and +4, as shown in Table 3 (column 1). The biased distribution of nucleotides bordering functional initiator codons is not observed when one examines AUG triplets that occur internally in the translated regions of eukaryotic mRNAs. Table 3, column 3, reveals a nearly random distribution of nucleotides flanking internal AUG codons. The third set of AUG triplets that we surveyed are those which occur in the 5′-noncoding region of the 15 mRNAs (see Table 2) in which translation begins at an AUG codon which is not first in line. Table 3, column 2, shows that most 'nonfunctional' upstream AUG triplets are flanked by sequences that rarely occur around functional initiator codons. Whereas 80% of functional initiator codons are preceded by A in position −3, no nonfunctional AUG triplet has an A in that position. No proven functional initiator codon is flanked by pyrimidines in both positions −3 and +4, although those are the most common sequences bordering nonfunctional AUG triplets that occur upstream from the initiation site. Thus, the data summarized in Table 3 raise the possibility that nucleotides bordering an AUG triplet influence its ability to serve as an initiator codon. Oligonucleotide binding studies support the idea that flanking nucleotides modulate the efficiency with which the AUG triplet is recognized. As shown in Table 4, binding of ^{32}P-labelled synthetic oligonucleotides to wheat germ ribosomes was enhanced 4-9-fold by placing a purine in position −3 or +4.

These results can be accommodated by a slightly modified version of the scanning model, which postulates that flanking nucleotides modulate the efficiency with which an AUG triplet is recognized as a stop signal by the migrating 40 S subunit. The idea is that all 40 S subunits will stop at the first AUG codon if the flanking sequences are optimal (ANNAUGN, GNNAUG$^{G}_{A}$). If the first AUG triplet is bordered by suboptimal sequences (GNNAUGY, YNNAUGR or YNNAUGY; R = purine, Y = pyrimidine), some 40 S subunits will stop and some will continue migrating, to be halted by an AUG triplet farther downstream. Whereas the original version of the scanning model (see the Introduction) predicted that AUG codons should never be found upstream from the functional initiation site, the modified scanning model admits that AUG triplets may occur within the 5′-noncoding region, provided that all of the upstream AUGs are flanked by suboptimal sequences such that (some)

Table 3. *Sequences flanking AUG triplets in eukaryotic mRNAs*

Sequence	Functional initiator AUG codons[a]	'Nonfunctional' upstream AUGs[b]	Internal AUG codons[c]
−3 +4			
ANNAUGG	74	1[d]	25
ANNAUGA	38	0	26
ANNAUGY	29	0	27
GNNAUGG	18	2[e]	39
GNNAUGA	4	0	18
GNNAUGY	4	4	38
YNNAUGG	11[f]	4	25
YNNAUGA	1(?)[g]	7	23
YNNAUGY	1	12	47
Total	180	30	268

[a] In addition to the 153 mRNAs listed previously (Kozak, 1981*b*), this tabulation of functional initiator codons includes mRNAs for duck β-globin (Hampe *et al.*, 1981), chicken embryonic ρ-globin (Roninson & Ingram, 1981), rabbit embryonic β_3-globin (Hardison, 1981), yeast mating type protein a1 (Astell *et al.*, 1981; Nasmyth *et al.*, 1981), reovirus s1 mRNA (Kozak, 1982*b*), rabies virus glycoprotein (Anilionis *et al.*, 1981), adenovirus-5 100 kDa protein from region L4 (Kruijer *et al.*, 1981), Moloney murine leukaemia virus *gag* protein (Shinnick *et al.*, 1981), Moloney murine sarcoma virus *gag* protein (Reddy *et al.*, 1981; Van Beveren *et al.*, 1981), mouse immunoglobulin V_H 441 gene segment (Ollo *et al.*, 1981), mouse immunoglobulin $V_\kappa(167)$ gene segment (Selsing & Storb, 1981), rat serum albumin (Sargent *et al.*, 1981*b*), adenovirus-12 protein IX (Bos *et al.*, 1981), adenovirus-7 early region e1a (van Ormondt *et al.*, 1980; Dijkema *et al.*, 1980), yeast his3 (Struhl & Davis, 1981), yeast mating type proteins α2 and α1 (Astell *et al.*, 1981), one of the large reovirus proteins (Kozak, 1982*a*), herpes simplex IE-mRNA-5 encoding a 12 kDa protein (Watson *et al.*, 1981), anglerfish pancreatic preprosomatostatin II (Hobart *et al.*, 1980), rat brain α-tubulin Lemischka *et al.*, 1981), mouse thyrotropin (α subunit) (Chin *et al.*, 1981), adenovirus-5 DNA binding protein (Kruijer *et al.*, 1981), 19 kDa protein from adenovirus-12 region e1b (Bos *et al.*, 1981), chicken histone H2A (D'Andrea *et al.*, 1981), the 21 kDa protein from adenovirus-7 region e1b (van Ormondt, personal communication), and hagfish preproinsulin (Chan *et al.*, 1981).

[b] The 15 mRNAs used for column 2 have one or more AUG triplets upstream from the functional initiator codon. (These messages are identified in the right side of Table 2). The functional initiator codon in each of these messages is included in column 1; only the nonfunctional upstream AUGs are represented in column 2.

[c] Internal AUG codons that occur in the reading frame used for translation were tabulated for nine viral and 15 cellular mRNAs. The messages surveyed were Semliki Forest virus 26S mRNA (Garoff *et al.*, 1980*a,b*), vesicular stomatitis virus N, NS, G and M messages (Rose & Gallione, 1981; Gallione *et al.*, 1981), the *src* gene of Rous sarcoma virus (Czernilofsky *et al.*, 1980), the poliovirus genome (Kitamura *et al.*, 1981), influenza virus haemagglutinin (Verhoeyen *et al.*, 1980), adenovirus-2 polypeptide VI (Akusjärvi & Persson, 1981*a*), chicken ovalbumin (McReynolds *et al.*, 1978), bovine preproparathyroid hormone (Kronenberg *et al.*, 1979), rat serum albumin (Sargent *et al.*, 1981*a*), rat calcitonin precursor (Jacobs *et al.*, 1981), chicken brain α- and β-tubulins (Valenzuela *et al.*, 1981), human leucocyte interferon (Taniguchi *et al.*, 1980), mouse α-amylase (Hagenbüchle *et al.*, 1980), rat pregrowth hormone (Seeburg *et al.*, 1977), rat α_{2u} globulin (Unterman *et al.*, 1981), yeast actin (Ng & Abelson, 1980), yeast iso-1-cytochrome *c* (Smith *et al.*, 1979), human chorionic gonadotropin (β-subunit) (Fiddes & Goodman, 1980), mouse thyrotropin (Chin *et al.*, 1981), and rat α_1-acid glycoprotein (Ricca & Taylor, 1981).

[d] The 21S mRNA of Rous sarcoma virus, which encodes the *src* protein, is spliced in a way that places the initiator codon of *gag* (AGCAUGG) 93 nucleotides upstream from the *src* initiation site (Swanstrom *et al.*, 1982).

[e] In SV40 late 16S mRNA, the sequence GNNAUGG occurs in two locations upstream from the VP1 initiation site. One of those AUG triplets is retained in the leader portion of late 19S mRNA.

[f] Some messenger RNAs in which the (first) functional initiator codon is flanked by YNNAUGG are unusual in that ribosomes also initiate at the next AUG downstream (see the text and Table 5).

[g] It is not known whether translation of SV40 VP1 *in vivo* begins at the first or the second AUG triplet within the sequence CUU*AUG*AAG*AUG*GCC. I have tentatively designated the first AUG as the initiator codon based on *in vitro* studies carried out by A. Mellor, R. Hewick & A. Smith (cited in Kozak, 1978).

Table 4. *Binding of ^{32}P-labelled oligonucleotides to wheat germ ribosomes*

^{32}P-labelled oligonucleotide	Bound (%)	^{32}P-labelled oligonucleotide	Bound (%)
A-U-G-C	3	A-A-U-G-U-C	6
C-C-C-A-U-G-C	3	A-A-U-G-C-C	12
A-C-C-A-U-G-C	5	A-A-U-G-A-C	18
G-C-C-A-U-G-C	6	A-A-U-G-G-C	23
C-C-C-C-C-C-A-U-G-C	1	A-A-U-G-U-U	1
C-C-C-A-C-C-A-U-G-C	7–9	A-A-U-G-U-C	5
C-C-C-G-C-C-A-U-G-C	9	A-A-U-G-U-A	1
		A-A-U-G-U-G	3
A-C-C-A-U-U-C	0.2	A-G-U-G-G-C	0.3
A-C-C-G-U-G-C	0.2	A-A-U-G-G-C	24
A-C-C-A-U-G-C	5		

The amount of ^{32}P-labelled oligonucleotide present in 80S initiation complexes was assayed by glycerol gradient centrifugation. The effect of varying the nucleotide in position −3 is shown on the left; positions +4 (and, as a control, +5) were varied in the oligonucleotides on the right. The control experiments in the lower portion of the table show that binding of oligonucleotides was dependent on an AUG codon. Data are from Kozak (1981*b*).

40 S ribosomes can get through. The data in Table 3, column 2, are consistent with this prediction, with only a few unexplained exceptions.

Table 3, column 1, lists several messages in which the functional initiator codon is flanked by suboptimal sequences: GNNAUGY (4 mRNAs) or YNNAUGG (11 mRNAs). By extending the argument presented in the preceding paragraph, we predicted that those mRNAs should have the unusual ability to direct synthesis of two proteins. In other words, if the first AUG codon occurs in an unfavourable context, only some 40 S ribosomes are expected to stop and initiate at that site; some 40 S subunits should bypass the first AUG and initiate at the next AUG downstream. As shown in Table 5, the expectation that two independently-initiated proteins can be translated from a single message has been verified for SV40 late 19 S mRNA, herpes thymidine kinase mRNA, reovirus s1 mRNA, and messages from early region e1b of adenovirus-5 and -7. In those five examples the leaky 5′-proximal initiator codon is flanked by suboptimal sequences. The lower portion of Table 5 lists a few other mRNAs which permit initiation at two nearby sites even though the first AUG codon is in a favourable context. In other words, we would have expected the first AUG triplet to be the exclusive initiation site in the e1b message of adenovirus-12, the N message of vesicular stomatitis virus, the message encoding MOPC-41 immunoglobulin, and SV40 late 16 S mRNA. It is curious that, in many cases, the 'leaky' 5′-proximal AUG codon lies very close to the cap. Thus, a possible explanation is that an AUG codon occurring within the first 10–12 nucleotides is inefficiently recognized irrespective of the flanking sequences. But this idea awaits testing. Among the eukaryotic mRNAs that display two functional initiation sites, the *src* message of Rous sarcoma virus is the only one for which no facile explanation comes to mind. It is noteworthy that, in all eukaryotic mRNAs that permit initiation at two distinct sites, the alternative initiation site lies just downstream

Table 5. *Eukaryotic mRNAs which appear to contain two functional initiation sites*

mRNA	First initiation site: Protein	First initiation site: Sequence	First initiation site: cap/AUG[a]	Second initiation site: Protein	Second initiation site: Sequence	Reading frame(s)[b]	References
SV40 late 19S	VP2	UCC*AUG*G	115	VP3[c]	GAA*AUG*G	1	Ghosh *et al.*, 1978*a*; Reddy *et al.*, 1978; Piatak *et al.*, 1979; Cepko *et al.*, 1981;
Herpes simplex	Thymidine kinase	CGU*AUG*G	110	39 kDa[c]	AAA*AUG*C (presumptive site)	1	McKnight, 1980; Preston & McGeoch, 1981
Poliovirus	5–10 kDa protein[d]	?[e]	?	High-M_r structural precursor	AUA*AUG*G	2	Celma & Ehrenfeld, 1975; Knauert & Ehrenfeld, 1979; Kitamura *et al.*, 1981
Reovirus s1	?[f]	CGG*AUG*G	13	?[d,f]	AUA*AUG*G	2	Kozak, 1982*b*
Adenovirus-5 region e1b	21 kDa tumour Ag	CUC*AUG*G	11–13	55 kDa tumour Ag[c]	UAA*AUG*G	2	Bos *et al.*, 1981
Adenovirus-7 region e1b	21 kDa tumour Ag	UCC*AUG*G	25–26	55 kDa tumour Ag[c]	UAA*AUG*G	2	H. van Ormondt, personal communication
Mouse immunoglobulin MOPC21	Ig	NNN*AUG*C	4–6	Ig[d]	AGC*AUG*G	1	Rose *et al.*, 1977 (Fig. 5); Hamlyn *et al.*, 1981
Adenovirus-12 region e1b	19 kDa tumour Ag	GUA*AUG*G	10–15	54 kDa tumour Ag[c]	UAA*AUG*G	2	Bos *et al.*, 1981
Vesicular stomatitis virus	N protein	AAA*AUG*U	14	?[d,f]	CAA*AUG*A	2	Gallione *et al.*, 1981
Mouse immunoglobulin MOPC41	Ig	AGC*AUG*G	?	Ig[d]	GAC*AUG*A	1	Zemell *et al.*, 1978; Seidman *et al.*, 1979
SV40 late 16S	Agnoprotein[c]	GCC*AUG*G	11	VP1	CUU*AUG*A or AAG*AUG*G	1[g]	Jay *et al.*, 1981
Rous sarcoma virus mutant OOO	*src* *N*-terminal fragment	ACC*AUG*G	~450	*src* *C*-terminal fragment	ACU*AUG*U		Oppermann *et al.*, 1981; H. Varmus & G. Mardon, personal communication

[a] This column indicates the number of nucleotides between the m^7G cap and the first initiator codon.

[b] A '1' in this column means that both proteins are translated from the same reading frame. A '2' means that the second functional AUG codon lies in a different reading frame from the first, thus permitting synthesis of two unrelated proteins.

[c] The second initiation site functions *in vivo* as well as *in vitro*.

[d] The second initiation site has been demonstrated only *in vitro*.

[e] The initiator codon for the 5–10 kDa protein has not yet been located on the poliovirus genome, although several candidate sites can be recognized upstream from the AUG triplet that initiates viral structural proteins. All of the upstream potential initiator codons are flanked by 'suboptimal' sequences (see the text).

[f] The two initiation sites were identified by ribosome-protection experiments. It has not been shown that this mRNA actually directs synthesis of two proteins.

[g] Although the two initiator codons in SV40 16S mRNA lie in the same reading frame, the two cistrons do not overlap. Thus, two unique proteins are made.

from the first.* This provides encouragement for the idea of scanning even though in some cases we do not quite understand why the first AUG is 'leaky'. Finally, it must be emphasized that initiation at two sites in a given mRNA is the exception, rather than the rule. *N*-Terminal heterogeneity was not detected in the products of translation *in vitro* directed by egg white ovomucoid mRNA (Thibodeau *et al.*, 1978), bovine preproparathyroid hormone mRNA (Habener *et al.*, 1978), adenovirus-2 hexon protein mRNA (Anderson & Lewis, 1980), or T7 phage gene 0.3, translated in a wheat germ extract (Anderson, 1982). In each of those messages ribosomes initiated exclusively at the first AUG codon, even though each mRNA has a second AUG triplet just downstream from, and in-phase with, the first. Unfortunately, precise identification of the *N*-terminal amino acid sequence of nascent polypeptide chains can only be carried out *in vitro*. The fidelity of initiation *in vivo* remains an unexplored question.

The mechanism by which flanking sequences (particularly in positions −3 and +4) modulate recognition of the AUG codon is open to speculation. It is conceivable that the sequence AUGG forms a four base-pair interaction with the anticodon loop of Met-$tRNA_i$ (CCAU), although that hypothesis is weakened somewhat by finding that oligonucleotides containing AUGA bind almost as efficiently as AUGG (Kozak, 1981*b*). The notion of context effects on codon recognition is hardly new, of course. A conserved pattern of nucleotides, apart from the Shine–Dalgarno sequence, has been discerned to the left and right of the initiator codon in bacterial ribosome binding sites (Scherer *et al.*, 1980). The nucleotide adjacent to the 3′-side of the AUG initiator codon is especially influential in modulating binding of *E. coli* ribosomes (Taniguchi & Weissmann, 1978; Schmitt *et al.*, 1980). In both prokaryotes and eukaryotes, the efficiency of readthrough of nonsense codons by suppressor tRNAs is dramatically affected by sequences flanking the nonsense codon (Akaboshi *et al.*, 1976; Feinstein & Altman, 1978; Bossi & Roth, 1980; Bienz *et al.*, 1981; Engelberg-Kulka, 1981). Although these examples do not shed much light on the mechanism, they affirm the ability of adjacent nucleotides to modulate codon–anticodon interactions.

Prospects for regulating translational initiation in eukaryotes

It is clear that some mRNAs initiate at higher rates than others (Lodish, 1976), although the structural features responsible for these differences have not yet been defined. The model just described leaves open many possibilities for modulating translational efficiency. Presence of the m^7G cap (Shatkin, 1982) does not guarantee efficient translation, since a few naturally occurring uncapped mRNAs can outcompete capped mRNAs (Herson *et al.*, 1979; Jen *et al.*, 1978). Moreover, among capped mRNAs the efficiency of translation varies widely. The ability of mRNAs to

* It is not known whether this generalization holds for poliovirus, because the location of the alternate initiation site detected by Celma & Ehrenfeld (1975) has not yet been mapped. Kitamura *et al.* (1981) have recently published the sequence of the poliovirus genome, and the reader is referred to their paper for speculation about the possible location of the second initiation site. Poliovirus is more complicated than the other messages under discussion here, since the major initiation site (encoding the structural proteins) is preceded by eight other AUG triplets.

compete for 40 S ribosomal subunits might depend on the extent to which the 5′-terminus (be it capped or uncapped) is exposed.

After the 40 S ribosome • initiation factor complex binds at the 5′-end of the message, subsequent steps in initiation might also be regulated. For example, by analogy with RNA polymerase complexes (Munson & Reznikoff, 1981), some 40 S ribosomal subunits might dissociate, either right at the 5′-end of the mRNA or at some point after they begin migrating. Another possibility is that migrating 40 S ribosomes sometimes halt prematurely, thereby allowing a 60 S subunit to join upstream from the functional initiation site. For example, when certain plant virus mRNAs are incubated with wheat germ ribosomes *in vitro*, 80 S ribosomes assemble at a point in the leader sequence that lacks AUG triplets (Ahlquist *et al.*, 1979; Filipowicz & Haenni, 1979; Pinck *et al.*, 1981). If the prematurely assembled 80 S ribosome is unable to migrate down to the AUG codon (a hypothesis which has yet to be tested), the mRNA template would be rendered non-functional. The explanation underlying this phenomenon might be that the rate of migration of 40 S subunits on mRNA is non-uniform. Certain primary or secondary structural features in a message might cause a 40 S subunit to pause, and temporizing might facilitate joining of the 60 S subunit.

The modified scanning model predicts that those unusual mRNAs which have AUG triplets upstream from the 'intended' initiator codon (see right side of Table 2)' should be translated inefficiently. Some 40 S subunits would be expected to stop (and join with a 60 S subunit) at each upstream AUG codon, thereby decreasing the number of 40 S ribosomes that reach the intended start-site for translation. It is interesting in this regard that SV40 19 S mRNA (the message encoding T antigen) can be transcribed from two different promoters, such that the 5′-noncoding region on the late form of the message is slightly longer than on the early form (Ghosh & Lebowitz, 1981). Whereas the early form of T antigen mRNA has no upstream AUG triplets, the late form (which is produced when the biological requirement for T antigen is low) has an AUG triplet preceding the functional initiator codon of T antigen. As yet, however, there is no evidence that the late form translates T antigen less efficiently than the early form.

A final opportunity to regulate formation of productive initiation complexes occurs after the 40 S subunit has halted at the AUG initiator codon. The processes whereby initiation factors are released and the 60 S subunit joins are not fully understood, but there is some evidence that the requirements for 60 S joining differ among various reovirus mRNAs (Kozak & Shatkin, 1977).

None of these prospects for regulating initiation is yet supported by good experimental evidence, although there are hints that can be fitted into the above scenarios. Even though eukaryotic ribosomes are quite flexible regarding the sequences required for initiation, in some messages the 5′-noncoding sequences are highly conserved (Efstratiadis *et al.*, 1980; Miyata *et al.*, 1980; Martin *et al.*, 1981). This suggests that the 5′-noncoding region of eukaryotic mRNAs might serve a useful, albeit dispensable, regulatory function. In support of the suggestion that late steps in initiation can be modulated (i.e. steps that occur after the message has successfully competed for a 40 S ribosome and initiation factors), Herson *et al.* (1979) have reported that the mRNA of satellite tobacco necrosis virus outcompetes most other messages, but is itself inefficient when assayed for ability to direct amino acid incorporation. The s1 mRNA of reovirus provides an interesting example of a message that is inherently inefficient; i.e. it is translated poorly even in the absence of competing mRNAs (Levin & Samuel, 1980). Although the structural features underlying differential utilization of eukaryotic mRNAs have not yet been identified, the question might soon be answered by focusing on mutant forms of mRNA which are translated less efficiently than are wild-type forms (Bendig *et al.*, 1980).

I am grateful to Dr. H. van Ormondt for sharing information about adenovirus early mRNA sequences, Dr. Henrik Garoff for a preprint of the Semliki Forest virus sequence, and Drs. Harold Varmus and J. Michael Bishop for preprints and helpful discussions about Rous sarcoma virus mRNAs. Work from my laboratory was supported by research grant AI-16634 and Career Development Award AI-00380 from the National Institutes of Health.

References

Ahlquist, P., Dasgupta, R., Shih, D.S., Zimmern, D. & Kaesberg, P. (1979) *Nature (London)* **281**, 277-282
Akaboshi, E., Inouye, M. & Tsugita, A. (1976) *Mol. Gen. Genet.* **149**, 1-4
Akusjärvi, G. & Persson, H. (1981*a*) *J. Virol.* **38**, 469-482
Akusjärvi, G. & Persson, H. (1981*b*) *Nature (London)* **292**, 420-426
Akusjärvi, G., Zabielski, J., Perricaudet, M. & Pettersson, U. (1981) *Nucleic Acids Res.* **9**, 1-17
Anderson, C.W. (1982) in *Genetic Engineering* (Setlow, J.K. & Hollaender, A., eds.), vol. 4, Plenum Press, New York and London, in press.
Anderson, C.W. & Lewis, J.B. (1980) *Virology* **104**, 27-41
Anilionis, A., Wunner, W.H. & Curtis, P.J. (1981) *Nature (London)* **294**, 275-278
Arrand, J.R., Soeda, E., Walsh, J.E., Smolar, N. & Griffin, B.E. (1980) *J. Virol.* **33**, 606-618
Astell, C.R., Ahlstrom-Jonasson, L. & Smith, M. (1981) *Cell* **27**, 15-23
Baker, C.C., Herisse, J., Courtois, G., Galibert, F. & Ziff, E. (1979) *Cell* **18**, 569-580
Barkan, A. & Mertz, J.E. (1981) *J. Virol.* **37**, 730-737
Beemon, K. & Hunter, T. (1978) *J. Virol.* **28**, 551-566
Bendig, M.M. & Folk, W.R. (1979) *J. Virol.* **32**, 530-535
Bendig, M.M., Thomas T. & Folk, W.R. (1980) *Cell* **20**, 401-409
Bienz, M., Kubli, E., Kohli, J., deHenau, S., Huez, G., Marbaix, G. & Grosjean, H. (1981) *Nucleic Acids Res.* **9**, 3835-3850
Bos, J.L., Polder, L.J., Bernards, R., Schrier, P.I., van den Elsen, P.J., van der Eb, A.J. & van Ormondt, H. (1981) *Cell* **27**, 121-131
Bossi, L. & Roth, J.R. (1980) *Nature (London)* **286**, 123-127
Celma, M.L. & Ehrenfeld, E. (1975) *J. Mol. Biol.* **98**, 761-780
Cepko, C.L., Hansen, U., Handa, H. & Sharp, P.A. (1981) *Mol. Cell. Biol.* **1**, 919-931
Chan, S.J., Emdin, S.O., Kwok, S.C.M., Kramer, J.M., Falkmer, S. & Steiner, D. (1981) *J. Biol. Chem.* **256**, 7595-7602
Chin, W., Kronenberg, H., Dee, P., Maloof, F. & Habener, J. (1981) *Proc. Natl. Acad. Sci. U.S.A.* **78**, 5329-5333
Cordell, B., Weiss, S.R., Varmus, H.E. & Bishop, J.M. (1978) *Cell* **15**, 79-91
Cremer, K., Reddy, E.P. & Aaronson, S.A. (1981) *J. Virol.* **38**, 704-711
Czernilofsky, A.P., Levinson, A.D., Varmus, H.E., Bishop, J.M., Tischer, E. & Goodman, H.M. (1980) *Nature (London)* **287**, 198-203
D'Andrea, R., Harvey, R. & Wells, J.R.E. (1981) *Nucleic Acids Res.* **9**, 3119-3128
Dijkema, R., Dekker, B.M.M., van Ormondt, H., de Waard, A., Maat, J. & Boyer, H.W. (1980) *Gene* **12**, 287-299
Dorner, A.J., Rothberg, P.G. & Wimmer, E. (1981) *FEBS Lett.* **132**, 219-223
Dunn, A.R., Mathews, M., Chow, L., Sambrook, J. & Keller, W. (1978) *Cell* **15**, 511-526
Dunn, J.J., Buzash-Pollert, E. & Studier, F.W. (1978) *Proc. Natl. Acad. Sci. U.S.A.* **75**, 2741-2745
Efstratiadis, A., Posakony, J.W., Maniatis, T., Lawn, R., O'Connell, C., Spritz, R., DeRiel, J., Forget, B., Weissman, S., Slightom, J., Blechl, A., Smithies, O., Baralle, F., Shoulders, C. & Proudfoot, N. (1980) *Cell* **21**, 653-668
Engelberg-Kulka, H. (1981) *Nucleic Acids Res.* **9**, 983-991

Faye, G., Leung, D., Tatchell, K., Hall, B.D. & Smith, M. (1981) *Proc. Natl. Acad. Sci. U.S.A.* **78**, 2258–2262
Feinstein, S. & Altman, S. (1978) *Genetics* **88**, 201–219
Fiddes, J.C. & Goodman, H.M. (1980) *Nature (London)* **286**, 684–687
Fields, S., Winter, G. & Brownlee, G.G. (1981) *Nature (London)* **290**, 213–217
Filipowicz, W. & Haenni, A.-L. (1979) *Proc. Natl. Acad. Sci. U.S.A.* **76**, 3111–3115
Frink, R.J., Draper, K.G. & Wagner, E.K. (1981) *Proc. Natl. Acad. Sci. U.S.A.* **78**, 6139–6143
Gallione, C.J., Greene, J.R., Iverson, L. & Rose, J. (1981) *J. Virol.* **39**, 529–535
Garoff, H., Frischauf, A.-M., Simons, K., Lehrach, H. & Delius, H. (1980*a*) *Proc. Natl. Acad. Sci. U.S.A.* **77**, 6376–6380
Garoff, H., Frischauf, A.-M., Simons, K., Lehrach, H. & Delius, H. (1980*b*) *Nature (London)* **288**, 236–241
Garoff, H., Kondor-Koch, C. & Riedel, H. (1982) *Curr. Top. Microbiol. Immunol.* in press
Ghosh, P.K. & Lebowitz, P. (1981) *J. Virol.* **40**, 224–240
Ghosh, P.K., Reddy, V.B., Swinscoe, J., Lebowitz, P. & Weissman, S.M. (1978*a*) *J. Mol. Biol.* **126**, 813–846
Ghosh, P.K., Reddy, V.B., Swinscoe, J., Choudary, P.V., Lebowitz, P. & Weissman, S.M. (1978*b*) *J. Biol. Chem.* **253**, 3643–3647
Ghosh, P.K., Roy, P., Barkan, A., Mertz, J., Weissman, S.M. & Lebowitz, P. (1981) *Proc. Natl. Acad. Sci. U.S.A.* **78**, 1386–1390
Glanville, N., Durnam, D. & Palmiter, R. (1981) *Nature (London)* **292**, 267–269
Goeddel, D., Leung, D., Dull, T., Gross, M., Lawn, R., McCandliss, R., Seeburg, P., Ullrich, A., Yelverton, E. & Gray, P. (1981) *Nature (London)* **290**, 20–26
Golini, F., Semler, B., Dorner, A. & Wimmer, E. (1980) *Nature (London)* **287**, 600–603
Guilley, H., Jonard, G., Kukla, B. & Richards, K.E. (1979) *Nucleic Acids Res.* **6**, 1287–1308
Habener, J., Rosenblatt, M., Kemper, B., Kronenberg, H., Rich, A. & Potts, J.T., Jr. (1978) *Proc. Natl. Acad. Sci. U.S.A.* **75**, 2616–2620
Hackett, P.B., Swanstrom, R., Varmus, H.E. & Bishop, J.M. (1982) *J. Virol.* **41**, 527–534
Hagenbuchle, O., Bovey, R. & Young, R. (1980) *Cell* **21**, 179–187
Hagenbuchle, O., Tosi, M., Schibler, U., Bovey, R., Wellauer, P.K. & Young, R.A. (1981) *Nature (London)* **289**, 643–646
Hamlyn, P., Gait, M. & Milstein, C. (1981) *Nucleic Acids Res.* **9**, 4485–4494
Hampe, A., Therwath, A., Soriano, P. & Galibert, F. (1981) *Gene* **14**, 11–21
Hardison, R.C. (1981) *J. Biol. Chem.* **256**, 11780–11786
Heintz, N., Zernik, M. & Roeder, R.G. (1981) *Cell* **24**, 661–668
Herson, D., Schmidt, A., Seal, S., Marcus, A. & van Vloten-Doting, L. (1979) *J. Biol. Chem.* **254**, 8245–8249
Hobart, P., Crawford, R., Shen, L.P., Pictet, R. & Rutter, W.J. (1980) *Nature (London)* **288**, 137–141
Hudson, P., Haley, J., Cronk, M., Shine, J. & Niall, H. (1981) *Nature (London)* **291**, 127–131
Jacobs, J., Goodman, R., Chin, W., Dee, P., Habener, J., Bell, N. & Potts, J. T., Jr. (1981) *Science* **213**, 457–459
Jay, G., Nomura, S., Anderson, C.W. & Khoury, G. (1981) *Nature (London)* **291**, 346–349
Jen, G., Birge, C.H. & Thach, R.E. (1978) *J. Virol.* **27**, 640–647
Kitamura, N., Semler, B., Rothberg, P., Larsen, G., Adler, C., Dorner, A., Emini, E., Hanecak, R., Lee, J., van der Werf, S., Anderson, C. & Wimmer, E. (1981) *Nature (London)* **291**, 547–553
Knauert, F. & Ehrenfeld, E. (1979) *Virology* **93**, 537–546
Koper-Zwarthoff, E., Brederode, F., Veeneman, G., van Boom, J. & Bol, J.F. (1980) *Nucleic Acids Res.* **8**, 5635–5647
Kozak, M. (1978) *Cell* **15**, 1109–1123
Kozak, M. (1979*a*) *Nature (London)* **280**, 82–85
Kozak, M. (1979*b*) *J. Biol. Chem.* **254**, 4731–4738
Kozak, M. (1980*a*) *Cell* **19**, 79–90
Kozak, M. (1980*b*) *J. Virol.* **35**, 748–756
Kozak, M. (1980*c*) *Cell* **22**, 459–467
Kozak, M. (1980*d*) *J. Mol. Biol.* **144**, 291–304
Kozak, M. (1981*a*) *Curr. Top. Microbiol. Immunol.* **93**, 81–123
Kozak, M. (1981*b*) *Nucleic Acids Res.* **9**, 5233–5252
Kozak, M. (1982*a*) *J. Virol.* **42**, 467–473
Kozak, M. (1982*b*) *J. Mol. Biol.* **156**, in press
Kozak, M. & Shatkin, A.J. (1977) *J. Biol. Chem.* **252**, 6895–6908
Kozak, M. & Shatkin, A.J. (1978) *J. Biol. Chem.* **253**, 6568–6577

Kronenberg, H., McDevitt, B., Majzoub, J., Nathans, J., Sharp, P., Potts, J.T., Jr. & Rich, A. (1979) *Proc. Natl. Acad. Sci. U.S.A.* **76**, 4981–4985

Kruijer, W., van Schaik, F. & Sussenbach, J.S. (1981) *Nucleic Acids Res.* **9**, 4439–4457

Law, S.W. & Dugaiczyk, A. (1981) *Nature (London)* **291**, 201–205

Lawrence, C., Hunter, T. & Walter, G. (1979) *J. Mol. Biol.* **134**, 95–108

Lemischka, I.R., Farmer, S., Racaniello, V. & Sharp, P. (1981) *J. Mol. Biol.* **151**, 101–120

Levin, K. & Samuel, C.E. (1980) *Virology* **106**, 1–13

Lodish, H.F. (1976) *Annu. Rev. Biochem.* **45**, 39–72

Martin, S.L., Zimmer, E.A., Davidson, W.S., Wilson, A.C. & Kan, Y.W. (1981) *Cell* **25**, 737–741

McKeown, M. & Firtel, R.A. (1981) *Cell* **24**, 799–807

McKnight, S.L. (1980) *Nucleic Acids Res.* **8**, 5949–5964

McReynolds, L., O'Malley, B.W., Nisbet, A.D., Fothergill, J.E., Givol, D., Fields, S., Robertson, M. & Brownlee, G.G. (1978) *Nature (London)* **273**, 723–728

Miller, J., Ricciardi, R., Roberts, B., Paterson, B. & Mathews, M. (1980) *J. Mol. Biol.* **142**, 455–488

Miyata, T., Yasunaga, T. & Nishida, T. (1980) *Proc. Natl. Acad. Sci. U.S.A.* **77**, 7328–7332

Mulligan, R.C. & Berg, P. (1981) *Mol. Cell. Biol.* **1**, 449–459

Munson, L.M. & Reznikoff, W.S. (1981) *Biochemistry* **20**, 2081–2085

Nakanishi, S., Teranishi, Y., Watanabe, Y., Notake, M., Noda, M., Kakidani, H., Jingami, H. & Numa, S. (1981) *Eur. J. Biochem.* **115**, 429–438

Nasmyth, K., Tatchell, K., Hall, B.D., Astell, C. & Smith, M. (1981) *Nature (London)* **289**, 244–250

Ng, R. & Abelson, J. (1980) *Proc. Natl. Acad. Sci. U.S.A.* **77**, 3912–3916

Ollo, R., Auffray, C., Sikorav, J.-L. & Rougeon, F. (1981) *Nucleic Acids Res.* **9**, 4099–4109

Oppermann, H., Levinson, A.D. & Varmus, H.E. (1981) *Virology* **108**, 47–70

Papkoff, J., Lai, M., Hunter, T. & Verma, I. (1981) *Cell* **27**, 109–119

Pelham, H. (1979) *FEBS Lett.* **100**, 195–199

Perler, F., Efstratiadis, A., Lomedico, P., Gilbert, W., Kolodner, R. & Dodgson, J. (1980) *Cell* **20**, 555–566

Perricaudet, M., LeMoullec, J.M. & Pettersson, U. (1980) *Proc. Natl. Acad. Sci. U.S.A.* **77**, 3778–3782

Piatak, M., Ghosh, P.K., Reddy, V.B., Lebowitz, P. & Weissman, S.M. (1979) in *Extrachromosomal DNA: ICN-UCLA Symposia on Molecular and Cellular Biology* (Cummings, D.J. *et al.*, eds.), vol. 15, pp. 199–215, Academic Press, New York

Pinck, M., Fritsch, C., Ravelonandro, M., Thivent, C. & Pinck, L. (1981) *Nucleic Acids Res.* **9**, 1087–1100

Porter, A.G., Barber, C., Carey, N.H., Hallewell, R.A., Threlfall, G. & Emtage, J.S. (1979) *Nature (London)* **282**, 471–477

Preston, C.M. & McGeoch, D.J. (1981) *J. Virol.* **38**, 593–605

Racaniello, V.R. & Baltimore, D. (1981) *Proc. Natl. Acad. Sci. U.S.A.* **78**, 4887–4891

Reddy, E.P., Smith, M.J. & Aaronson, S.A. (1981) *Science* **214**, 445–450

Reddy, V.B., Dhar, R. & Weissman, S.M. (1978) *J. Biol. Chem.* **253**, 621–630

Ricca, G.A. & Taylor, J.M. (1981) *J. Biol. Chem.* **256**, 11199–11202

Roninson, I.B. & Ingram, V.M. (1981) *Proc. Natl. Acad. Sci. U.S.A.* **78**, 4782–4785

Rose, J.K. & Gallione, C.J. (1981) *J. Virol.* **39**, 519–528

Rose, S.M., Kuehl, W.M. & Smith, G.P. (1977) *Cell* **12**, 453–462

Rosenberg, M. & Paterson, B.M. (1979) *Nature (London)* **279**, 696–701

Sangar, D.V., Black, D.N., Rowlands, D.J., Harris, T. & Brown, F. (1980) *J. Virol.* **33**, 59–68

Sargent, T.D., Yang, M. & Bonner, J. (1981*a*) *Proc. Natl. Acad. Sci. U.S.A.* **78**, 243–246

Sargent, T.D., Jagodzinski, L.L., Yang, M. & Bonner, J. (1981*b*) *Mol. Cell. Biol.* **1**, 871–883

Scherer, G., Walkinshaw, M., Arnott, S. & Morre, D.J. (1980) *Nucleic Acids Res.* **8**, 3895–3907

Schmitt, M., Manderschied, U., Kyriatsoulis, A., Brinckmann, U. & Gassen, H.G. (1980) *Eur. J. Biochem.* **109**, 291–299

Seeburg, P.H., Shine, J., Martial, J.A., Baxter, J.D. & Goodman, H.M. (1977) *Nature (London)* **270**, 486–494

Seidman, J.G., Max, E.E. & Leder, P. (1979) *Nature (London)* **280**, 370–375

Selsing, E. & Storb, U. (1981) *Cell* **25**, 47–58

Shatkin, A.J. (1982) *Biochem. Soc. Symp.* **47**, 129–143

Sherman, F., Stewart, J.W. & Schweingruber, A.M. (1980) *Cell* **20**, 215–222

Shih, D.S. & Kaesberg, P. (1973) *Proc. Natl. Acad. Sci. U.S.A.* **70**, 1799–1803

Shine, J. & Dalgarno, L. (1974) *Proc. Natl. Acad. Sci. U.S.A.* **71**, 1342–1346

Shinnick, T.M., Lerner, R.A. & Sutcliffe, J.G. (1981) *Nature (London)* **293**, 543–548

Smith, A.E. (1977) *FEBS Symp.* **43,** 37–46
Smith, M., Leung, D.W., Gillam, S., Astell, C.R., Montgomery, D.L. & Hall, B.D. (1979) *Cell* **16,** 753–761
Solnick, D. (1981) *Cell* **24,** 135–143
Steitz, J.A. & Jakes, K. (1975) *Proc. Natl. Acad. Sci. U.S.A.* **72,** 4734–4738
Stiles, J.I., Szostak, J.W., Young, A.T., Wu, R., Consaul, S. & Sherman, F. (1981) *Cell* **25,** 277–284
Struhl, K. & Davis, R.W. (1981) *J. Mol. Biol.* **152,** 553–568
Subramani, S., Mulligan, R. & Berg, P. (1981) *Mol. Cell Biol.* **1,** 854–864
Swanstrom, R., Varmus, H.E. & Bishop, J.M. (1982) *J. Virol.* **41,** 535–541
Taniguchi, T. & Weissmann, C. (1978) *J. Mol. Biol.* **118,** 533–565
Taniguchi, T., Mantei, N., Schwarzstein, M., Nagata, S., Muramatsu, M. & Weissmann, C. (1980) *Nature (London)* **285,** 547–549
Thibodeau, S.N., Palmiter, R.D. & Walsh, K.A. (1978) *J. Biol. Chem.* **253,** 9018–9023
Unterman, R.D., Lynch, K.R., Nakhasi, H.L., Dolan, K.P., Hamilton, J.W., Cohn, D.V. & Feigelson, P. (1981) *Proc. Natl. Acad. Sci. U.S.A.* **78,** 3478–3482
Valenzuela, P., Quiroga, M., Zaldivar, J., Rutter, W.J., Kirschner, M.W. & Cleveland, D.W. (1981) *Nature (London)* **289,** 650–655
Van Beveren, C., van Straaten, F., Galleshaw, J.A. & Verma, I.M. (1981) *Cell* **27,** 97–108
van Ormondt, H., Maat, J. & Dijkema, R. (1980) *Gene* **12,** 63–76
Verhoeyen, M., Fang, R., MinJou, W., Devos, R., Huylebroeck, D., Saman, E. & Fiers, W. (1980) *Nature (London)* **286,** 771–776
Vogeli, G., Ohkubo, H., Sobel, M.E., Yamada, Y., Pastan, I. & de Crombrugghe, B. (1981) *Proc. Natl. Acad. Sci. U.S.A.* **78,** 5334–5338
Watson, R.J., Umene, K. & Enquist, L.W. (1981) *Nucleic Acids Res.* **9,** 4189–4199
Weaver, C.A., Gordon, D.F. & Kemper, B. (1981) *Proc. Natl. Acad. Sci. U.S.A.* **78,** 4073–4077
Wengler, G., Wengler, G. & Gross, H.J. (1979) *Nature (London)* **282,** 754–756
Wieringa, B., Geert, A.B. & Gruber, M. (1981) *Nucleic Acids Res.* **9,** 489–501
Winter, G. & Fields, S. (1981) *Virology* **114,** 423–428
Winter, G., Fields, S. & Brownlee, G.G. (1981) *Nature (London)* **292,** 72–75
Woodworth-Gutai, M. (1981) *Virology* **109,** 353–365
Zain, S., Sambrook, J., Roberts, R.J., Keller, W., Fried, M. & Dunn, A.R. (1979) *Cell* **16,** 851–861
Zemell, R., Burstein, Y. & Schechter, I. (1978) *Eur. J. Biochem.* **89,** 187–193

Biochem. Soc. Symp. 47, 129–143
Printed in Great Britain

5′-Terminal Caps, Cap-Binding Proteins and Eukaryotic mRNA Function

AARON J. SHATKIN, EDWARD DARZYNKIEWICZ*, YASUHIRO FURUICHI, HANS KROATH, MAUREEN A. MORGAN, STANLEY M. TAHARA and MINORU YAMAKAWA

Roche Institute of Molecular Biology, Nutley, NJ 07110, U.S.A.

Introductory Overview of mRNA Modification

Eukaryotic mRNAs are easily distinguished from other RNAs, including prokaryotic messengers, by the presence of one or more unique structural features. They include methylated adenosine in internal positions, 3′-terminal polyadenylic acid and 5′ ends that are 'capped' by 7-methylguanosine. All three kinds of modifications occur in the nucleus on RNA polymerase II pre-mRNA transcripts of both cellular and viral origin (for reviews, see Revel & Groner, 1978; Darnell, 1979; Shenk, 1981). By contrast, animal viruses that replicate in the cytoplasm of infected cells produce mRNAs that usually are not methylated at internal sites (reviewed by Banerjee, 1980). Furthermore, some functional viral mRNAs (reovirus, frog virus 3) are not 3′-polyadenylated (Stoltzfus *et al.*, 1973; Willis & Granoff, 1976), and others (picornaviruses) are devoid of a 5′-terminal cap (Flanegan *et al.*, 1977; Lee *et al.*, 1977). From these diverse observations, the biological significance of mRNA methylation and terminal modification for eukaryotic genetic expression is not readily apparent.

Some eukaryotic genes, e.g. those coding for cellular histones (Kedes, 1979) and vaccinia virus proteins (Wittek *et al.*, 1980), are notable for the absence of intron sequences. The corresponding transcripts also are not methylated internally (Kedes, 1979; Moss *et al.*, 1977), and it therefore seemed reasonable to suggest that methylated adenosine acts as a recognition signal during mRNA splicing. However, studies with methylation inhibitors have indicated that processing and intracellular transport of mRNA, like translation, are probably not strictly dependent on internal methylation (Kaehler *et al.*, 1977; Dimock & Stoltzfus, 1978, 1979; Bachellerie *et al.*, 1978). Thus a biochemical role for methyladenosine in mRNA remains to be demonstrated.

The importance of 3′-polyadenylation for eukaryotic mRNA formation and function is somewhat clearer. Addition of poly(A) to cleaved nascent transcripts is a primary maturation event that defines mRNA 3′ ends (Hofer & Darnell, 1981; Fitzgerald & Shenk, 1981). A stabilizing effect of 3′-poly(A) on cytoplasmic mRNA has been described recently (Zeevi *et al.*, 1981). In addition, the presence

* Present address: Institute of Experimental Physics, University of Warsaw, Warsaw, Poland

of 3′-poly(A) is apparently important for the long-term translation of globin mRNA in microinjected *Xenopus* oocytes (Huez *et al.*, 1975) and for reinitiation of ovalbumin synthesis in reticulocyte lysate (Doel & Carey, 1976).

One of the recently discovered and intensely studied eukaryotic mRNA modification events is 5′-terminal capping (Shatkin, 1976). Cap structures of the type $m^7G(5')ppp(5')N^mpNp\ldots$ (Fig. 1) have been detected on short pre-mRNA transcripts. This strongly suggests that capping is an early event in cellular mRNA synthesis (Salditt-Georgieff *et al.*, 1980). Studies of viral systems also indicate that capping probably occurs as part of the initiation of transcription. Caps are retained during transcript processing, and in various systems their presence has been found to influence gene expression at the levels of transcription, translation and mRNA stability.

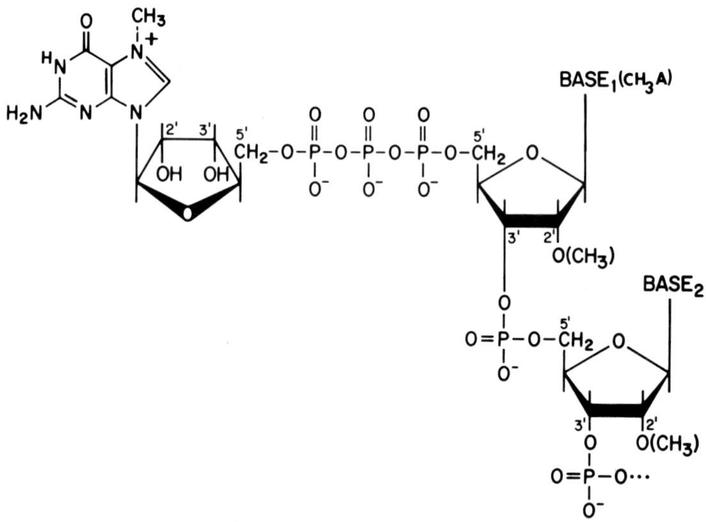

Fig. 1. *General structure of 5′-terminal cap of eukaryotic mRNAs*

The terminus invariably is m^7G. Caps of some mRNAs (e.g. from plants, plant viruses, lower eukaryotes including yeast) have no additional methyl groups and are referred to as cap zero. Most others contain a 2′-*O*-methylated penultimate residue, i.e. a cap one structure. If the base 1 is adenine, it may be ring-methylated in the *N*-6 position. A cytoplasmic 2′-*O*-methylating activity can also modify base 2 to form a cap two structure.

Cap Synthesis and Transcription Initiation

The biochemical mechanism of cap formation has been elucidated by studying the complex of enzymes associated with purified reovirus (Furuichi *et al.*, 1976) and vaccinia virus (Moss *et al.*, 1976). Capping enzymes isolated from HeLa cell nuclei (Venkatesan *et al.*, 1980*a*) and from purified vaccinia virus (Venkatesan *et al.*, 1980*b*; Shuman *et al.*, 1980) catalyse the same series of reactions. They include:

(1) RNA polymerase $pppN + pppN \rightarrow pppNpN\ldots + PP_i$
(2) Nucleotide phosphohydrolase $pppNpN\ldots \rightarrow ppNpN\ldots + P_i$

(3) Guanylyltransferase ppNpN...+ ppp$\overset{*}{\text{p}}$G ⇄ G$\overset{*}{\text{p}}$ppNpN...+ PP_i
(4) Methyltransferase 1
G$\overset{*}{\text{p}}$ppNpN...+ S-adenosylmethionine → m^7G$\overset{*}{\text{p}}$ppNpN...+ S-adenosylhomocysteine
(5) Methyltransferase 2
m^7G$\overset{*}{\text{p}}$ppNpN...+ S-adenosylmethionine → m^7G$\overset{*}{\text{p}}$ppN^{m}pN...+ S-adenosylhomocysteine

The reovirus-associated guanylyltransferase and the corresponding activities purified from HeLa cells and vaccinia virus are capable of capping dinucleotides (reaction 3 above). The enzymes can also use polynucleotides containing 5′-diphosphate ends as guanylate acceptors. Consequently the exact time of cap addition to pre-mRNA transcripts *in vivo* is difficult to establish. What seems clear from recent analyses of the reovirion capping enzymes is that all five of the above reactions can be carried out in the absence of chain elongation. This is illustrated by the results shown in Fig. 2. Reovirus cores derived from virions by chymotrypsin digestion

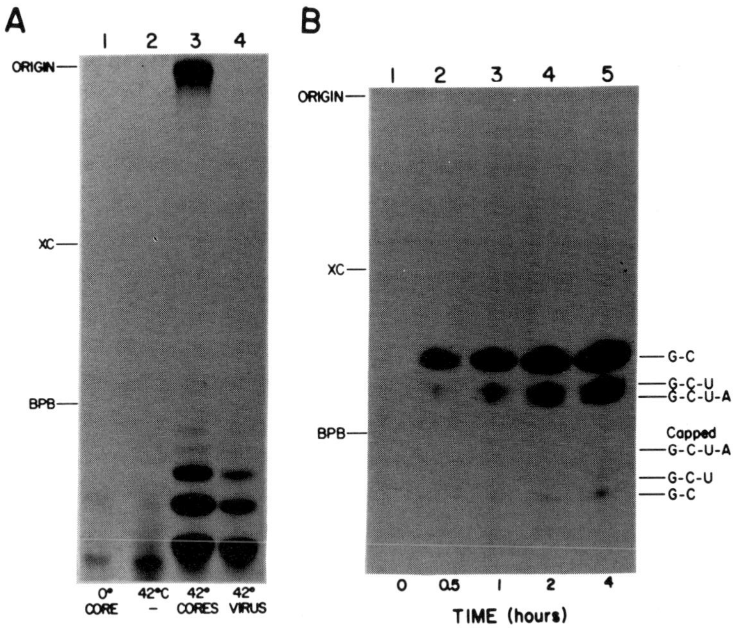

Fig. 2. *Synthesis of initiator oligonucleotides and viral mRNA by reovirus-associated enzymes*

A. Transcription reaction mixtures (25 μl) containing [α-^{32}P]CTP as the radioactive precursor and reovirions (25 μg) or viral cores derived from them by chymotrypsin digestion were incubated for 30 min, and the resulting products were analysed by electrophoresis in a 20% polyacrylamide gel as described previously (Yamakawa *et al.*, 1981). mRNA synthesized by cores remained at the origin. Initiator oligonucleotides that migrated faster than the marker dyes xylene cyanol (XC) and bromphenol blue (BPB) were eluted and analysed further by chromatography and electrophoresis after enzymic digestions. Capped and uncapped structures corresponding to the 5′-terminal sequence of viral mRNA were identified, e.g. m^7GpppG^{m}pC, m^7GpppGpC, GpppGpC and ppGpC. B. Virion transcriptase products labelled with [α-^{32}P]CTP and S-adenosyl[*methyl*-^{3}H]methionine were collected after 0, 0.5, 1, 2, and 4 h of incubation (lanes 1–5, respectively). Samples were digested with alkaline phosphatase before gel analysis. Material migrating in the positions of the capped oligonucleotides was eluted and analysed further as described in the legend to Table 1.

synthesize capped viral mRNAs *in vitro*. They also produce a high molar excess of short 'initiator' oligonucleotides that correspond in sequence to mRNA 5′-termini (panel A, lane 3). Intact reovirions produce the same capped and uncapped oligonucleotides, indicating that they contain the capping enzymes in an active form (lane 4). Although initiation and capping of nascent oligonucleotides as short as dinucleotides are effectively carried out by whole virions, no mRNA is produced even during an incubation period of several hours (panel B). Apparently elongation by the virion RNA polymerase is prevented by structural constraints on the intact particles. Analyses of the relative proportions of GpppG, m^7GpppG and m^7GpppG^m at the 5′-ends of the capped oligonucleotides synthesized by the reovirion enzymes is consistent with the order of reactions shown above (Table 1). These results indicate that capping accompanies initiation of transcription *in vitro*. Studies on adenovirus mRNA formation in infected cells have also demonstrated that caps are added early to initiated 5′-ends of incomplete transcripts (Babich *et al.*, 1980). Thus capping occurs before extensive chain growth and presumably precedes other modifications including internal methylation and polyadenylation as well as splicing. The early addition of caps to nuclear transcripts and their retention during mRNA maturation and passage to the cytoplasm is consistent with a protective effect of blocked 5′-termini on eukayotic mRNAs (Furuichi *et al.*, 1977).

A functional relationship between initiation of transcription and capping has been described for the virion-associated RNA polymerases in purified insect cytoplasmic polyhedrosis virus (Furuichi, 1978) and influenza virus (Krug, 1981). The number of starts made by the polyhedrosis virus transcriptase is markedly increased by the presence of *S*-adenosylmethionine. The methyl donor as well as some of its structural analogues are positive allosteric effectors of transcription initiation. They act by lowering the K_m for ATP, the initiating nucleotide for the RNA polymerase. Apparently the methyltransferase and RNA polymerase are structurally coupled in the virion nucleoprotein complex, and binding of methyl donor by the transferase results in activation of the polymerase.

The mechanism of influenza virus mRNA synthesis is unusual. The virion polymerase uses capped heterologous polynucleotides, i.e. nascent cellular pre-mRNAs in infected cells, to prime viral mRNA formation. An endonuclease that is present

Table 1. *Composition of capped virion oligonucleotide products*

Capped oligonucleotides eluted in water from the gel shown in Fig. 2B were counted, desalted and digested with P_1 nuclease (Furuichi *et al.*, 1976). Cap structures were separated by paper chromatography in isobutyric acid/0.5 M-NH_4OH (10 : 6, v/v) and localized by counting the paper cut into 1 cm strips. The caps were extracted and digested with venom phosphodiesterase followed by alkaline phosphatase; the resulting 7-methylguanosine and 2′-*O*-methylguanosine were separated by high voltage paper electrophoresis. Total and relative amounts of each of the capped oligonucleotides were calculated on the basis of the specific activities of the [α-^{32}P]CTP and *S*-adenosyl[*methyl*-^{3}H]methionine used in the reaction mixture.

		Composition (%)		
Oligonucleotide	Total (pmol)	GpppG	m^7GpppG	m^7GpppG^m
Capped G-C	0.093	78.5	17.6	3.9
Capped G-C-U	0.061	63.9	31.7	4.4
Capped G-C-U-A	0.053	92.4	6.1	1.5

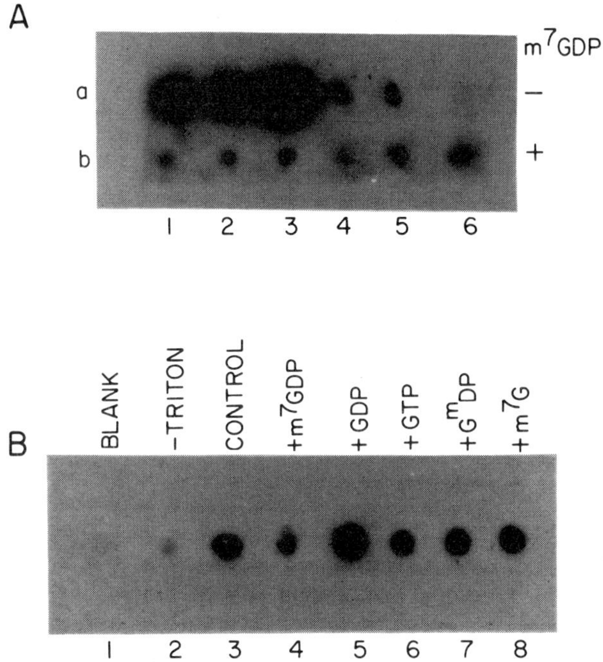

Fig. 3. *Cap binding activity of cellular (A) and influenza virus (B) proteins detected by filter blotting*

m^{7}GpppG^{m}pC was obtained from capped reovirus mRNA by RNAase T2 and alkaline phosphatase digestion followed by high voltage paper electrophoresis. The purified caps were radiolabelled to a specific activity of 1–2 μCi/pmol by incubation with [^{32}P]$\overset{*}{p}$Cp and T4 RNA ligase. After phosphatase treatment and repurification, the m^{7}GpppG^{m}pC$\overset{*}{p}$C was used as an affinity probe for cap binding activity. Protein samples in 2 μl of binding buffer [20 mM-Hepes (pH 7.4)/70 mM-KCl/3 mM-magnesium acetate/1 mM-dithiothreitol) were spotted on strips of cellulose nitrate paper (BA85, S & S) which had been soaked for 10 min in the buffer. To each protein spot was applied 2 μl of m^{7}GpppG^{m}pC$\overset{*}{p}$C in water with or without 0.5 mM-m^{7}GDP. The strip was soaked for 5 min at room temperature, washed three times in 100 ml each of the buffer, dried and the spots were located by autoradiography (Kroath & Shatkin, 1982). A. Samples of rabbit erythrocyte cap binding protein eluted from m^{7}GDP-Sepharose affinity resin (Sonenberg *et al.*, 1979) with m^{7}GDP (spots 1–3) and subsequently with 1 M-KCl (spots 4–6) were tested with 0.04 pmol (27000 c.p.m.) of m^{7}GpppG^{m}pC$\overset{*}{p}$C. Protein applied (in ng): spot 1, 360; 2, 140; 3, 340; 4, 560; 5, 590; 6, 580. Exposure 20 h; radioactivity bound in spot 3a = 6300 c.p.m. and in 3b = 140 c.p.m. B. Purified influenza A virus strain WSN (kindly provided by Dr. Robert Krug) was incubated in a transcription reaction mixture (11 μg in 25 μl) containing 50 mM-Tris/HCl (pH 7.8)/1 mM-dithiothreitol/120 mM-potassium acetate/5 mM-magnesium acetate/1 mM of each rNTP and 0.4% Triton N101. After 10 min at 31°C, 2 μl aliquots were assayed as in A but with 0.11 pmol (46000 c.p.m.) of m^{7}GpppG^{m}pC$\overset{*}{p}$C and 0.1 mM of the indicated compounds. Spot 1 is + Triton, − virus; 2 is − Triton, + virus; 3–8 are + Triton, + virus. Exposure was 24 h.

in virions cleaves the foreign RNAs at a measured distance (approx. 10–15 nucleotides) from the cap. The polymerase utilizes the capped fragments as primers to transcribe the influenza RNA templates. Thus the resulting mRNAs are chimaeric and consist of a newly formed influenza sequence joined to a derived, heterologous 5′-capped leader. The dependence of this series of unique events on a preformed cap suggests that influenza virus contains cap recognizing protein(s). This is further indicated by the ability of influenza virus, like cellular cap binding protein, to bind ^{32}P-labelled cap structures as detected by a filter blotting assay (Fig. 3). Binding depends on detergent activation of the virus and is inhibited by cap analogues. It is not a general property of viruses that contain transcriptases since vaccinia and reoviruses synthesize capped viral mRNAs entirely *de novo* but do not have detectable cap binding activity (Table 2). Association of cap binding activity with an influenza virion P protein has recently been demonstrated by using radiolabelled cap analogues (Blaas *et al.*, 1981) and capped polynucleotides as photoaffinity probes (Ulmanen *et al.*, 1981). It remains to be determined if this influenza protein is related to cellular cap binding proteins, but influenza transcription primed by capped mRNA *in vitro* is inhibited by cap binding protein purified from rabbit erythrocytes (Kroath & Shatkin, 1982).

Table 2. *Binding of* $m^7GpppG^m p\overset{*}{C}pC$ *to purified viruses*

Purified influenza virus, vaccinia virus, and reovirus (11 μg, 18 μg and 25 μg of protein, respectively), were incubated for 10 min at 31°C in transcription mixtures as in Fig. 3B. Aliquots of 2 μl were removed and tested with 0.06 pmol (23000 c.p.m.) of ^{32}P-labelled caps in the presence and absence of 0.4 mM-m^{7}GDP. Spots located by autoradiography were cut out and counted in toluene-based scintillant.

Additions		Binding (c.p.m.)		
		Influenza	Vaccinia	Reo
Triton	m^{7}GDP			
–	–	46	33	47
–	+	24	23	11
+	–	333	17	28
+	+	13	10	6

Fascinating problems remain to be solved concerning the enzymic mechanisms of cap formation, the spatial configurations assumed by caps, and the functional interactions of caps during mRNA maturation and function. One important functional effect of caps is its facilitating effect on protein synthesis. The enhancement appears to occur at the level of initiation. Recent findings suggest that cap binding protein(s) are involved in promoting stable interactions between 40 S ribosomal subunits and 5′-cap-vicinal sequences in eukaryotic mRNAs.

Role of 5′-Cap in the Initiation of Translation: Reversible Inhibition by Cap Analogues

Many different kinds of experiments have provided evidence that the presence of a 5′-terminal cap enhances eukaryotic mRNA translation (Shatkin, 1976; Filipowicz, 1978; Banerjee, 1980). For example, removal of caps by chemical or enzymic treatment of viral and cellular mRNAs diminished their capacity to bind to ribosomes in cell-free systems and to direct protein synthesis in microinjected *Xenopus* oocytes. Furthermore, capped reovirus mRNAs effectively formed stable initiation complexes *in vitro* while the corresponding uncapped RNAs did not. Ribosomal subunits initially attached to reovirus mRNAs at or near the cap, as shown by 40 S subunit protection of mRNA 5′-sequences, including the 5′-proximal AUG, against digestion by RNAase. In most other viral and cellular mRNAs the 5′-proximal AUG also serves as the initiator codon for polypeptide synthesis (Kozak, 1981).

Involvement of the cap in the initiation of translation is probably mediated or at least influenced by cap binding protein(s). A protein of M_r approx. 24000 that binds to caps (24K-CBP) has been purified from rabbit reticulocyte ribosomal salt wash by affinity column chromatography on m^{7}GDP-Sepharose (Sonenberg *et al.*, 1979). It stimulated the translation *in vitro* of capped but not of uncapped mRNAs (Sonenberg *et al.*, 1980). Binding of the 24K-CBP to mRNA 5′-ends was competed out by the presence of an excess of cap analogues such as m^{7}GMP and m^{7}GDP. The same kinds of compounds added to cell-free translating systems prevented ribosome binding to capped mRNAs. This is illustrated in Fig. 4. Reovirus mRNA formed 80 S initiation complexes in wheat germ extract (panel A), and the presence of 1 mM-m^{7}GMP inhibited binding by 83% (panel B). One interpretation of these results is that cap binding protein(s) interact with mRNA 5′-termini and promote conformation(s) favourable for attachment of ribosomal subunits.

It was suggested earlier from a comparison of three-dimensional structure and translation inhibitory activity that active cap analogues (and by inference capped mRNA 5′ ends) assume a 'rigid' conformation that is preferred for initiation of

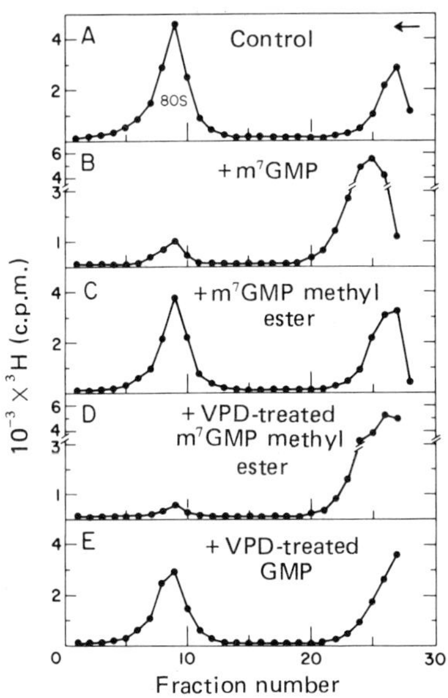

Fig. 4. *Cap analogue effects on ribosome binding*

Capped, 5′-[^{3}H]methyl-labelled reovirus mRNA was synthesized *in vitro* by incubating viral cores in transcription reaction mixtures containing *S*-adenosyl[*methyl*-^{3}H]methionine (Muthukrishnan *et al.*, 1976). The purified mRNA was incubated in wheat germ 'S23' extract in the presence of 0.2 mM-sparsomycin under conditions of 80S initiation complex formation but not polypeptide chain elongation. After 10 min at 25°C complexes were analysed by glycerol gradient centrifugation. Incubation mixtures contained 1 mM-m^{7}GMP (B) or the m^{7}GMP methyl ester without treatment (C) or after digestion as in Fig. 6 with venom phosphodiesterase (D) or venom plus calf intestine alkaline phosphatase (E).

protein synthesis (Hickey *et al.*, 1977; Adams *et al.*, 1978). The rigidity of m^7Gp-(p). . . apparently results from an electrostatic interaction between the positively charged imidazole portion of the N-7 alkylated guanosine and the negative phosphate group(s) (see Fig. 1). To test the proposed correlation between preferred conformation and cap function, the methyl ester of m^7GMP was chemically synthesized. This cap analogue was predicted to have a decreased negativity due to the availability of only one, rather than two, hydroxyls on the phosphate group (Fig. 5). N.m.r. spectroscopy studies indicated that the conformations of m^7GMP and m^7GMP methyl ester were very similar (Darzynkiewicz *et al.*, 1981). However, other properties of the two compounds were strikingly different. For example, the methyl ester was resistant to attack by alkaline phosphatase while m^7GMP was converted to the positively charged nucleoside by phosphatase digestion (Fig. 6, lanes 5 and 3). Treatment with venom phosphodiesterase hydrolysed the phosphate-

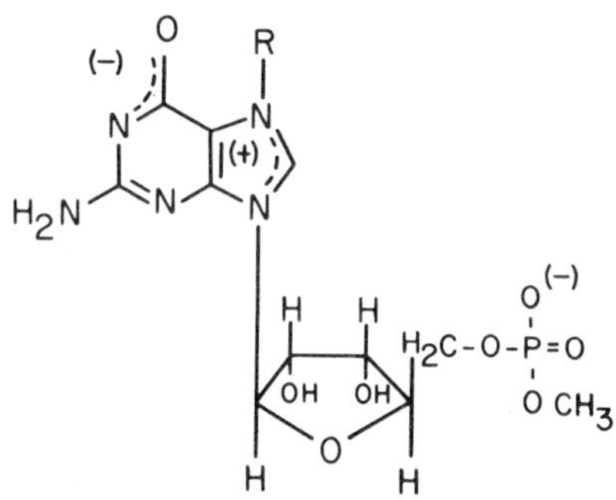

Fig. 5. *Structure of 7-methylguanosine-5′-phosphate methyl ester*

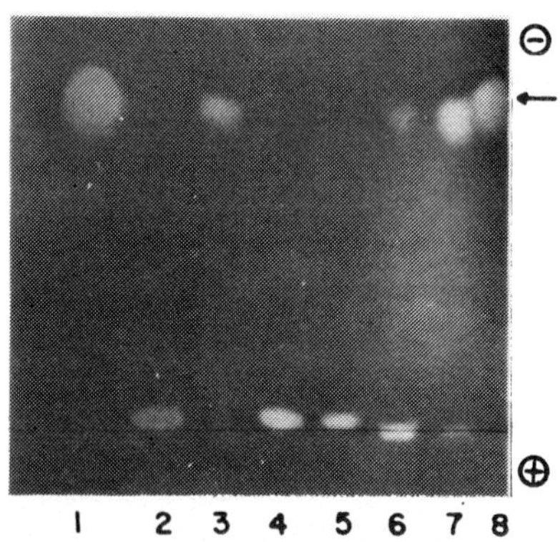

Fig. 6. *Susceptibility of m^7GMP methyl ester to enzyme digestion*

After paper electrophoresis in pyridine/acetate buffer (pH 3.5) the fluorescent samples were photographed under u.v. light. Lanes 1 and 8 are 10 nmol of m^7G. Lanes 2, 3 are 50 nmol of m^7GMP before and after digestion with calf intestine alkaline phosphatase [20 units/ml in 0.02 M-Tris/HCl (pH 8), 37°C, 30 min]. Other lanes are 50 nmol of m^7GMP methyl ester analysed directly (4) or (5) treated with phosphatase as in lane 3 or (6) digested with venom phosphodiesterase [2 units/ml in 0.02 M-Tris/HCl (pH 8), 10 mM-magnesium acetate, 0.1 M-NaCl, 37°C, 30 min] and repeated, followed by 2 min at 50°C to inactivate the enzyme and (7) treated sequentially with phosphodiesterase and phosphatase but at 30 units of calf phosphatase/ml.

methyl ester, and the resulting m^7GMP could then be converted to m^7G by incubation with phosphatase (lanes 6 and 7).

These enzyme susceptibility results were reflected in functional tests of cap analogue activity as shown in Fig. 4. In contrast to the strong inhibitory activity of 1 mM-m^7GMP on ribosome binding (panel B), the same concentration of methyl ester had little effect on translation initiation as measured by 80 S initiation complex formation (panel C). Conversion of the m^7GMP methyl ester to m^7GMP by venom phosphodiesterase digestion was accompanied by acquisition of inhibitory activity (panel D). Restoration of cap analogue activity was not due to a non-specific effect (e.g. the presence of an inhibitory contaminant) because sequential digestion of m^7GMP methyl ester with phosphodiesterase and phosphatase yielded the non-inhibitory nucleoside product, m^7G (panel E). These findings indicate that although the rigid conformation of m^7GMP (as determined from n.m.r. measurements) was retained following methyl esterification, the cap analogue activity of the esterified m^7GMP was lost. Thus a rigid 5′-terminal conformation may be necessary, but clearly it is not sufficient for cap recognition during initiation of translation.

The inhibitory activity of m^7GMP on ribosome binding may be due to competition with mRNA for protein(s) that recognize and bind to the 5′-cap during initiation. To test this possiblity a chemical cross-linking procedure was used as described in detail previously (Sonenberg *et al.*, 1978). In this method affinity purified 24K-CBP was radiolabelled by reductive cross-linking to periodate-oxidized, 5′-^{3}H-labelled mRNA. Cap-specific interaction of the 24K-CBP with reovirus mRNA was progressively inhibited by increasing concentrations of m^7GMP. At 0.25 mM the extent of inhibition was 85% (Fig. 7, lane 3). As also observed in the ribosome binding assay (Fig. 4C), the methyl ester was far less inhibitory, and a level of 0.25 mM decreased 24K-CBP radiolabelling by only 11% (Fig. 7, lane 7). The results provide evidence for a direct effect of cap analogues on the interaction of 24K-CBP with mRNA 5′ ends. They suggest further that the presence of an extra methyl group on the phosphate moiety of m^7GMP perturbs the 'fit' between cap and the binding protein(s) involved in initiation complex formation.

Capped mRNA translation *in vitro* is increased by addition of purified 24K-CBP, and it was of interest to determine if this enhancement was affected by the presence of cap analogues. HeLa cell-free translating extracts synthesized mainly viral capsid protein in response to Sindbis virus mRNA. The yield of this M_r 33000 polypeptide was increased several-fold by addition of affinity purified 24K-CBP (Fig. 8, lanes 1–3). The increase was partially prevented by 0.1 mM m^7GMP (Fig. 8, lane 5), and a higher concentration was almost completely inhibitory (lane 4). Consistent with results obtained in the other assays for cap analogue activity, 0.25 mM-m^7GMP methyl ester had little effect on translation of Sindbis mRNA in HeLa cell extracts (Fig. 8, lanes 7–9). These and other data taken together provide strong evidence for a functional interaction between mRNA 5′-caps and 24K-CBP during initiation of translation.

Translation Inhibition by Antibody to 24K-CBP

A functional role of 24K-CBP in protein synthesis was also suggested by the results of studies with a monoclonal antibody to this polypeptide. Antibody was

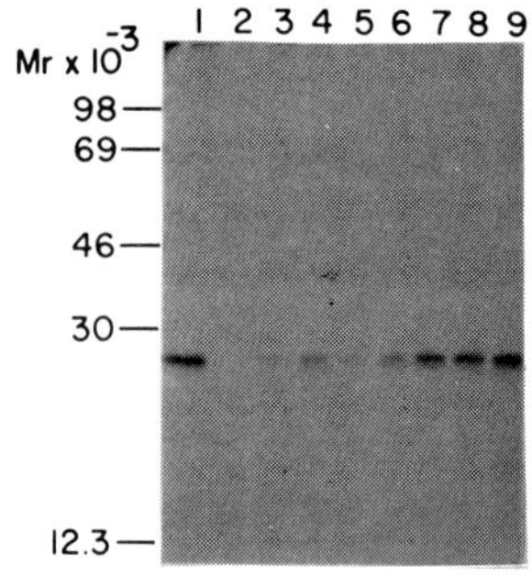

Fig. 7. *Cap analogue inhibition of cap binding protein interaction with mRNA as measured by chemical cross-linking*

Each sample contained 2.3 μg of [^{3}H]methyl-labelled, periodate-oxidized reovirus mRNA (specific radioactivity 10^5 c.p.m./μg) and 1.8 μg of 24K-CBP purified from rabbit erythrocytes by affinity chromatography. Binding was done (Sonenberg *et al.*, 1979) in the absence (lane 1) or presence of 0.5, 0.25, 0.1, or 0.05 mM-m^{7}GMP (lanes 2–5, respectively) or the same concentration of m^{7}GMP methyl ester (lanes 6–9, respectively). After $NaBH_3CN$ reduction and RNAase digestion, proteins radiolabelled by covalently bound [^{3}H]caps transferred from oxidized mRNA were detected by polyacrylamide gel electrophoresis and fluorography. Films were traced, and the normalized levels of crosslinking based on relative densities of the approx. 24 kDa band in lanes 1–9 were: 1.0, 0, 0.15, 0.31, 0.15, 0.38, 0.89, 0.78, and 1.12, respectively. Values obtained with control samples that contained GMP and GMP methyl ester (0.5 mM) were 1.19 and 1.25, respectively.

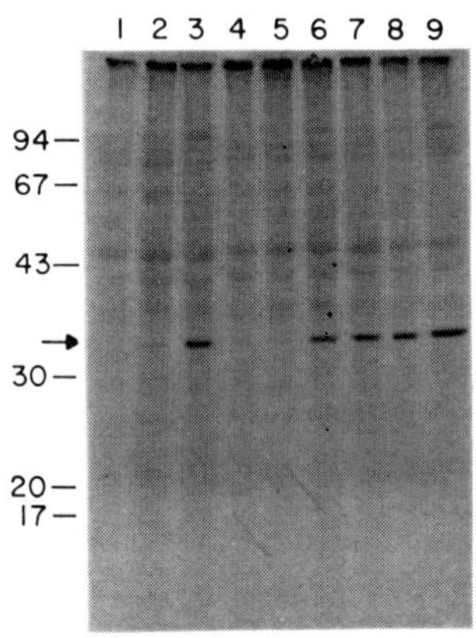

Fig. 8. *Effect of cap analogues on cell-free translation*

HeLa cell-free protein synthesizing extracts were incubated with Sindbis virus mRNA and [^{35}S]methionine for 60 min at 37°C (Sonenberg *et al.*, 1980). Products were analysed by polyacrylamide-gel electrophoresis and autoradiography. Lane 1, no added mRNA; lane 2 approx. 0.3 μg of Sindbis virus mRNA; lanes 3–9 mRNA plus 1.4 μg of affinity purified 24K-CBP. Samples in lanes 4, 5 and 6 also contained m^{7}GMP and in lanes 7, 8 and 9 m^{7}GMP methyl ester at 0.25, 0.1, and 0.05 mM, respectively. Normalized levels of viral capsid protein (arrow, M_r approx. 33000) determined from densitometry tracings of lanes 2–9 were 1.0, 5.5, 0, 0.6, 4.1, 5.3, 4.7 and 6.7, respectively.

prepared by fusing non-producer, P3X63Ag8.653 myeloma cells with spleen cells from mice that had been immunized with 24K-CBP. The antigen was purified by m^{7}GDP–Sepharose affinity chromatography from rabbit erythrocytes rather than from reticulocyte ribosomal high salt wash. This was done in an effort to avoid the

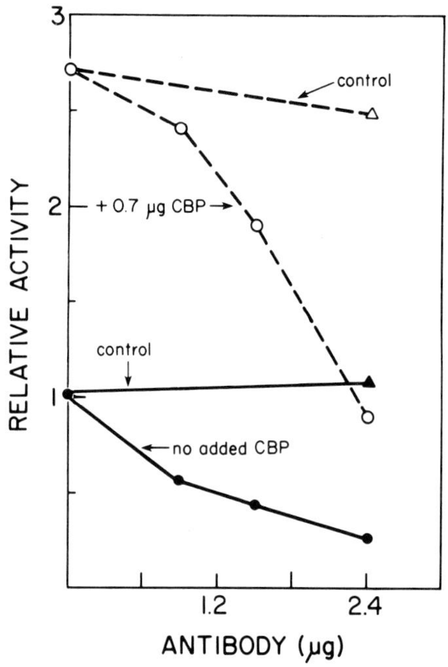

Fig. 9. *Inhibition of capped mRNA translation by anti-(24K-CBP) monoclonal antibody*

Sindbis virus mRNA (approx. 0.3 μg) was translated in 25 μl incubation mixtures (Rose *et al.*, 1978) that included 9 μl of micrococcal nuclease-treated HeLa cell extract, 140 mM-K^+, 0.5 mM-magnesium acetate, 10 μCi of [^{35}S]methionine (Amersham Searle, approx. 1200 Ci/mmol), and, where indicated, 0.7 μg of 24K-CBP purified from rabbit erythrocytes. Hybridoma cells from a clone selected for production of antibody to 24K-CBP were injected intraperitoneally into mice, and $1\frac{1}{2}$ to 2 weeks later the ascites fluid was collected, clarified and made to 40% saturation with solid $(NH_4)_2SO_4$. The precipitated protein was applied to a tandem column (0.7 cm × 8 cm) consisting of equal layers of CM-52 above DE-52 resin equilibrated with phosphate-buffered saline containing 1 mM-EGTA (Palacios *et al.*, 1972). Peak A_{280} fractions were pooled and dialysed exhaustively against 0.1 M-potassium phosphate buffer (pH 8) containing 0.02% NaN_3. The dialysed protein solution (160 A_{280} units) was applied to a column of protein A-Sepharose (1.5 cm × 4 cm) equilibrated with 0.1 M-potassium phosphate (pH 8) containing 0.02% NaN_3. After washing with 30 ml of the equilibration buffer, samples were eluted sequentially with 40 ml of 0.1 M-sodium citrate (pH 6) followed by the same volume of 0.1 M-sodium citrate (pH 4.5), both containing 0.2% NaN_3. Peak fractions from the pH 6 eluate (1.3 A_{280} units) and pH 4.5 eluate (0.8 A_{280} units) were pooled separately and dialysed against 20 mM-Tris/HCl (pH 7.5)/150 mM-KCl. The pH 6 eluate, which was active by enzyme-linked immunoadsorbent (ELISA) assay, was used after concentration to 0.32 mg of protein/ml. The control sample was P3X63 IgG_1 of unknown specificity from ascites fluid of Balb/c mice injected with P3X63 myeloma cells. Protein was precipitated with $(NH_4)_2SO_4$ as above, and after dialysis 12 A_{280} units were applied to protein A-Sepharose. The pH 6 eluate (1.7 A_{280} units) was dialysed and concentrated to 0.6 mg/ml. Both antibody preparations just before use were treated for 30 min at 37°C with 0.04% diethyl pyrocarbonate (by dilution from a 1% solution in ethanol) in order to eliminate residual ribonuclease. Recently, however, we found that this treatment inactivated the antibody as measured by radioimmunoassay and ELISA assays but apparently not as determined by inhibition of capped mRNA translation. Affinity purified 24K-CBP was pre-incubated for 10 min at room temperature with the purified anti-(24K-CBP) or control immunoglobulin and then added to translation assay mixtures. After incubation at 37°C for 60 min, samples were precipitated (approx. 3 h) with 1 ml of acetone, pelleted by centrifugation, air-dried, dissolved in 20 μl of Laemmli sample buffer (Laemmli, 1970), heated for 5 min at 100°C and analysed by electrophoresis in a 12.5% modified Laemmli slab gel containing a 3% stacking gel. The autoradiogram (Kodak X-Omat film) of the dried gel was traced, and the relative translational activities were determined from densitometry readings.

high molecular weight polypeptides that copurify with 24K-CBP in some conditions (Tahara *et al.*, 1981). The hybridoma clone was grown in the ascites form, and immunoglobulin was purified from the fluid by $(NH_4)_2SO_4$ precipitation and protein A–Sepharose chromatography. High-titre antibodies were not obtained, possibly because of tolerance to a rabbit protein that presumably is related to a putative cap binding protein present in the mouse (and perhaps conserved among other mammals). However, the anti-(24K-CBP) inhibited translation of Sindbis virus capped mRNA in extracts of HeLa cells (Fig. 9). The stimulation of capped mRNA translation obtained by addition of purified 24K-CBP was also antibody sensitive; 2.4 μg of purified antibody/25 μl assay diminished translation to below the unstimulated level of protein synthesis. Control antibodies, e.g. a monoclonal antibody to reovirus cores or an immunoglobulin fraction from mouse ascites cells, had little or no effect. It should be noted that tissue culture fluids from hybridoma cells grown *in vitro* were not used because the growth medium from non-producer parental cells yielded protein fractions that were inhibitory for cell-free translation.

The antibody to cap binding protein was similarly active in mouse L cell and HeLa protein synthesizing extracts. It inhibited [^{35}S]methionine incorporation into Sindbis virus capsid protein in both the unstimulated and 24K-CBP stimulated reactions (Fig. 10). By contrast, a naturally uncapped RNA, encephalomyocarditis (EMC) virus RNA, was translated equally well in HeLa cell extracts in the absence or presence of the antibody. This finding is consistent with the 24K-CBP specificity of the monoclonal antibody and cap-independent expression of EMC virus RNA.

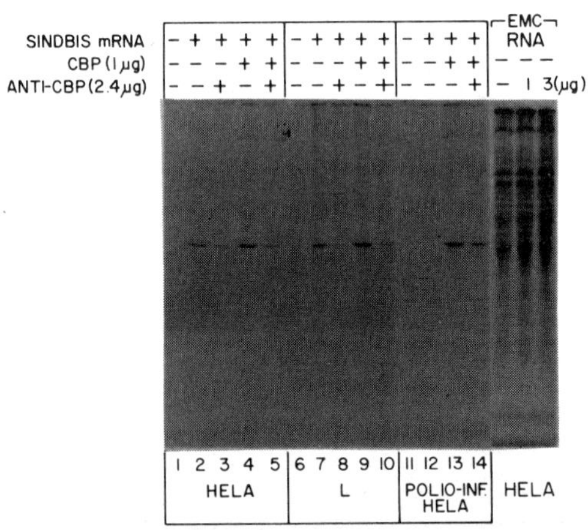

Fig. 10. *Inhibition of mouse and HeLa cell-free translation by anti-CBP antibody*

Gel analysis by autoradiography of [^{35}S]methionine-labelled products directed by Sindbis virus capped mRNA and EMC virus uncapped RNA in uninfected HeLa and L cell extracts and in extracts of poliovirus-infected HeLa cells. Incubation mixtures included, as indicated, approx. 0.1–0.3 μg of Sindbis mRNA or 0.25 μg of EMC virus RNA, 24K-CBP or in the case of extracts of polio-infected cells, CBP II (restoring activity, see Tahara *et al.*, 1981) and/or anti-CBP immunoglobulin. Conditions were as described in the legend to Fig. 9, except that for translation by L cell extracts, the K^+ and magnesium acetate levels were 100 mM and 2 mM, respectively.

Poliovirus, a human picornavirus relative of mouse EMC virus, also contains an uncapped RNA that functions as viral messenger. Infection of HeLa cells with poliovirus results in the shut-off of host (capped) mRNA translation in favour of viral protein formation. The switch to translation of uncapped RNAs is maintained in extracts prepared from poliovirus-infected cells. In such extracts capped mRNA translation can be restored by addition of protein preparations that contain 24K-CBP (Trachsel *et al.*, 1980; Tahara *et al.*, 1981). Restoring activity was decreased by exposure to anti-(24K-CBP), indicative of a functionally important role of cap binding proteins in the initiation of capped mRNA translation (Fig. 10).

Conclusion

Most viral and cellular mRNAs of eukaryotes are 5′-terminally modified by the addition of a cap structure, $m^7G^{5'}ppp^{5'}N$. The presence of a cap has important consequences, most strikingly with respect to the ability of mRNA to initiate protein synthesis effectively. The facilitating effect of cap on translation initiation appears to be mediated by cap binding protein(s) such as a 24000 dalton cap binding protein (24K-CBP) purified from rabbit blood cells. It stimulated capped mRNA translation *in vitro* and could be chemically cross-linked to the 5′ end of oxidized, capped mRNA. Both cap recognition processes were inhibited by an excess of the cap analogue, m^7GMP. By contrast, the methyl ester of m^7GMP was inactive as a cap analogue, suggesting that the interaction of cap binding protein with cap is a highly specific and discriminatory one. Monoclonal antibody prepared against affinity purified 24K-CBP inhibited translation of Sindbis virus capped mRNA but not naturally uncapped EMC virus RNA in extracts of HeLa and mouse L cells. Stimulation of capped mRNA function by 24K-CBP was also prevented by the antibody, as was the restoration of capped mRNA translation by addition of a cap binding protein complex to extract of poliovirus-infected HeLa cells.

The possible regulatory roles of caps and cap binding proteins in gene expression is being explored. Of some interest is the mechanism of enhancement of capped mRNA translation by 24K-CBP. Presumably the effect is related to the finding that 40S ribosomal subunits attach to the 5′ end of mRNA before positioning on the initiator (usually 5′-proximal) AUG triplet during initiation complex formation (Kozak, 1981). Ribosome binding to capped mRNA (unlike naturally uncapped messengers: Kozak, 1980*b*; R. Jackson, personal communication) is an ATP-dependent process (Marcus, 1970; Trachsel *et al.*, 1977; Kozak, 1980*a, b*). However, binding of capped mRNA is ATP-independent if the mRNA is unfolded, e.g. substituted with IMP in place of GMP during synthesis (Kozak, 1980*a*; Morgan & Shatkin, 1980) or treated with bisulphite to convert CMP residues to UMP (Kozak, 1980*a*) and thus prevent G·C base-pairing. Furthermore, unlike native mRNAs, binding of the modified, denatured mRNAs is relatively insensitive to inhibition by cap analogues. The combined findings that capped mRNAs depleted of secondary structure have a lower ATP requirement and a decreased dependence on the cap suggests that cap binding protein facilitates ribosome attachment to (and correct positioning on) capped mRNA by an ATP-dependent process. Recently we observed that eukaryotic initiation factors eIF-4A and eIF-4B, two factors that are involved in mRNA binding to ribosomal subunits, can be cross-linked to the 5′ end of oxidized, capped mRNA in the presence of ATP, Mg^{2+} and 24K-CBP (Grifo *et al.*, 1982). Thus, interaction of mRNA with the cap binding protein may direct binding of other protein factors that are involved in the early stages of protein synthesis initiation.

We thank the copyright holders, Academic Press Inc. (London) Ltd., for permission to reproduce Figs. 2, 3 and 4 from Darzynkiewicz *et al.* (1981).

References

Adams, B.L., Morgan, M., Muthukrishnan, S., Hecht, S.M. & Shatkin, A.J. (1978) *J. Biol. Chem.* **253**, 2589–2595
Babich, A., Nevins, J.R. & Darnell, Jr., J.E. (1980) *Nature (London)* **287**, 246–248
Bachellerie, J.-P., Amalric, F. & Caboche, M. (1978) *Nucleic Acids Res.* **5**, 2927–2943
Banerjee, A.K. (1980) *Microbiol. Rev.* **44**, 175–205
Blaas, D., Patzelt, E. & Kuchler, E. (1981) in *Proc. Int. Congr. Virol. 5th*, p. 393, no. P43/01
Darnell, J.E., Jr. (1979) *Prog. Nucleic Acid Res. Mol. Biol.* **22**, 327–353
Darzynkiewicz, E., Antosiewicz, J., Ekiel, I., Morgan, M.A., Tahara, S.M. & Shatkin, A.J. (1981) *J. Mol. Biol.* **153**, 451–458
Dimock, K. & Stoltzfus, C.M. (1978) *Biochemistry* **17**, 3627–3632
Dimock, K. & Stoltzfus, C.M. (1979) *J. Biol. Chem.* **254**, 5591–5594
Doel, M.T. & Carey, N.H. (1976) *Cell* **8**, 51–58
Filipowicz, W. (1978) *FEBS Lett.* **96**, 1–11
Fitzgerald, M. & Shenk, T. (1981) *Cell* **24**, 251–260
Flanegan, J.B., Pettersson, R.F., Ambros, V., Hewlett, M. & Baltimore, D. (1977) *Proc. Natl. Acad. Sci. U.S.A.* **74**, 961–965
Furuichi, Y. (1978) *Proc. Natl. Acad. Sci. U.S.A.* **75**, 1086–1090
Furuichi, Y., Muthukrishnan, S., Tomasz, J. & Shatkin, A.J. (1976) *J. Biol. Chem.* **251**, 5043–5053
Furuichi, Y., LaFiandra, A. & Shatkin, A.J. (1977) *Nature (London)* **266**, 235–239
Grifo, J.A., Tahara, S.M., Leis, J.P., Morgan, M.A., Shatkin, A.J. & Merrick, W.C. (1982) *J. Biol. Chem.* **257**, 5246–5252
Hickey, E.D., Weber, L.A., Baglioni, C., Kim, C.H. & Sarma, R.H. (1977) *J. Mol. Biol.* **109**, 173–183
Hofer, E. & Darnell, Jr., J.E. (1981) *Cell* **23**, 585–593
Huez, G., Marbaix, G., Hubert, E., Cleuter, Y., Leclercq, M., Chantrenne, H., Soreq, H., Nudel, U. & Littauer, U.Z. (1975) *Eur. J. Biochem.* **59**, 589–592
Kaehler, M., Coward, J. & Rottman, F. (1977) *Biochemistry* **16**, 5770–5775
Kedes, L.H. (1979) *Annu. Rev. Biochem.* **48**, 837–870
Kozak, M. (1980*a*) *Cell* **19**, 79–90
Kozak, M. (1980*b*) *Cell* **22**, 459–467
Kozak, M. (1981) *Curr. Top. Microbiol. Immunol.* **93**, 81–123
Kroath, H. & Shatkin, A.J. (1982) *J. Virol.* **41**, 1105–1108
Krug, R.M. (1981) *Curr. Top. Microbiol. Immunol.* **93**, 125–149
Laemmli, U.K. (1970) *Nature (London)* **227**, 680–685
Lee, Y.F., Nomoto, A., Detjen, B.M. & Wimmer, E. (1977) *Proc. Natl. Acad. Sci. U.S.A.* **74**, 59–63
Marcus, A. (1970) *J. Biol. Chem.* **245**, 962–966
Morgan, M.A. & Shatkin, A.J. (1980) *Biochemistry* **19**, 5960–5966
Moss, B., Gershowitz, A., Wei, C.-M. & Boone, R.F. (1976) *Virology* **72**, 341–351
Moss, B., Gershowitz, A., Weber, L.A. & Baglioni, C. (1977) *Cell* **10**, 113–120
Muthukrishnan, S., Morgan, M., Banerjee, A.K. & Shatkin, A.J. (1976) *Biochemistry* **15**, 5761–5768
Palacios, R., Palmiter, R.D. & Schimke, R.T. (1972) *J. Biol. Chem.* **247**, 2316–2321
Revel, M. & Groner, Y. (1978) *Annu. Rev. Biochem.* **47**, 1079–1126
Rose, J.K., Trachsel, H., Leong, K. & Baltimore, D. (1978) *Proc. Natl. Acad. Sci. U.S.A.* **75**, 2732–2736
Salditt-Georgieff, M., Harpold, M., Chen-Kiang, S. & Darnell, J.E. (1980) *Cell* **19**, 69–78
Shatkin, A.J. (1976) *Cell* **9**, 645–653
Shenk, T. (1981) *Curr. Top. Microbiol. Immunol.* **93**, 25–46
Shuman, S., Surks, M., Furneaux, H. & Hurwitz, J. (1980) *J. Biol. Chem.* **255**, 11588–11598

Sonenberg, N., Morgan, M.A., Merrick, W.C. & Shatkin, A.J. (1978) *Proc. Natl. Acad. Sci. U.S.A.* **75**, 4843–4847
Sonenberg, N., Rupprecht, K.M., Hecht, S.M. & Shatkin, A.J. (1979) *Proc. Natl. Acad. Sci. U.S.A.* **76**, 4345–4349
Sonenberg, N., Trachsel, H., Hecht, S. & Shatkin, A.J. (1980) *Nature (London)* **285**, 331–333
Stoltzfus, C.M., Shatkin, A.J. & Banerjee, A.K. (1973) *J. Biol. Chem.* **248**, 7993–7998
Tahara, S.M., Morgan, M.A. & Shatkin, A.J. (1981) *J. Biol. Chem.* **256**, 7691–7694
Trachsel, H., Erni, B., Schreier, M. & Staehelin, T. (1977) *J. Mol. Biol.* **116**, 755–767
Trachsel, H., Sonenberg, N., Shatkin, A.J., Rose, J.K., Leong, K., Bergmann, J.E., Gordon, J. & Baltimore, D. (1980) *Proc. Natl. Acad. Sci. U.S.A.* **77**, 770–774
Ulmanen, I., Broni, B.A. & Krug, R.M. (1981) *Proc. Natl. Acad. Sci. U.S.A.* **78**, 7355–7359
Venkatesan, S., Gershowitz, A. & Moss, B. (1980*a*) *J. Biol. Chem.* **255**, 903–908
Venkatesan, S., Gershowitz, A. & Moss, B. (1980*b*) *J. Biol. Chem.* **255**, 2829–2834
Willis, D. & Granoff, A. (1976) *Virology* **73**, 543–547
Wittek, R., Cooper, J.A., Barbosa, E. & Moss, B. (1980) *Cell* **21**, 487–493
Yamakawa, M., Furuichi, Y., Nakashima, K., LaFiandra, A.J. & Shatkin, A.J. (1981) *J. Biol. Chem.* **256**, 6507–6514
Zeevi, M., Nevins, J.R. & Darnell, Jr., J.E. (1981) *Cell* **26**, 39–46

Biochem. Soc. Symp. **47**, 145–161
Printed in Great Britain

Association of an M_r 50 000 Cap Binding Protein with the Cytoskeleton in BHK Cells

HANS TRACHSEL,* ALBERT ZUMBÉ,† CHRISTIAN STÄHLI,‡ MARKUS HÜMBELIN*
and NAHUM SONENBERG §

* *Biocenter, University of Basel, CH-4056 Basel, Switzerland,* † *Friedrich-Miescher Institut, Postfach 273, CH-4002 Basel, Switzerland,* ‡ *Hoffmann-La Roche & Co. AG, Grenzacherstrasse, CH-4002 Basel, Switzerland, and* § *Department of Biochemistry, McGill University, Montreal, Canada*

Synopsis

A monoclonal antibody (anti-CBP antibody) is shown to be directed against cap binding protein(s) (CBP) by virtue of its ability to (*a*) inhibit the translation of capped reovirus mRNA in a cell-free system derived from L-cells and (*b*) inhibit the specific (cap analogue-inhibited) cross-linking of proteins to the oxidized 5′ terminal cap structure of reovirus mRNA.

Anti-CBP antibody reacts with an M_r 50000 polypeptide in rabbit reticulocyte polysomes and this polypeptide appears to be associated with the 5′ cap structure of mRNA. In BHK-21 cells immunofluorescence microscopy reveals that the antibody reacts with a fibrous network extending through the cytoplasm in a radial arrangement. The network behaves like intermediate filaments in colchicine-treated cells suggesting a direct or indirect linkage of CBP with intermediate filaments. The association of CBP with a cytoskeletal element is further confirmed by isolation of proteins from Triton X-100-extracted cells and identification of CBP in the cytoskeletal fraction with anti-CBP antibody. The major polypeptide reacting with anti-CBP antibody is an M_r 50000 component. Tryptic peptide mapping shows that this polypeptide is related to an M_r 24000 polypeptide identified as cap binding protein in earlier experiments [Sonenberg, Morgan, Merrick & Shatkin (1978) *Proc. Natl. Acad. Sci. U.S.A.* **75**, 4843–4847].

Introduction

Most of our knowledge about the mechanism of eukayotic protein synthesis initiation stems from experiments with cell-free systems reconstituted from individual purified components (for a review, see [1]). Among these components are seven initiation factors designated eIF-1, -2, -3, -4A, -4B, -4C and -5 [2–9] and at least one cofactor [10–12]. Their function in initiation complex formation in reconstituted systems has been described [13–15] and is summarized in Fig. 1. The initiation factor eIF-3 and probably also eIF-4C [16] bind to the 40S ribosomal subunit and inhibit the joining of 40S and 60S ribosomal subunits [15]. In the next step the ternary complex (eIF-2, Met-tRNA$_f$, GTP) binds to the 40S ribosomal subunit. The subsequent binding of mRNA is dependent on ATP hydrolysis, mRNA

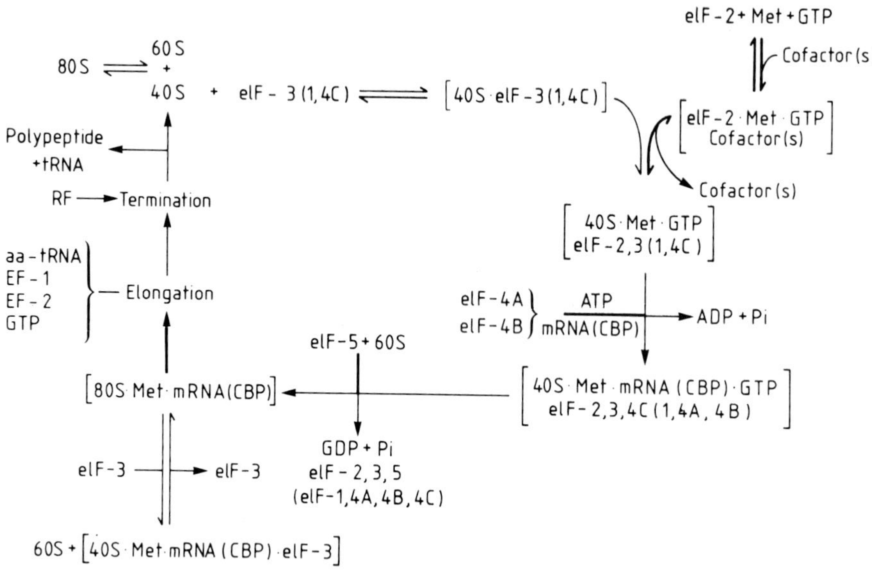

Fig. 1. *Scheme of protein synthesis initiation*

Abbreviations: 80S, 80S ribosome; 60S, 60S ribosomal subunit; 40S, 40S ribosomal subunit; eIF-, eukaryotic initiation factor; Met, initiator Met-tRNA; CBP, cap binding protein; EF-, elongation factor; aa-tRNA, aminoacyl-tRNA; RF, release factors; [ ], complex; ( ), components in parentheses have not directly been shown to be associated with the complex.

binding factors (eIF-4A and 4B) and cap binding protein. The latter is most probably associated with mRNA (see below). In the last step, the 60S ribosomal subunit joins the 40S initiation complex to form the 80S initiation complex. Protein synthesis initiation is facilitated by the presence in eukaryotic mRNAs of the 5′ terminal cap structure $m^7G(5')ppp(5')N$ [17]. This structure mediates its effect(s) in initiation through proteins which specifically bind to it. In crude protein-synthesizing systems several polypeptides (cap binding proteins) were shown to cross-link specifically to the cap structure [18]. In addition, an M_r 24000 polypeptide (24K-CBP) that binds to the cap structure has been identified in partially purified initiation factor fractions [19] and purified by affinity chromatography on m^7GDP-Sepharose [20].

Recently, cross-linking experiments done in the presence of ATP/Mg^{2+} revealed the presence in partially purified initiation factor preparations of cap binding proteins of M_r 28000, 50000 and 80000 in addition to 24K-CBP [21]. These findings suggest the existence of a whole family of cap binding proteins. So far, it is not known what the relative contributions to cap recognition of the different cap binding proteins are, but it is clear that they play an important role in the initiation of translation of capped mRNAs [22, 23] and the regulation of translation in poliovirus-infected HeLa cells [24, 25].

We prepared monoclonal antibodies directed against cap binding proteins using as antigen a partially purified preparation of rabbit reticulocyte eIF-3 which was shown previously to contain cap binding protein(s) [25]. One of the antibodies (anti-CBP antibody) was found to react with several, structurally related polypeptides which share tryptic and chromotryptic peptides with 24K-CBP [26]. Here we show that one of these polypeptides, an M_r 50000 component, is associated with polyribosomes in rabbit reticulocyte lysate and the cytoskeleton in BHK-21 cells.

Materials and methods

1. Preparation of the monoclonal antibody (anti-CBP antibody)

(a) Immunization. BALB c/J female mice were injected with rabbit reticulocyte eIF-3 that had been purified through steps 1–4 as described [27]. The immunization schedule of Stähli *et al.* [28] was used with 70 μg of eIF-3 per mouse for the two initial injections and 100–500 μg of eIF-3 per mouse for the booster injections just before fusion.

(b) Cell fusion and cell culture. Spleen cells from an immunized mouse were fused with SP-2-01 myeloma cells as described [28]. Culture supernatants were tested using a solid-phase radioimmunoassay [28] with 5–10 μg of eIF-3/ml as antigen. Culture supernatants containing antibodies directed against eIF-3 were further tested by radioimmunoassay with 1 μg of affinity-purified 24K-CBP/ml [20] as antigen. Positive cultures (anti-CBP clones) were recloned by limiting dilution as described [28].

(c) Purification of the monoclonal 19S (IgM) anti-CBP antibody. Cells (1×10^7–2×10^7) of anti-CBP clone were injected intraperitoneally into BALB c/J mice which had received an injection of 0.15 ml of complete Freund's adjuvant 7 days before; 2–3 weeks later the mice were killed and the ascites fluid was collected. Antibodies were purified from ascites fluid by $(NH_4)_2SO_4$ precipitation (1.75 M final concn.) and DEAE-cellulose chromatography [29]. In the final purification step, anti-CBP antibodies were separated from IgG and lower molecular weight components by centrifugation in 12 ml 10–40% convex exponential sucrose gradients in 20 mM-Hepes (pH 7.6)/150 mM-NaCl. Centrifugation was in the Spinco SW 40 Ti rotor at 40000 rev./min for 16 h at 4°C.

(d) Labelling of anti-CBP antibody. Anti-CBP antibody was purified from culture medium (hybridoma supernatant) by affinity chromatography on (sheep anti-mouse IgG)–Sepharose eluted with 10 mM-HCl. The pH of the collected fractions was immediately adjusted to neutrality with KOH. Fractions containing anti-CBP antibody were concentrated by lyophilization and labelled with ^{125}I by using the chloramine-T method [30].

2. Viral mRNA

Reovirus mRNA containing 5′-terminal m^7GpppG^mp was synthesized with viral cores in the presence of 10 μM-*S*-adenosylmethionine as previously described [31]. For cross-linking experiments, cores were incubated in the presence of *S*-adenosyl-[^3H]methionine (sp. radioactivity 70 Ci/mmol; New England Nuclear). Periodate oxidation of [^3H]methylated mRNA (80000 c.p.m./μg) was done as described before [31].

3. Cell-free translation assay

Cell-free extracts (25 μl) prepared from L-cells were incubated under translation conditions [32] with 5 μg of reovirus mRNA. Mixtures were incubated for 60 min at 30°C, at which time 5 μl of the reaction mixture was withdrawn, mixed with sample buffer, heated for 5 min at 100°C and analysed by SDS/polyacrylamide-gel electrophoresis and fluorography as described [33].

4. Cross-linking of proteins to mRNA

[^{3}H]Methyl-labelled oxidized reovirus mRNA (1 μg) was incubated with 90 μg of reticulocyte ribosomal salt wash [34] in 20 mM-Hepes (pH 7.3)/70 mM-KCl/1 mM-magnesium acetate/2 mM-dithiothreitol/1 mM-ATP/0.2 mM-GTP for 10 min at 30°C. Complexes formed between proteins and mRNA were stabilized by reduction with $NaBH_3CN$, followed by treatment with RNAase [33] and detected after analysis by SDS/polyacrylamide-gel electrophoresis and fluorography [33].

5. Isolation of polysomes from rabbit reticulocyte lysate

The procedures for the preparation of rabbit reticulocyte lysate were as described [35]. Incubation mixtures (30 μl) contained 10 mM-Tris/HCl (pH 7.6), 75 mM-KCl, 1 mM-magnesium acetate, 0.2 mM-GTP, 40 μM each of 20 unlabelled amino acids, 10 mM-creatine phosphate, 1.5 μg of creatine kinase, 20 μM-haemin and 15 μl of reticulocyte lysate. The incubation mixtures were incubated for 10 min at 30°C, diluted to 100 μl with 10 mM-Tris/HCl (pH 7.6)/75 mM-KCl/1 mM-magnesium acetate and fractionated on 5–40% sucrose gradients (3.8 ml) in 20 mM-Tris/HCl (pH 7.6)/75 mM-KCl/1 mM-magnesium acetate. Centrifugation was in the Spinco SW56 Ti rotor at 50000 rev./min for 75 min at 4°C.

6. SDS/polyacrylamide-gel electrophoresis

Electrophoresis in slab gels was according to Laemmli [36]. The separating gel contained 15% (w/v) acrylamide and 0.086% (w/v) bisacrylamide.

7. SDS/polyacrylamide-gel/nitrocellulose paper blotting technique (protein blotting)

The procedure of Towbin *et al.* [37] was followed. After the electrophoretic transfer of the proteins, the nitrocellulose sheet (Millipore) was incubated for 1 h at room temperature with 2.5% (w/v) bovine serum albumin and 5% (v/v) horse serum in Tris-buffered saline [20 mM-Tris/HCl (pH 7.6)/150 mM-NaCl]. After washing with Tris-buffered saline, the sheet was incubated overnight with a solution containing 1–2 μg of purified anti-CBP antibody/ml, 0.5% bovine serum albumin and 1% (v/v) horse serum in Tris-buffered saline. Bound antibody was detected by reaction with rabbit anti-mouse IgG conjugated to horseradish peroxidase, followed by incubation in *o*-dianisidine and H_2O_2, which gives a red-brown colour reaction [37].

8. Cross-linking of protein to mRNA in rabbit reticulocyte lysate

Rabbit reticulocyte lysate incubation mixture (500 μl) was incubated in the presence of [^{3}H]methyl-labelled reovirus mRNA (10 μg, 7.5×10^5 c.p.m.) as described [33] except that sparsomycin was omitted. Cross-linking of protein to reovirus mRNA was carried out with 30 mM-$NaBH_3CN$ for 3 h at 0°C. The reaction mixture was fractioned by glycerol gradient centrifugation [33] and fractions containing polysomes and monosomes were precipitated with 3 vol. of ethanol in 10 mM-magnesium acetate overnight. The precipitates were pelleted, dried, digested with pancreatic RNAase and fractionated by SDS/polyacrylamide-gel electrophoresis as described [33]. Polypeptides in the gel were then electrophoretically transferred to a nitrocellulose sheet [37], the sheet was cut in 3 mm slices and the slices counted for ^{3}H radioactivity.

9. Cell culture

Subconfluent cultures of BHK-21 cells were maintained in Dulbecco's modified Eagle's medium (Gibco). For immunofluorescence, cells were seeded on glass coverslips in Petri-dishes.

10. Antisera

(a) Anti-tubulin antibody. Porcine brain tubulin was isolated by using the polymerization–depolymerization method [38] and further purified by preparative SDS/polyacrylamide-gel electrophoresis. The tubulin band was cut out from the gel and eluted electrophoretically. Rabbits were immunized by injecting 1 mg of tubulin in incomplete Freund's adjuvant. The injection was repeated twice with 0.5 mg of tubulin in incomplete Freund's adjuvant at weekly intervals. Rabbit anti-tubulin antibodies were isolated by affinity chromatography as described [39, 40].

(b) Anti-decamin antibody. Rabbit antisera directed against decamin, the intermediate filament protein of BHK cells [41], was generously given by Dr. R. D. Goldman, Carnegie-Mellon University, Pittsburgh, PA, U.S.A.

(c) Anti-actin antibody. Antisera from rabbits directed against actin were kindly provided by Dr. G. Gabbiani, University of Geneva, Geneva, Switzerland.

(d) Fluorochrome-coupled antibodies. The IgG fraction from goat serum containing antibodies directed against mouse IgM (Nordic) was purified by precipitation with $(NH_4)_2SO_4$ and DEAE-cellulose chromatography [29] and then conjugated with FITC (fluorescein isothiocyanate) using the procedure of Cebra & Goldstein [42]. The reaction mixture was passed through a Sephadex G-25 column (Pharmacia) to separate the unconjugated fluorochrome from the conjugated IgG, to re-establish a pH of 7.5 and to replace the salt. Finally, the IgG was chromatographed on a DEAE cellulose column to remove under- and over-conjugated IgG [43]. Fluorescein-labelled goat anti-rabbit IgG and rhodamine-labelled goat anti-rabbit IgG were purchased from Nordic.

11. Immunofluorescence microscopy

(a) Preparation of cells. For fixation of microtubules, cells grown on coverslips were rinsed with phosphate-buffered saline and treated with freshly prepared 3.7% formaldehyde in phosphate-buffered saline for 20 min at 4°C. Coverslips were then rinsed quickly with phosphate-buffered saline, treated with acetone at −20°C for 5 min and air-dried. For fixation of intermediate filaments the formaldehyde step was omitted and acetone-fixation done as described above. For fixation of microfilaments, coverslips were rinsed in phosphate-buffered saline, treated with ethanol at 4°C for 30 s and rinsed again in phosphate-buffered saline.

(b) Immunofluorescent staining of fixed cells. Coverslips with fixed cells were incubated with the first antibody for 45 min at 37°C. The coverslips were rinsed with phosphate-buffered saline and then the appropriate fluorescein- or rhodamine-labelled second antibody was applied. After a further incubation for 45 min at 37°C and subsequent washing with phosphate-buffered saline the coverslips were mounted with phosphate-buffered saline/glycerol (1:1, v/v) on microscope slides. Antibody dilutions were made in phosphate-buffered saline containing 0.5% bovine serum albumin. Purified anti-CBP antibody was diluted to about 50 μg/ml. Antisera (rabbit anti-decamin antibody and rabbit anti-actin antibody), rabbit anti-tubulin antibody and commercial fluorochrome-labelled antibodies were diluted 1:20–1:50. Stained cells were viewed with a Zeiss microscope equipped with epifluorescent optics.

Photographs were taken with ×40, or oil immersion ×63 objectives using Ilford HP5 film. Control experiments in which the specific (first) antibody was omitted gave no staining and there was no detectable reaction of fluorochrome-coupled goat anti-rabbit IgG with mouse IgM or fluorochrome-coupled goat anti-mouse IgM with rabbit IgG (results not shown).

12. Preparation of extracted cytoskeleton

BHK-21 cells were grown in Petri dishes (four dishes, approx. 4×10^6–8×10^6 cells), rinsed twice in buffer A [20 mM-Tris/HCl (pH 7.6)/50 mM-NaCl/0.3 M-sucrose/2 mM-magnesium acetate/1 mM-phenylmethane sulphonyl fluoride], lysed by a 30 s treatment with buffer A plus 0.5% Triton X-100 and washed in buffer A minus sucrose. At this stage the cells were detached from the dish with a rubber policeman, centrifuged (Spinco SS-34 at 2000 rev./min for 10 min at 4°C) and the pellet was treated with DNAase [10 μg of DNAase I (Worthington)/ml] in buffer A minus sucrose for 2 h at 4°C with occasional stirring (volume 3 ml). After a second centrifugation as above, the sediment was taken up in 200 μl of SDS sample buffer.

13. Tryptic peptide analysis

Polypeptide bands were cut out from fixed and stained SDS/polyacrylamide gels, the protein was labelled in the slice with ^{125}I (0.5 mCi per slice), digested with trypsin and the peptides were analysed [44].

Results

We immunized mice with unwashed eIF-3 to obtain monoclonal antibodies directed against cap binding protein [25] and the subunits of eIF-3 (Materials and Methods, sections 1*a* and 1*b*). Hybridoma clones were selected by screening with eIF-3 and affinity-purified 24K-CBP [20] in a solid-phase radioimmunoassay (Materials and Methods, section 1). Monoclonal antibodies were judged to be anti-CBP antibodies by virtue of their ability to (*a*) interfere with the translation of capped (but not of uncapped) mRNAs in cell-free systems and (*b*) inhibit cross-linking of cap binding protein(s) to the oxidized 5′ cap structure of reovirus mRNAs.

Antibodies produced by one such clone (anti-CBP antibody) were further analysed. They belong to the IgM class with H-chains of the μ-type and L-chains of the κ-type as determined by reaction with rabbit anti-mouse μ and κ antisera. Fig. 2 shows the inhibition of translation of reovirus capped mRNA by anti-CBP antibody in a cell-free extract from L-cells. We have shown previously that the same antibody interferes only with the translation of capped reovirus mRNA but not of naturally uncapped EMC RNA in this system or in HeLa cell extracts [23]. Furthermore, we showed that anti-CBP antibody does not inhibit the translation of uncapped reovirus mRNA in cell-free extracts from reovirus-infected L-cells where

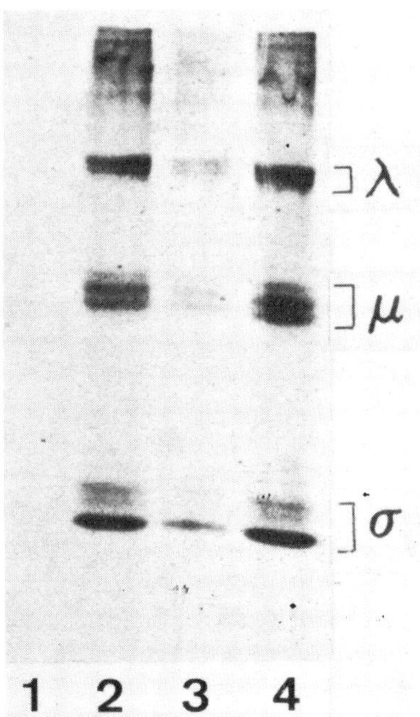

Fig. 2. *Effect of anti-CBP antibody on translation of reovirus mRNA in cell-free extract from L-cells*

The experiment was done as described in Materials and Methods, sections 2 and 3. The autoradiograph is shown. Track 1, no mRNA added; track 2, reovirus mRNA; track 3, reovirus mRNA + 8 μg of anti-CBP antibody purified from ascites fluid (Materials and Methods, section 1*c*); track 4, reovirus mRNA + 8 μg of immunoglobulin fraction from Ehrlich ascites fluid [23]. λ, μ and σ show the positions in the gel of the three size classes of reovirus-coded proteins.

initiation presumably proceeds via a cap-independent mechanism [32]. The inhibition of cross-linking of proteins to oxidized reovirus mRNA by anti-CBP antibody is demonstrated in Fig. 3. A number of proteins in the crude fraction are cross-linked to the 5′ cap structure (track 1) but only three proteins bind specifically, that means their binding is inhibited in the presence of the cap analogue m^7GDP (track 2). These proteins are the 24K-CBP (plus an M_r 28000 polypeptide which is not resolved in this gel) and two additional proteins of M_r 50000 and M_r 80000. The cross-linking of the latter proteins has recently been shown to depend on the presence of Mg^{2+} and ATP [21]. Anti-CBP antibody inhibits the cross-linking of the three proteins which bind specifically to the cap structure (tracks 4 and 5). These experiments define the monoclonal antibody as anti-cap binding protein antibody.

In order to identify the polypeptide(s) reacting with anti-CBP antibody in a crude protein-synthesizing system, rabbit reticulocyte lysate was incubated under protein synthesis conditions and then fractionated by sucrose density gradient sedimentation (Fig. 4). Anti-CBP antibody reacts with an M_r 50000 polypeptide in polysomes and monosomes (Fig. 4, tracks 1 and 2) and an M_r 210000 polypeptide in post-polysomal supernatant (Fig. 4, track 3). The presence of the M_r 50000 polypeptide in fractions containing polysomes and monosomes, but not in post-polysomal supernatant, suggests that it is associated with translated mRNA, with ribosomes actively translating mRNA or with all ribosomes. To distinguish between

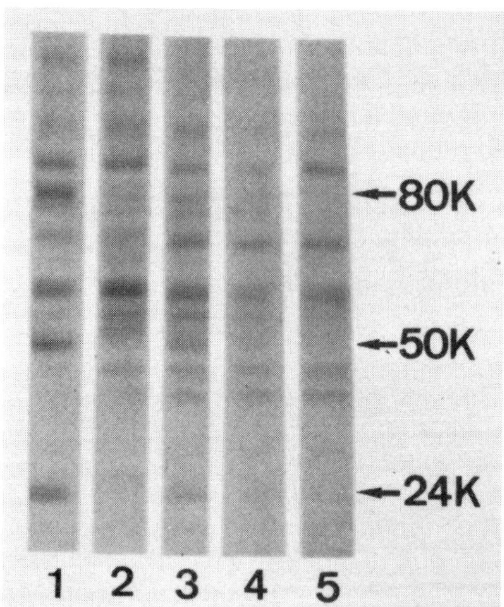

Fig. 3. *SDS/polyacrylamide-gel electrophoresis of proteins cross-linked to [³H]methyl-labelled-reovirus mRNA*

The experiment was done as described in Materials and Methods, sections 2 and 4. The autoradiograph is shown. Track 1, no addition; track 2, 0.4 mM-m^7GDP; track 3, 2 μg of anti-CBP antibody purified from ascites fluid (Materials and Methods, section 1*c*); track 4, 4 μg of anti-CBP antibody; track 5, 8 μg of anti-CBP antibody. Tracks 1–4 also contained the additional Hepes and KCl contributed by the anti-CBP antibody in track 5.

these possibilities, ^{125}I-labelled anti-CBP antibody was added to reitculocyte lysates which had been preincubated under conditions for protein synthesis in the presence or absence of haemin. Incubation mixtures incubated in the presence of haemin were then partially digested with RNAase and the amount of anti-CBP antibody bound to disomes was compared with the amount of anti-CBP bound to disomes in the incubation mixture preincubated in the absence of haemin. The rationale of this experiment is the following. If the M_r 50000 polypeptide is associated with ribosomes, then disomes formed under the two sets of conditions should bind the same amount of antibody. However, if the M_r 50000 polypeptide is associated with the 5′ cap structure of mRNA, then disomes formed under conditions of limiting initiation (in the absence of haemin) should bind two to three times more antibody than do disomes generated by RNAase degradation. The results of this experiment are shown in Fig. 5. Polysomes in rabbit reticulocyte lysate bind anti-CBP antibody (Fig. 5*a*). The distribution of ^{125}I radioactivity (anti-CBP antibody) over the polysome region in the sucrose density gradient suggests an association of the antigen with mRNA rather than ribosomes. Under conditions of limiting initiation (Fig. 5*b*), disomes carry roughly three times as much bound antibody as do disomes generated

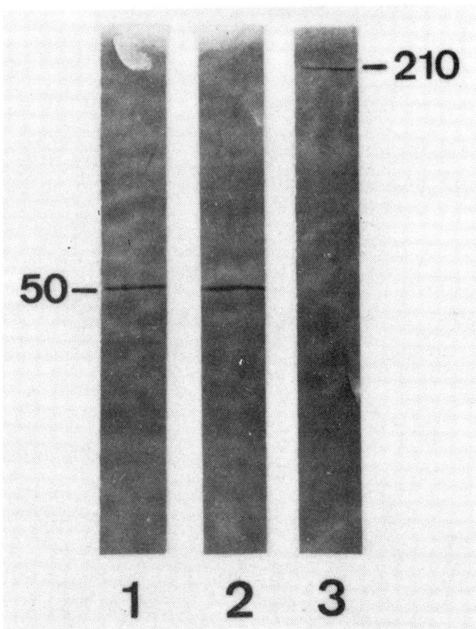

Fig. 4. *Polypeptides in rabbit reticulocytes lysate reacting with anti-CBP antibody*

Rabbit reticulocyte lysate was incubated for protein synthesis and then fractionated by sucrose density gradient sedimentation as described in Materials and Methods (section 5). Fractions containing polysomes and monosomes were pooled separately, collected by centrifugation (Spinco SW 60 Ti, 58000 rev./min, 4 h, 2°C) and further fractionated by SDS/polyacrylamide gel electrophoresis (Materials and Methods, section 6). An aliquot (10 μl) of pooled post-polysomal supernatant was run on the gel. Polypeptides were blotted on to a nitrocellulose sheet, reacted with anti-CBP hybridoma culture supernatant and then with peroxidase-coupled second antibody (Materials and Methods, section 7). The *o*-dianisidine-stained blots are shown. Track 1, polysomes; track 2, monosomes; track 3, post-polysomal supernatant. The positions on the blots of M_r 210000 and M_r 50000 are indicated.

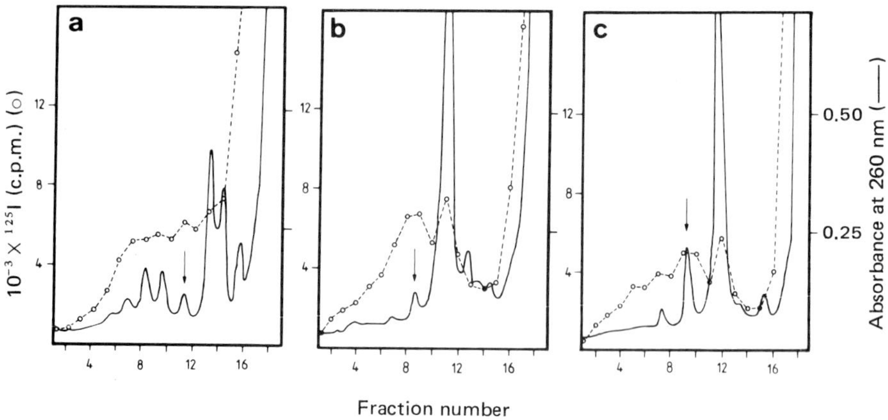

Fig. 5. *Binding of anti-CBP antibody to polysomes*

Rabbit reticulocyte lysate was incubated in the presence or absence of haemin for 10 min at 30°C (Materials and Methods, section 5). Then 5 pmol of ^{125}I-labelled anti-CBP antibody (150000 c.p.m./pmol; Materials and Methods, section 1*d*) was added to every incubation mixture. Incubation was continued for 3 min at 30°C with (a, b) no addition or (c) in the presence of 1 mM-$CaCl_2$ and 24 units of micrococcal nuclease (Worthington). Incubation mixtures were cooled on ice and (c) was made 2 mM in EGTA. Finally, the incubation mixtures were fractioned by sucrose density gradient sedimentation (Materials and Methods, section 5). (a) Plus haemin; (b) minus haemin; (c) plus haemin and then digested with RNAase. Arrows point to the disome peaks.

by RNAase digestion from polysomes formed under standard incubation conditions (Fig. 5*c*) Furthermore, the bulk of 80S monosomes has no antibody bound. These data favour an association of the antigen with either the 3′ or 5′ end of mRNA. Most likely the antigen is the M_r 50000 polypeptide (Fig. 4), and it is bound to the 5′ cap structure of mRNA since the antibody interferes specifically with the translation of capped mRNAs. These conclusions are supported by the experiment shown in Fig. 6. When [^{3}H]methyl-labelled reovirus mRNA is incubated with rabbit reticulocyte lysate under translation conditions, an M_r 50000 polypeptide is found associated with the 5′ cap structure of this mRNA in polysomes and monosomes (Fig. 6). This polypeptide comigrates in SDS/polyacrylamide gels with the M_r 50000 polypeptide of endogenous polysomes which reacts with anti-CBP antibody (Fig. 6, arrow). In accordance with Fig. 4, little or no M_r 210000 polypeptide is found associated with mRNA in polysomes.

Preliminary experiments indicated that anti-CBP antibody also reacts with protein from HeLa cells, Balb C/3T3 cells and BHK cells. In order to localize CBP in cells by indirect immunofluorescence microscopy with anti-CBP antibody we chose BHK cells. Fig. 7 shows typical patterns obtained by immunofluorescence microscopy with anti-actin antibody (Fig. 7*a*), anti-tubulin antibody (Fig. 7*b*) and anti-decamin antibody (Fig. 7*c*) on fixed BHK-21 cells. Observation of BHK-21 cells by this technique using anti-CBP antibody reveals that CBP is distributed along a very fine network of fibres (Fig. 7*d*). These fibres extend through the cytoplasm in a radial arrangement and are more abundant around the nucleus. Identical results were obtained with antibody purified from mouse ascites fluid and hybridoma culture supernatant and a typical staining pattern could be obtained with about 5–10 μg

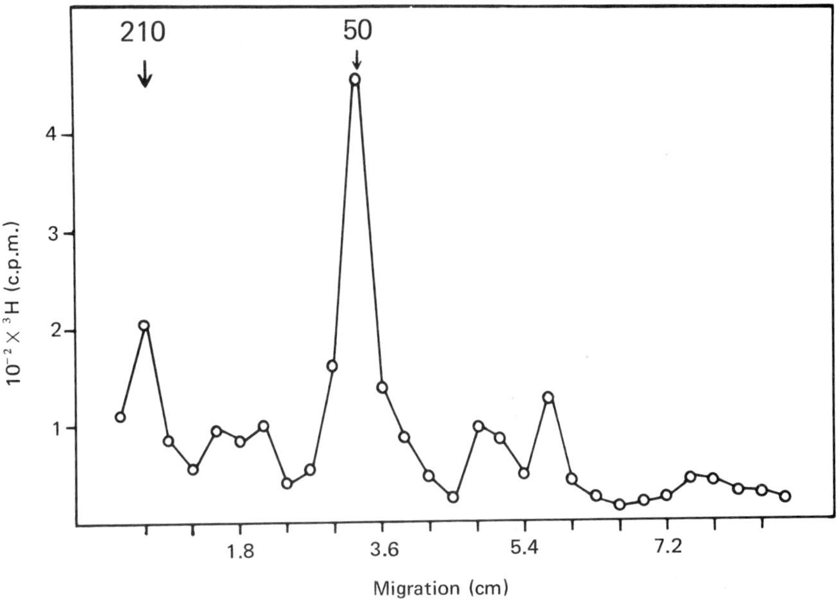

Fig. 6. *Cross-linking of protein to reovirus mRNA*

The experiment was done as described in Materials and Methods, section 8. The arrows point to the positions on the protein blot of the M_r 50000 polypeptide in rabbit reticulocyte polysomes and the M_r 210000 polypeptide in post-polysomal supernatant which react with anti-CBP antibody.

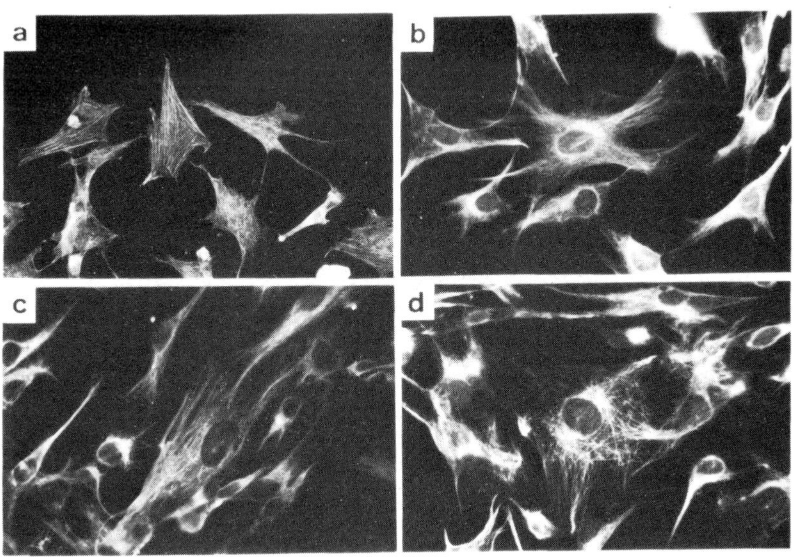

Fig. 7. *Immunofluorescence microscopy (single labelling)*

BHK-21 cells were prepared for immunofluorescence microscopy (Materials and Methods, section 11) and reacted with: (a) rabbit anti-actin antibody and goat anti-rabbit IgG-FITC; (b) rabbit anti-tubulin antibody and goat anti-rabbit IgG-FITC; (c) rabbit anti-decamin antibody and goat anti-rabbit IgG-FITC; (d) mouse anti-CBP antibody and goat anti-mouse IgM-FITC. Magnification × 200.

of antibody from hybridoma culture supernatant/ml. Furthermore, the same staining pattern was obtained using a completely different preparation method in which cells are extracted with Triton X-100 and the resulting cytoskeleton preparations reacted with anti-CBP antibody without prior fixation (results not shown). The fibrous network reacting with anti-CBP antibody resembles the network stained by either anti-tubulin antibody (microtubules, Fig. 7*b*) or anti-decamin antibody (intermediate filaments, Fig. 7*c*) but has no similarities with the patterns stained by anti-acting antibody (microfilaments, Fig. 7*a*).

In order to distinguish between reaction of anti-CBP antibody with microtubules or intermediate filaments, BHK-21 cells were double-stained for immunofluorescence microscopy using anti-CBP and anti-decamin antibodies (Figs. 8*a* and 8*b*) or anti-CBP and anti-tubulin antibodies (Figs. 9*a* and 9*b*). At the resolution of the light microscope, use of the double-staining techniques with interphase cells shows complete superposition of the labelling patterns obtained with anti-CBP and anti-decamin antibodies (Figs. 8*a* and 8*b*) and significant but not complete superposition of the patterns obtained with anti-CBP and anti-tubulin antibodies (Figs. 9*a* and 9*b*). After treatment of BHK-21 cells with 20 μg of colchicine/ml for 2 h the distribution of CBP as revealed by immunofluorescence microscopy begins to change. The filaments form thicker bundles and aggregate into perinuclear bundles in the form of a ring or blob (Figs. 10*c* and 10*d*). This process continues over several hours and reaches completion within 10 h (results not shown). When cells are double-labelled for immunofluorescence microscopy using anti-CBP and anti-decamin antibodies (Figs. 10*c* and 10*d*), similar staining patterns are observed. Perinuclear filament bundles are seen and they are similar to structures that have already been described for intermediate filaments in colchicine-treated cells [45–48]. When the double-labelling technique is used with anti-CBP and anti-tubulin antibodies (Figs. 10*a* and 10*b*), it is evident that microtubules are depolymerized (as

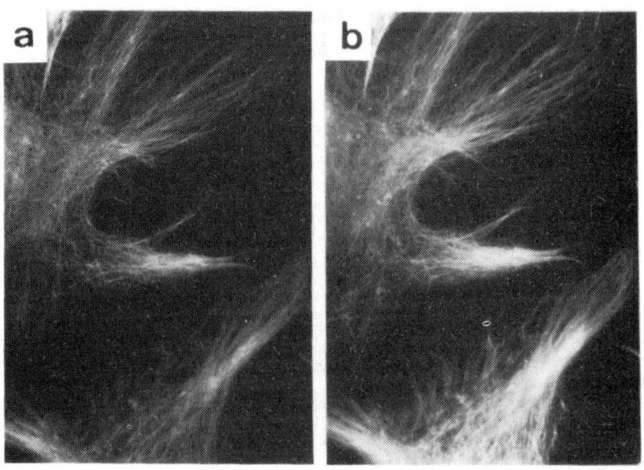

Fig. 8. *Immunofluorescence microscopy (double labelling)*

BHK-21 cells were prepared for immunofluorescence microscopy (Materials and Methods, section 11) and reacted with mouse anti-CBP antibody followed by rabbit anti-decamin antibody and then goat anti-mouse IgM-FITC followed by goat anti-rabbit IgG-rhodamine. (a) Fluorescein pattern; (b) rhodamine pattern of the same cell. Magnification ×400.

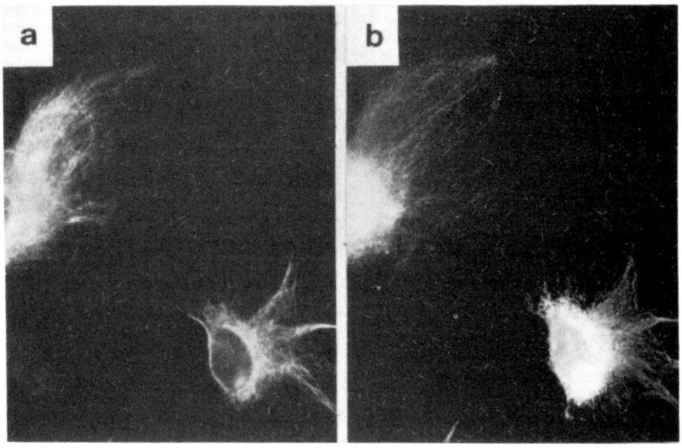

Fig. 9. *Immunofluorescence microscopy (double labelling)*

BHK-21 cells were prepared for immunofluorescence microscopy (Materials and Methods, section 11) and reacted with mouse anti-CBP antibody followed by rabbit anti-tubulin antibody and then goat anti-mouse IgM-FITC followed by goat anti-rabbit IgG-rhodamine. (a) Fluorescein pattern; (b) rhodamine pattern of the same cell. Magnification ×400.

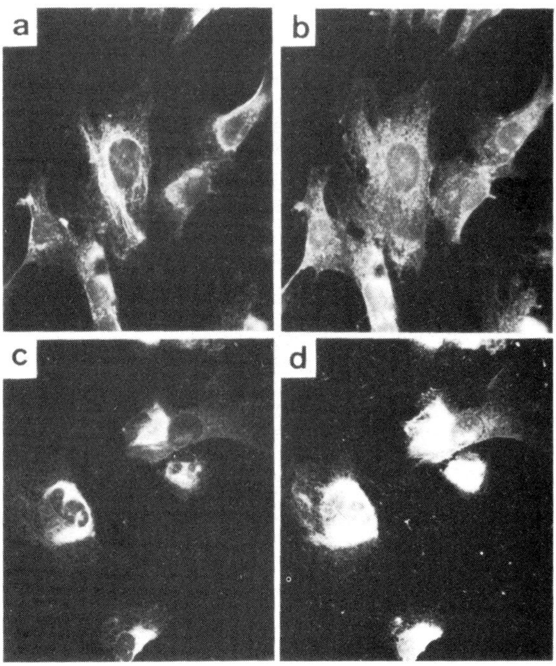

Fig. 10. *Immunofluorescence microscopy (double labelling)*

BHK-21 cells were pretreated with 20 μg of colchicine/ml for 2 h, prepared for immunofluorescence microscopy (Materials and Methods, section 11) and reacted: (a, b) with mouse anti-CBP antibody followed by rabbit anti-tubulin antibody; (c, d) with mouse anti-CBP antibody followed by rabbit anti-decamin antibody, and then (a, c) goat anti-mouse IgM-FITC and (b, d) goat anti-rabbit IgG-rhodamine. (a, c) Fluorescein patterns; (b, d) rhodamine patterns. (a and b) and (c and d) are the same cells. Magnification × 200.

indicated by diffuse staining with anti-tubulin antibody, Fig. 10*b*) while CBP is present on filaments coiled around the nuclear membrane (Fig. 10*a*). These data suggest that CBP is associated directly or indirectly with intermediate filaments in BHK-21 cells.

The association of CBP with the cytoskeletal fraction of BHK cells was also demonstrated by another method (Fig. 11). BHK-21 cells grown on glass coverslips were either dissolved directly in SDS sample buffer or extracted on the coverslip with Triton X-100. Part of the extracted cells were treated with anti-CBP antibody for immunofluorescence microscopy. The characteristic fibrous network described above (e.g. Fig. 7*d*) was observed (results not shown). The rest of the extracted cells were dissolved in SDS sample buffer. Total cell protein (Fig. 11, track 1) and cytoskeletal proteins (Fig. 11, track 2) were separated by SDS/polyacrylamide-gel elctrophoresis and blotted electrophoretically on to a nitrocellulose sheet. Anti-CBP antibody was then used to identify CBP on the protein blots. The experiment shows that a major polypeptide of M_r 50000 and a minor component of M_r 93000 remain associated with the cytoskeletal fraction (Fig. 11, track 2). The most abundant polypeptide in this cytoskeletal fraction as revealed by staining with Coomassie Blue is decamin (results not shown). This M_r 55000 polypeptide does not react with anti-CBP antibody.

Earlier experiments [26] demontrated that anti-CBP antibody reacts with an M_r 50000 polypeptide in rabbit reticulocyte polysomes and an M_r 210000 poly-

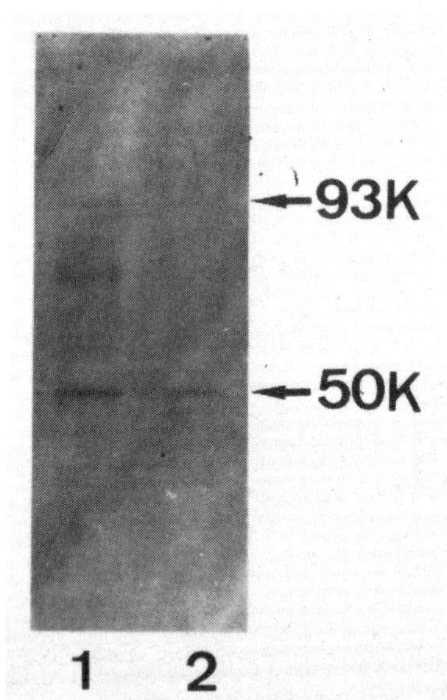

Fig. 11. *SDS/polyacrylamide-gel-nitrocellulose blot*

BHK-21 cells (approx. 10^5 cells) were either dissolved directly in SDS sample buffer or first extracted with Triton X-100 (Materials and Methods, section 12) and then dissolved in sample buffer. Proteins were separated by SDS/polyacrylamide-gel electrophoresis, transferred electrophoretically from the gel on to a nitrocellulose sheet and identified by reaction with mouse anti-CBP antibody (Materials and Methods, sections 6 and 7). Track 1, intact cells; track 2, Triton X-100-extracted cells.

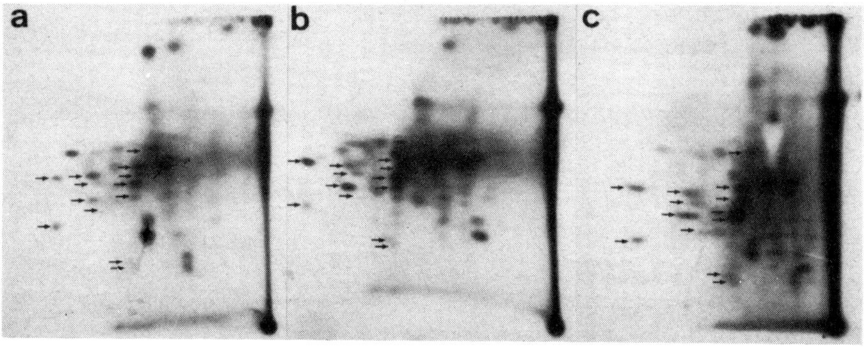

Fig. 12. *Tryptic peptide patterns*

The M_r 50000 polypeptide from Triton X-100-extracted cells which was shown to react with mouse anti-CBP antibody (Fig. 11) and the M_r 24000 cap binding protein (24K-CBP [20]) were analysed as described in Materials and Methods, section 13. The autoradiograph is shown. (a) 24K-CBP; (b) M_r 50000 polypeptide; (c) 24K-CBP plus M_r 50000 polypeptide. Electrophoresis was from right to left (first dimension) and chromatography from bottom to top (second dimension).

peptide in post-polysomal supernatant. Furthermore, polypeptides ranging in molecular weight from M_r 24000 to M_r 210000, including components of M_r 50000 and M_r 93000 were found in partially purified initiation factor preparations [26]. It was shown by peptide mapping that these polypeptides are structurally related and furthermore, share trypic and chymotryptic peptides with an M_r 24000 cap binding protein (24K-CBP) isolated by affinity chromatography on m^7GDP–Sepharose [20]. Fig. 12 shows that the M_r 50000 polypeptide in the BHK cytoskeletal fraction (Fig. 12*b*) shares tryptic peptides with 24K-CBP from rabbit reticulocytes (Fig. 12*a*). The mixing experiment confirms the identity of peptides indicated by arrows (Fig. 12*c*). These data, together with the data shown in Fig. 11, demonstrate directly the presence of CBP on BHK-21 cytoskeletons, as suggested above by the immunofluorescence microscopy data.

Discussion

To extend our knowledge about the mechanism and regulation of eukaryotic protein synthesis initiation it is necessary to study the role of individual initiation factors in crude cell lysates or even in intact cells. As a first step towards this end, we began to produce monoclonal antibodies directed against initiation factors and initiation factor subunits.

One of the monoclonal antibodies produced so far was defined as anti-CBP antibody by its specific inhibitory effects on capped mRNA translation *in vitro* (Fig. 2 [23]) and cross-linking of proteins to oxidized mRNA (Fig. 3). This antibody reacts with an M_r 50000 polypeptide in rabbit reticulocyte polysomes and an M_r 210000 polypeptide in post-polysomal supernatant (Fig. 4). The M_r 50000 polypeptide is a functional cap binding protein inasmuch as it appears to be bound to mRNA (Fig. 5) and is able to cross-link to the oxidized 5′ cap structure of reovirus mRNA (Fig. 6). The M_r 210000 polypeptide is structurally related to the M_r

50000 polypeptide as suggested by comparison of partial chymotryptic peptides of the two polypeptides [26], but apparently is not a functional CBP (Figs. 5 and 6). However, it could be a precursor form of active CBP [26].

Immunofluorescence microscopy data (Figs. 7–10) and biochemical experiments (Figs. 11 and 12) demonstrate that the M_r 50000 CBP is associated with the cytoskeleton in BHK-21 cells. The cytoskeleton in animal cells contains three major fibrous structures known as microtubules, microfilaments and intermediate filaments (for a review see [49]). In addition, an extensive cytoplasmic meshwork termed microtrabeculae [50, 51] was found.

The M_r 50000 polypeptide appears to be bound to intermediate filaments; however, due to the low resolution of immunofluorescence microscopy we cannot distinguish whether this polypeptide is associated directly with intermediate filaments or bound to an independent structure which is linked to intermediate filaments. A likely candidate for such a structure is the microtrabecular network [50, 51]. Electron microscopical evidence suggests that this network carries polysomes [5] and consequently we might expect to find CBP bound to it as well. The association of polysomes with some element of the cytoskeleton was also demonstrated by using biochemical methods [52]. Lenk *et al.* [53] repared cytoskeletons by detergent extraction of suspension-grown HeLa cells and presented evidence for the specific retention of polysomes with the cytoskeletal fraction. Degradation of mRNA with low levels of ribonuclease released ribosomes from the cytoskeleton implying an attachment of polysomes via mRNA. More recent experiments indicate that binding of mRNA to the cytoskeleton might be a requirement for translation [54]. This is further supported by the finding that shut-off of host protein synthesis in poliovirus-infected HeLa cells is accompanied by the release of host mRNAs from the cytoskeleton [55].

These findings [52–55] are consistent with the observation that CBP is functionally inactivated in poliovirus-infected HeLa cells [13, 14] and indicate that CBP may be involved in linking host mRNAs to the cytoskeleton. Experiments to test this hypothesis are in progress.

We thank Ruth Fessler and Heinz Kuster for excellent technical assistance, Dr. H. Towbin for help with the protein blotting technique, Dr. F. Köhler for providing the myeloma cell line, Drs. R.D. Goldman and G. Gabbiani for providing the anti-decamin and anti-actin antibodies, and Drs. A.J. Shatkin, R. Bürk and T. Staehelin for many helpful discussions. Supported in part by the M.R.C. (Canada) grant MA-7214 to N.S. We thank Academic Press Inc. for permission to reproduce Figs. 5 and 6 from [26].

References

1 Jagus, R., Anderson, W.F. & Safer, B. (1981) *Progr. Nucleic Acid Res. Mol. Biol.* **25**, 127–185
2 Schreier, M.H., Erni, B. & Staehelin, T. (1977) *J. Mol. Biol.* **116**, 727–753
3 Safer, B., Anderson, W.F. & Merrick, W.C. (1975) *J. Biol. Chem.* **250**, 9067–9075
4 Safer, B., Adams, S.L., Kemper, W.M., Berry, K.W., Lloyd, M. & Merrick, W.C. (1976) *Proc. Natl. Acad. Sci. U.S.A.* **73**, 2584–2588.
5 Kemper, W.M., Berry, K.W. & Merrick, W.C. (1976) *J. Biol. Chem.* **251**, 5551–5557
6 Merrick, W.C., Kemper, W.M. & Anderson, W.F. (1975) *J. Biol. Chem.* **250**, 5556–5562
7 Benne, R. & Hershey, J.W.B. (1976) *Proc. Natl. Acad. Sci. U.S.A.* **73**, 3005–3009
8 Benne, R., Wong, C., Luedi, M. & Hershey, J.W.B. (1976) *J. Biol. Chem.* **251**, 7675–7681
9 Benne, R., Luedi, M. & Hershey, J.W.B. (1977) *J. Biol. Chem.* **252**, 5798–5803
10 Amesz, H., Goumans, H., Haubrich-Morree, T., Voorma, H. O. & Benne, R. (1979) *Eur. J. Biochem.* **98**, 513–520

11 Siekirka, J., Ichiro-Mitsui, K. & Ochoa, S. (1981) *Proc. Natl. Acad. Sci. U.S.A.* **78**, 220–223
12 Ghosh-Dastidar, P., Yagmai, B., Das, A., Das, H.K. & Gupta, N.K. (1980) *J. Biol. Chem.* **255**, 365–368
13 Trachsel, H., Erni, B., Schreier, M.H. & Staehelin, T. (1977) *J. Mol. Biol.* **116**, 754–767
14 Benne, R., Luedi, M. & Hershey, J.W.B. (1977) *J. Biol. Chem.* **252**, 5788–5805
15 Trachsel, H. & Staehelin, T. (1979) *Biochim. Biophys. Acta* **565**, 305–314
16 Thomas, A., Goumans, H., Voorma, H.O. & Benne, R. (1980) *Eur. J. Biochem.* **107**, 39–45
17 Shatkin, A.J. (1976) *Cell* **9**, 645–653
18 Sonenberg, N., Morgan, M.A., Testa, D., Colonna, R.J. & Shatkin, A.J. (1979) *Nucleic Acids Res.* **7**, 15–29
19 Sonenberg, N., Morgan, M.A., Merrick, W.C. & Shatkin, A.J. (1978) *Proc. Natl. Acad. Sci. U.S.A.* **75**, 4843–4847
20 Sonenberg, N., Rupprecht, K.M., Hecht, S.M. & Shatkin, A.J. (1979) *Proc. Natl. Acad. Sci. U.S.A.* **76**, 4345–4349
21 Sonenberg, N. (1981) *Nucleic Acids Res.* **9**, 1643–1656
22 Sonenberg, N., Trachsel, H., Hecht, S.M. & Shatkin, A.J. (1980) *Nature (London)* **285**, 331–333
23 Sonenberg, N., Skup, D., Trachsel, H. & Millward, S. (1981) *J. Biol. Chem.* **256**, 4138–4141
24 Rose, J.K., Trachsel, H., Leong, K. & Baltimore, D. (1978) *Proc. Natl. Acad. Sci. U.S.A.* **75**, 2732–2736
25 Trachsel, H., Sonenberg, N., Shatkin, A.J., Rose, J.K., Leong, K., Bergmann, J.E., Gordon, J. & Baltimore, D. (1980) *Proc. Natl. Acad. Sci. U.S.A.* **77**, 770–774
26 Sonenberg, N. & Trachsel, H. (1982) *Curr. Topics Cell. Regul.* **21**, in the press
27 Schreier, M.H., Erni, B. & Staehelin, T. (1977) *J. Mol. Biol.* **116**, 727–753
28 Stähli, C., Staehelin, T., Miggiano, V. & Haering, P. (1980) *J. Immunol. Methods* **32**, 237–304
29 Peterson, E.A. & Soker, H.A. (1962) *Methods Enzymol.* **5**, 3–27
30 Hunter, W.M. & Greenwood, F.C. (1964) *Biochem. J.* **91**, 43–46
31 Muthukrishnan, S., Morgan, M.A., Banerjee, A.K. & Shatkin, A.J. (1976) *Biochemistry* **15**, 5761–5768
32 Skup, D. & Millward, S. (1980) *Proc. Natl. Acad. Sci. U.S.A.* **77**, 152–156
33 Sonenberg, N. & Shatkin, A.J. (1977) *Proc. Natl. Acad. Sci. U.S.A.* **74**, 4288–4292
34 Schreier, M.H. & Staehelin, T. (1973) *J. Mol. Biol.* **73**, 329–349
35 Hunt, T., Vanderhoff, G. & London, I.M. (1972) *J. Mol. Biol.* **66**, 471–481
36 Laemmli, U.K. (1970) *Nature (London)* **227**, 680–685
37 Towbin, H., Staehelin, T. & Gordon, J. (1979) *Proc. Natl. Acad. Sci. U.S.A.* **76**, 4350–4354
38 Shelanski, M.L., Gaskin, F. & Cantor, C.R. (1973) *Proc. Natl. Acad. Sci. U.S.A.* **70**, 765–768
39 Fuller, G.M., Brinkley, B.R. & Boughter, J.M. (1975) *Science* **187**, 948–950
40 Weingarten, M.D., Lockwood, A.H., Hwo, S. & Kirschner, W. (1975) *Proc. Natl. Acad. Sci. U.S.A.* **72**, 1858–1862
41 Starger, J.M. & Goldman, R.D. (1977) *Proc. Natl. Acad. Sci. U.S.A.* **78**, 2422–2426
42 Cebra, J.J. & Goldstein, G. (1965) *J. Immunol.* **95**, 230–245
43 Fujiwara, K. & Pollard, T.D. (1976) *J. Cell Biol.* **71**, 848–875
44 Elder, J.H., Pickett II, R.A., Hampton, J. & Lerner, R.A. (1977) *J. Biol. Chem.* **252**, 6510–6515
45 Goldman, R.D. & Knipe, D. (1973) *Cold Spring Harbor Symp. Quant. Biol.* **37**, 528–534
46 Granger, B.L., Gard, D.L. & Lazarides, E. (1979) *J. Cell Biol.* **83**, 314a
47 Granger, B.L. & Lazarides, E. (1979) *Cell* **18**, 1053–1063
48 Holtzer, H., Sanger, J.W., Ishikawa, H. & Strahs, K. (1973) *Cold Spring Harbor Symp. Quant. Biol.* **37**, 549–566
49 Small, J.V., Celis, J.E. & Isenberg, G. (1980) in *Transfer of Cell Constituents into Eukaryotic Cells* (Celis, J.E., Graessmann, A. & Loyter, A., eds.), pp. 75–111, Plenum, New York
50 Wolosewick, J.J. & Porter, K.R. (1976) *Am. J. Anat.* **147**, 303–324
51 Wolosewick, J.J. & Porter, K.R. (1979) *J. Cell Biol.* **82**, 114–139
52 Herman, R., Zieve, G., Williams, J., Lenk, R. & Penman, S. (1976) *Prog. Nucleic Acid Res. Mol. Biol.* **19**, 379–401
53 Lenk, R., Ransom, L., Kaufmann, Y. & Penman, S. (1977) *Cell* **10**, 67–78
54 Cervera, M., Dreyfuss, G. & Penman, S. (1981) *Cell* **23**, 113–120
55 Lenk, R. & Penman, S. (1979) *Cell* **16** 289–301

Biochem. Soc. Symp. 47, 163-182
Printed in Great Britain

Messenger Ribonucleoprotein Complexes in Gene Expression

H.R.V. ARNSTEIN

Department of Biochemistry, King's College, Strand, London WC2R 2LS, U.K.

Introduction

In eukaryotic cells both nascent mRNA transcripts and the final products after post-transcriptional processing exist in association with a variety of different proteins. It is the purpose of this review to consider the various protein-RNA complexes which contain either mRNA or its precursor (pre-mRNA or heterogeneous nuclear RNA, hnRNA) with particular reference to their structural organization and possible function in gene expression. For convenience the discussion has been divided into three sections: nuclear RNP containing mRNA sequences, cytoplasmic mRNP associated with polyribosomes and mRNA-protein complexes occurring free in the cytoplasm, which have also been called informosomes (for reviews see Greenberg, 1975; Preobrazhensky & Spirin, 1978; Heinrich *et al.*, 1978; van Venrooij & Janssen, 1978; Greenberg & Setyono, 1981). Typically, these RNA-protein complexes are characterized by sedimentation coefficients 2.0-2.5 times that of the RNA component and by a low buoyant density in CsCl of approx. 1.4 $g \cdot ml^{-1}$ indicating an RNA : protein ratio of about 1 : 3. Despite the large amount of protein, however, these complexes are very sensitive to ribonuclease and differ in this respect from ribosomes.

Nuclear pre-mRNP (hnRNP)

General structure of hnRNP

Biochemical and electron microscopic evidence (for reviews see Malcolm & Sommerville, 1977; van Venrooij & Janssen, 1978; Martin *et al.*, 1978; Heinrich & Northemann, 1981) shows that both nascent and completed RNA transcripts occur in the nucleus in association with proteins in the form of condensed repeating units (monomer particles) approximately 25 nm in diameter interspersed by more open-structured heterogeneous complexes (Fig. 1). Monoparticles and heterogeneous complexes differ from each other in protein composition and susceptibility to attack by nucleases and proteolytic enzymes (Fig. 2). By using a chromatin spreading technique to examine individual transcription units of 2-4 h old *Drosophila melanogaster* embryos it has been demonstrated (Beyer *et al.*, 1980) that RNP particles containing nascent hnRNA are non-random. Thus, the number of RNP particles on the nascent RNA fibrils varies greatly for different transcription units,

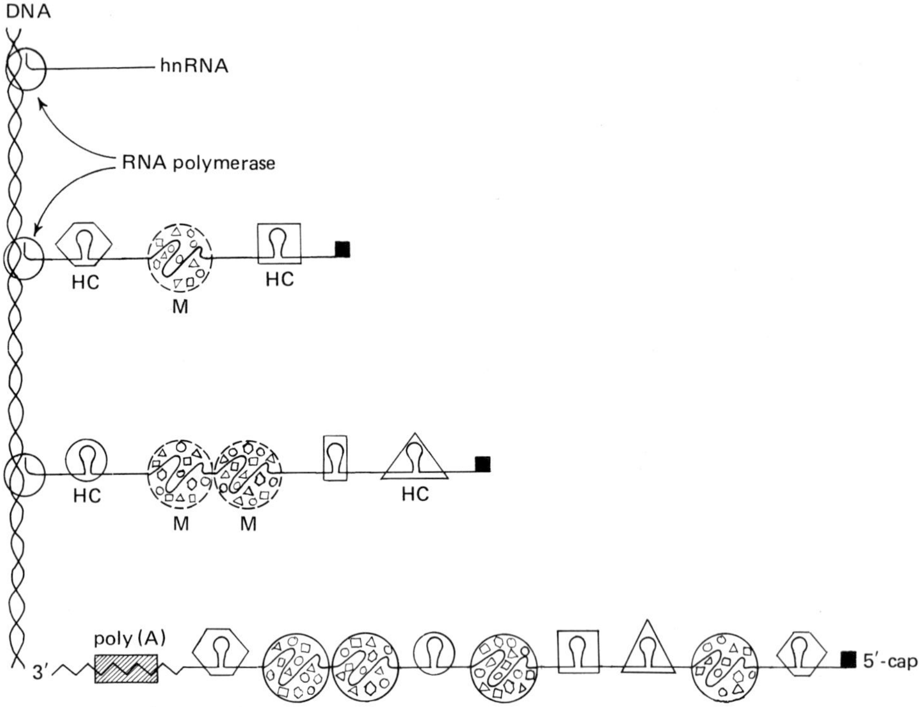

Fig. 1. *General structure of eukaryotic hnRNA-protein complexes*

During transcription, proteins (□, △, ⬡ and ○) bind to nascent hnRNA giving rise to compact 40 S monoparticles (M) and heterogeneous complexes (HC). Complete transcripts contain additional protein(s) (▨) on the poly(A) tract.

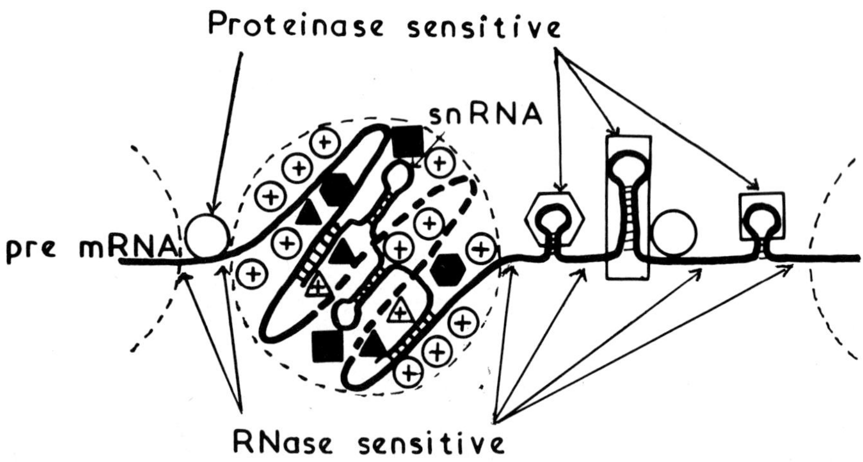

Fig. 2. *Hypothetical structure of monoparticles*

The major protein constituents comprise three pairs of proteins (A, B and C). Several low molecular weight stable RNA species (small nuclear RNA, snRNA) with some complementary sequences to hnRNA are also known to be present, probably in different monomers. Monoparticles are relatively resistant to attack by proteinases and ribonucleases and are left intact when hnRNP is partially degraded. Proteins of monomer particles (small symbols): ○, group A; △, ▲, group B; ■, group C; ⬢, others; Basic proteins are shown by +. Proteins of heterogeneous complexes are indicated by large symbols.

a few containing no obvious particles at all. The distribution of particles along the length of individual nascent RNA molecules is non-uniform and there seems to be no obvious relationship to size or activity of the transcription unit, but on average there is a particle-stabilizing sequence every 2000 nucleotides. From measurements of the distance between midpoints of adjacent particles it has been estimated that approx. 400 nucleotides are actually associated with an RNP particle. The non-random distribution of RNP particles is in agreement with the protection in the RNP of specific sequences such as AGC, GGC, AGGC and GAGC against digestion by staphylococcal nuclease (Augenlicht, 1979).

Isolation and chemical composition of hnRNP

The isolation of hnRNP usually involves extraction of purified nuclei with salt. Low (0.01 M) or moderately high (0.1 M) ionic strength buffers at pH 8 (Samarina *et al.*, 1968) in the presence of a ribonuclease inhibitor yield a mixture of hnRNP mono- and poly-particles sedimenting in sucrose gradients within the range of 30–200 S. Omission of the ribonuclease inhibitor results in partial degradation of hnRNP and recovery of only the 30–40 S monoparticles. The structure of hnRNP is, however, less regular than that of nucleosomes and digestion with staphylococcal nuclease generates a complex pattern of transient RNA bands ranging in molecular weight from 1.5×10^3 to 6×10^5 without any regular repeating structures (Wahrman & Augenlicht, 1979). If Mg^{2+} is chelated by ATP, pyrophosphate or EDTA in the extraction buffer, hnRNP can be removed from nuclei at pH 7 (Ishikawa *et al.*, 1974). Other methods include sonication of the nuclei to facilitate extraction of hnRNP with a pH 7 buffer in the presence of a ribonuclease inhibitor to preserve the polyparticles sedimenting at greater than 50 S (Parsons & McCarty, 1968; Bhorjee & Pederson, 1972).

Nuclear RNP complexes containing pre-mRNA have been isolated from a wide variety of eukaryotic cells (Table 1). Although differing in detail all complexes have certain features in common, including a buoyant density of 1.4 $g \cdot ml^{-1}$ in CsCl, the presence of both rapidly labelled RNA containing mRNA sequences, several stable low molecular weight RNA species (snRNA) and a set of proteins which may differ not only according to the cells from which the RNP originated but also as a result of the isolation procedure used. To obtain reproducible protein patterns the addition of PMSF is required before isolation of hnRNP particles in order to inhibit cellular serine proteinases. In addition, the particles themselves contain a different proteolytic enzyme activity, which is not inhibited by PMSF, iodoacetamide, or inhibitors of trypsin and chymotrypsin, but is inactivated by bisulphite and *p*-hydroxymercuribenzoate (Stevenin *et al.*, 1977).

The RNA components of hnRNP particles. Two distinct classes of RNA are present in hnRNP particles: a major fraction containing rapidly labelled heterodisperse poly(A)$^+$ RNA having sedimentation coefficients ranging from 25 S to 150 S or more and a minor component (about 10% of the total) comprising several metabolically stable low M_r RNA species, usually termed small nuclear RNA (snRNA), approx. 100–300 nucleotides in length (Zieve, 1981). In addition, 15 S nucleoprotein particles containing poly(A) may be present (Maxwell *et al.*, 1981)

Table 1. *Nuclear hnRNA-protein complexes from eukaryotic cells*

Source	Method of isolation from nuclei	Characteristics of RNP complexes: Size or density	Proteins ($10^{-3} \times M_r$)	RNA	References
HeLa cells	Extraction with 0.1 M-salt, pH 8	40 S monoparticles	Major constituents are three doublets (32/34, 36/37 and 42/44); also a (110) protein		Beyer *et al.* (1977)
HeLa cells	Extraction with 0.01 M-salt, pH 7, with sonication	40–150 S polyparticles	Many proteins (39–180)	20–80 S hnRNA	Pederson (1974)
HeLa cells	Salt extraction with sonication, then release of snRNP from hnRNP		Four proteins (10–14)	Five species of snRNA	Brunel *et al.* (1981)
Rat liver	Salt extraction, pH 8, with sonication	50–200 S polyparticles	Six to ten basic proteins (30–45); many proteins of higher M_r	hnRNA and six species of snRNA	Heinrich & Northemann (1981)
Rat brain	Extraction with 0.1 M-salt, pH 7.4, and sonication	30–50 S monoparticles; up to 200 S polyparticles	Basic proteins (28–38); acidic proteins (40–110)	hnRNA (up to 30 S) and five species of snRNA	Fuchs *et al.* (1980) Fuchs & Jacob (1979)
Mouse L cells	Extraction with 0.01 M-salt, pH 7, and sonication		Many proteins (30–> 90) crosslinked to RNA by u.v. irradiation. A protein (60) was bound to the poly(A) tract	hnRNA	Setyono & Greenberg (1981)
Mouse C11D LM(TK⁻) cells	Extraction with 0.1 M-salt and sonication; fractionation in metrizamide gradient	1.31 g · ml⁻¹ 1.18 g · ml⁻¹	Major proteins (40–60) Major proteins (30–40)	hnRNA hnRNA	Houssais (1977)
Duck erythroblasts	Extraction with 0.1 M-salt, pH 8, and 1 mM-ATP	40 S monoparticles	90 different polypeptides	hnRNA and eight species of snRNA	Maxwell *et al.* (1981) Maundrell & Scherrer (1979)

presumably as a result of the degradation of hnRNA by nuclease (Quinlan *et al.*, 1977).

Analysis of the hnRNA from various cells has shown that some molecular species contain sequences homologous with cytoplasmic poly(A)$^+$ mRNA, whereas others appear to be confined to the nucleus. Thus, 30 S hnRNP from mouse ascites cells contains a mixture of poly(A)$^+$ sequences with M_r 6×10^5 and a complexity of 3×10^4, approx. 10–15% being homologous to poly(A)$^+$ mRNA and the remainder restricted to the nucleus (Kinniburgh & Martin, 1976). Moreover, RNA from 40 S RNP obtained from chick erythroblast nuclei was found to have messenger activity in the wheat germ system, some of the cell-free translation product being identified as globin by chromatography on carboxymethylcellulose (Hendrick & de Kloet, 1975) suggesting the presence of some fully processed mRNA in the RNP. Both pre-β-globin mRNA (15 S) and spliced β-globin mRNA (9 S) have been identified in nuclear RNP by hybridization with ^{32}P-labelled nick-translated β-globin DNA (Pederson & Davis, 1980) and β-globin mRNA sequences in hnRNP from Friend erythroleukaemic cells have been shown to be closely associated with proteins by using cross-linking by u.v. irradiation (Mayrand & Pederson, 1981). Ovalbumin mRNA sequences have been detected in hnRNP from nuclei of oestrogen-stimulated chick oviduct (Thomas *et al.*, 1981).

The proteins of hnRNP complexes. Despite the great complexity of the proteins which have been found to occur in hnRNP particles, a pattern is beginning to emerge which makes it possible to begin to suggest possible structural models for the arrangement of proteins in the RNP complexes. In the case of the 40 S monoparticles there is now good evidence for the existence of a comparatively small number of so-called packaging proteins with molecular weights in the 30000–40000 range. Thus, in HeLa cells three pairs of proteins have been described (Beyer *et al.*, 1977): group A (M_r 32000 and 24000), group B (M_r 36000 and 37000) and group C (M_r 42000 and 44000). Group A constitutes 60% of the total particle protein mass. Both the group A proteins are basic (pI 9.2 and 9.4) and have a characteristic amino acid composition including a very high content of glycine (25%) and the unusual modified amino acid, N^G-dimethylarginine. Similar proteins have been found in hnRNP particles from brain nuclei (Fuchs *et al.*, 1980), hamster fibroblasts and mouse ascites cells (Beyer *et al.*, 1977). Both group A and group B proteins dissociate from the 40 S RNP particle at 0.3 M-NaCl, whereas the group C proteins remain in salt resistant RNP complexes sedimenting at 20 S and 30 S. These results are not consistent with the simple ‘informofer’ model (Lukanidin *et al.*, 1972) which has a self-contained 30 S protein core and external RNA, but support an interspersed RNP structure in which some proteins (groups A and B) interact with the RNA electrostatically and others are more firmly associated. In HeLa cells, only two proteins (M_r 41500 and 43000) were found to be present in nuclear hnRNP complexes in sufficiently close association with the RNA to become cross-linked efficiently by u.v. irradiation of the intact cells, two other polypeptides (M_r 36000 and 37000) reacting less efficiently. After infection of cells with adenovirus, hnRNP complexes containing specific viral mRNA sequences and the same two major proteins were isolated, indicating that both cellular and viral hnRNP are similar in general structure (van Eekelen *et al.*, 1981*a*).

Additional proteins are associated with heterogeneous RNP complexes larger than 40 S, although it is not clear to what extent they may be only minor components. Most of these proteins are acidic and have molecular weights larger than 40000 (Fuchs & Jacob, 1979; Maundrell & Scherrer, 1979). Amongst them a poly(A)-binding protein of M_r 70000–80000 has been reported to be present in nuclear poly(A)-protein complexes from HeLa cells (Kish & Pederson, 1975; Kumar & Pederson, 1975) and mouse ascites cells (Martin *et al.*, 1973; Quinlan *et al.*, 1974). At one time it was thought that this protein might be identical with the M_r 78000 protein associated with the poly(A) segment of cytoplasmic mRNP, although the evidence was far from clear cut (Preobrazhensky & Spirin, 1978). Recently, however, it has been shown by an examination of proteins cross-linked to RNA by u.v. irradiation that the protein associated with the poly(A) tract of hnRNA of mouse L cells has an M_r of 60000 and is clearly different from the protein present on the poly(A) of cytoplasmic mRNA, which has an M_r of 78000 (Setyono & Grenberg, 1981).

Some enzyme activities have been detected in hnRNP, including a protein kinase which phosphorylates the M_r 28000 and 37000 proteins of HeLa cell core particles (Blanchard *et al.*, 1977) and ribonuclease D (Rech *et al.*, 1979).

At the present time, the available evidence indicates that completely different sets of proteins are associated with nuclear hnRNP and cytoplasmic mRNP complexes.

Function of hnRNP proteins

Several functions, including packaging of hnRNA, processing of pre-mRNA involving splicing and transport of mRNA from the nucleus to the cytoplasm have been suggested, but the present evidence is mainly circumstantial.

Packaging. The presence of 400–2000 nucleotides (Kinniburgh & Martin, 1976) in the hnRNP monomers having a diameter of approx. 25 nm (Beyer *et al.*, 1980) indicates a 10–50-fold condensation in the length of transcripts within the RNP particles on the assumption that the internucleotide distance of extended RNA is 0.7 nm. It is reasonable, therefore, to conclude that one function of the hnRNP proteins may be concerned with the packaging of the primary RNA transcripts in order to accommodate them within the nucleus. The similarity in protein composition of the core hnRNP particles from different cells (Beyer *et al.*, 1977) is in agreement with such a function.

Processing of hnRNA transcripts. As shown by electron microscopy of spread chromatin from *Drosophila melanogaster* and *Calliphora erythrocephala*, nascent hnRNP fibrils form loops ranging in length between 1000 and 15000 bases, which occur at non-random points along the transcripts and correspond to segments released during RNA processing (Beyer *et al.*, 1981). Some small nuclear RNAs occur in hnRNP complexes and it has been suggested that they may be involved in the splicing of RNA transcripts (Murray & Holliday, 1980; Rogers & Wall, 1980; Lerner *et al.*, 1980). This idea is based on the complementarity of some snRNAs to splice junctions in pre-mRNA, as well as on the high degree of conservation of both the RNA sequences and the associated proteins (Lerner *et al.*, 1980; Lang *et al.*, 1981). An endoribonuclease active against double-stranded RNA (ribonuclease D) (Rech *et al.*, 1979) is present in hnRNP and may be related to a function in RNA processing, and the presence of an endogenous protein kinase activity which phosphorylates M_r 28000 and 37000 proteins of hnRNP may be connected with a regulatory function, although the enzyme is insensitive to cyclic AMP, cyclic GMP or calcium (Blanchard *et al.*, 1977).

Transport of mRNA from the nucleus. The transfer of mRNA from the nucleus into the cytoplasm appears to be dependent not only on the processing of mRNA precursors in the nucleus (Lerner *et al.*, 1980) but also on the presence of mRNA-binding proteins in the cytosol (Moffett & Webb, 1981; Roy *et al.*, 1979). These transport proteins are present as minor components in the cytosol as well as in the cytoplasmic mRNP fraction and their activity in releasing mRNA from rat liver nuclei is markedly enhanced by cyclic AMP, this effect being presumably mediated by a cyclic AMP-dependent protein kinase (Moffett & Webb, 1981). A comparison of the proteins present in the released mRNP particles with those of free, polysomal and heterogeneous nuclear mRNP proteins indicates that, with the possible exception of an M_r 78000 protein, most if not all proteins associated with hnRNA are removed before or at the time of exit of the RNA from the nucleus (Roy *et al.*, 1979). Since in mouse L cells the protein associated with the poly(A) tract of hnRNA has an M_r of 60000 and thus differs from the protein of M_r 78000 in the poly(A) of cytoplasmic mRNP it is likely that the hnRNP proteins are replaced by a completely new set of proteins during the transport of mRNA from the nucleus (Setyono & Greenberg, 1981). Since cordycepin and α_3-amanitin inhibit this transport process, it has been suggested that the poly(A) segment is involved either directly or indirectly by providing binding sites for specific proteins (Chisick *et al.*, 1979).

Free Cytoplasmic RNP Particles

Free RNP particles have been detected in the postribosomal supernatant fraction from many eukaryotic cells. They are characterized by a high protein : RNA ratio, usually 4 : 1, and the absence of ribosomal RNA and proteins. The proteins present in these complexes are diverse and no uniform pattern has emerged, probably because different laboratories have investigated RNPs from different cells and have used different methods for their isolation and purification.

Two main subgroups of free cytoplasmic RNPs may be distinguished: mRNA-containing nucleoproteins and complexes containing RNA species other than mRNA. This distinction is, however, not always clear cut because in some cases at least both mRNA and non-messenger RNAs may be present in the same complex (Vincent *et al.*, 1980).

Free mRNP particles

These mRNA–protein complexes include stored untranslatable mRNP as well as translatable mRNP particles containing mRNA freshly exported from the nucleus or temporarily released during the ribosome cycle following peptide chain termination on a monoribosome (Fig. 3).

mRNP may be untranslatable because the mRNA component has been temporarily or permanently modified in some way even though all or part of the coding sequence is still intact and one may include in this group mRNA which is in the early stages

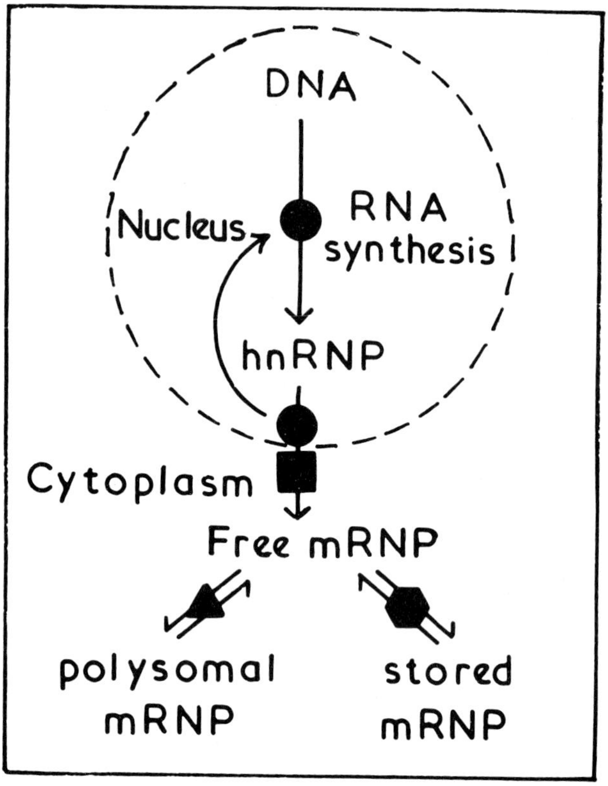

Fig. 3. *Inter-relationships between hnRNP and cytoplasmic free stored and polysomal mRNP*
Transport of mRNA from the nucleus into the cytoplasm is associated with an exchange of the proteins in the hnRNP complexes. Free mRNA and polysomal mRNA in the cytoplasm are interconverted by processes involving associations with different proteins (●, ■, ⬢, ▲) in mRNP complexes.

of degradation. Usually, however, untranslatable mRNP contains normal mRNA in association with translational inhibitors such as particular proteins or other RNAs. In this case the mRNA extracted from the mRNP complexes is fully active in cell-free protein synthesis systems. Alternatively, accumulation of certain mRNAs in non-polysomal cytosolic mRNP may be due to selective translational control involving initiation factors or other components needed for protein synthesis.

Stored mRNP. Unfertilized eggs of a number of species, such as sea urchins and amphibia, contain a store of translationally inactive maternal mRNA. After fertilization the mRNA becomes available for protein synthesis, although not necessarily immediately, and the timing depends both on the particular mRNA and the species. For example, histone H1 mRNA is accumulated in *Xenopus* during oogenesis, but is translated inefficiently until late cleavage when there is a rapid increase in the rate of synthesis of H1 histone (Woodland *et al.*, 1979). The exact mechanism of this process is not known, but the mRNA itself seems to be translatable (Ruderman *et al.*, 1979). In sea urchin eggs and embryos all species of mRNA are represented in both polysomes and free RNPs, but there are differences in their distribution, some being relatively concentrated in the free RNP fraction, others occurring predominantly on polysomes, and activation of stored mRNP is a selective process (Infante & Heilmann, 1981).

It is likely that the translational repression of stored messengers is due to the association of the mRNA with inhibitory protein(s). Thus, it has been shown that the template activity of mRNP from sea urchin eggs depends on the protein content, which can be varied by using different buffers in the isolation procedure. After extraction with 0.35 M-Na^+ the mRNP contained fewer proteins and was much more active in cell-free translation in a wheat-germ system than when 0.35 M-K^+ was used (Jenkins *et al.*, 1978). Other examples of the storage of mRNAs as translationally repressed cytoplasmic RNP include myosin heavy chain mRNA (Doetschman *et al.*, 1980) and mRNAs in the dormant stages of wheat germ embryogenesis (Ajtkhozhin *et al.*, 1976) and *Artemia salina* embryos (Grosfeld & Littauer, 1975; Slegers & Kondo, 1977; Slegers *et al.*, 1981), mouse sarcoma-180 ascites cells (Geoghegan *et al.*, 1979) rat liver (Northemann *et al.*, 1980), duck erythroblasts (Vincent *et al.*, 1980), rabbit brain (Mahony & Brown, 1979) and rabbit reticulocytes (Buhl *et al.*, 1981).

Isolation and structure of free cytoplasmic mRNP

RNP complexes have been found in the post-ribosomal supernatant fraction of many eukaryotic cells (Table 2). These complexes contain mRNAs, which have been identified by translation of the extracted RNA in cell-free systems, and a large assortment of different proteins. In some cases, low-M_r RNA species have also been demonstrated in addition to the poly(A)$^+$ mRNAs.

The proteins have molecular weights in the range 20000 to over 100000 and are clearly different from ribosomal proteins. Variations in protein composition of the mRNP complexes depend not only on the cell from which they are derived, but also on the method of isolation (Bag & Sells, 1979), particularly on the ionic strength and Mg^{2+} in the buffers used during fractionation of cell extracts in sucrose gradients or subsequently. There is evidence to show that purification of mRNP by washing with high-salt (0.5 M) buffers, by affinity chromatography on oligo(dT)-cellulose or poly(U)-glass and elution by formamide or low ionic strength buffers at elevated temperatures, or electrophoresis in agarose gels, gives rise to differences in the proteins retained in the mRNP.

The complexity of the free cytoplasmic mRNA-containing particles is illustrated by the recent detailed analysis of globin mRNP isolated from the cytoplasm of duck erythroblasts (Vincent *et al.*, 1980, 1981). A 20 S RNP complex obtained by centrifugation in a low-salt sucrose gradient was found to contain about 20% RNA of which 36% was identified as globin mRNA. No other mRNA was found, but two low-M_r RNAs (7 S and 4 S) were also present. Treatment of the 20 S particle with high salt gave rise to two globin mRNA-containing core particles with sedimentation coefficients 13 S and 16 S, which differed considerably in protein composition. Although the two RNPs had several proteins in common, the 13 S core particles contained five or six proteins (M_r 25000, 33000, 35000, a 48000 doublet and 125000) which were absent from the 16 S core. Differences have been reported in the nucleotide sequences protected against nuclease digestion of the two free

Table 2. *Isolation and properties of some free cytoplasmic mRNA–protein complexes*

Origin of free mRNP	Method of preparation	Properties of mRNA–protein complexes: Translational activity	Sedimentation coefficient and density	Protein/RNA ratio	RNA component	Proteins ($10^{-3} \times M_r$)	References
Duck erythroblasts	Low and high salt sucrose gradient fractionation	Repressed	20 S 1.39 g · ml^{-1}	4	Globin mRNA (36%), 7 S and 4 S RNA		Vincent *et al.* (1981) Vincent *et al.* (1980)
			16 S		Globin mRNA	14 proteins (40–155)	
			13 S		Globin mRNA, 7 S RNA	19 proteins (25–155)	
Rabbit reticulocytes	Low salt sucrose gradient fractionation	Stimulatory	25–30 S		Globin mRNA	Approx. 16 proteins (20–100)	Buhl *et al.* (1981)
Rat liver	Low and high salt sucrose gradient fractionation	Repressed	20 S		Preproalbumin mRNA and 243-nucleotide inhibitory RNA	Six major proteins (36–117) and several minor proteins in both mRNP particles	Northemann *et al.* (1980)
		Repressed	40 S		Preproalbumin mRNA and four small RNA species		
Rat liver	Affinity chromatography on oligo(dT)–cellulose and elution by (*a*) low salt at 2°C, (*b*) low salt at 37°C and (*c*) 50% formamide		(*a*) 1.55 g · ml^{-1}		Poly(A)$^+$ mRNA	Eight proteins (51–90)	Bag & Sells (1979)
			(*b*) 1.47 g · ml^{-1}		Poly(A)$^+$ mRNA	Five proteins additional to above (44–64)	
			(*c*) 1.38 g · ml^{-1}		Poly(A)$^+$ mRNA	Only four proteins (49–66)	
Mouse ascites 180 cells	Low salt sucrose gradient fractionation	Repressed	20–70 S		Poly(A)$^+$ mRNA		Geoghegan *et al.* (1979)

Mouse L cells	High salt sucrose gradient fractionation and (*a*) adsorption on poly(U)–glass filters, (*b*) chromatography on oligo(dT)–cellulose and elution with 50% formamide or (*c*) isopycnic Cs_2SO_4 gradient centrifugation		$1.38–1.47\ g \cdot ml^{-1}$			(*a*) Six major proteins (32–74), one minor protein (48) (*b*) Two major proteins (33, 77) (*c*) Five major proteins (32–77), three minor proteins (47–67)	Obrig *et al.* (1979)
Rabbit brain	High salt sucrose gradient fractionation	Inhibitory	<40 S $1.28\ g \cdot ml^{-1}$ (major fraction)	6	Translatable	Six major proteins (60–150), several minor proteins (30–150)	Mahony & Brown (1979)
		Stimulatory	>40 S				
Chick embryo muscle	High salt sucrose gradient fractionation and oligo(dT)–cellulose affinity chromatography with elution by 50% formamide		60 S $1.42\ g \cdot ml^{-1}$		Myosin heavy chain and actin mRNAs	Approx. ten proteins (40–100)	Jain & Sarkar (1979)
	Low salt sucrose gradient fractionation		120 S $1.42\ g \cdot ml^{-1}$			Seven major proteins (44–98)	Bag & Sarkar (1976)
Artemia salina	Low salt sucrose gradient fractionation; oligo(dT)–cellulose chromatography with elution at 36°C	Repressed or stimulatory	5–30S $1.39\ g \cdot ml^{-1}$ $1.27\ g \cdot ml^{-1}$		Poly(A)$^+$ mRNA and inhibitory oligonucleotide	Seven proteins (24–87)	Slegers *et al.* (1981)
	Low salt sucrose gradient zonal centrifugation	Stimulatory	22 S		Poly(A)$^-$ mRNA		De Herdt *et al.* (1979)

cytoplasmic duck globin mRNP particles and the 15 S polysomal mRNP (Goldenberg & Scherrer, 1981), indicating specific interactions of the different proteins with the mRNA which may in addition alter the secondary structure of the latter.

Free cytoplasmic RNP complexes lacking mRNA

An RNP particle sedimenting at 10–15 S and devoid of mRNA has been isolated recently from the postribosomal fraction of chick embryonic muscle (Bag *et al.*, 1980; Sarkar *et al.*, 1981). This particle contains a 4 S RNA giving three major 70–90-nucleotide bands on polyacrylamide-gel electrophoresis in formamide and a large number of polypeptides with M_r values in the range 12000–220000, but there is uncertainty about the similarity of these proteins and those present in the free cytosolic mRNP complexes. Both the 10 S RNP and the 4 S RNA inhibit cell-free protein synthesis in the nuclease-treated reticulocyte lysate.

Function of free cytoplasmic mRNP complexes

Although it is well established that mRNA accumulates in certain processes such as oogenesis as RNP complexes which are stored untranslated and made available for protein synthesis at a subsequent stage of development, little is known about the molecular mechanism(s) involved in this translational control. Moreover, even tissues such as liver, where there is no obvious requirement for the storage of mRNA, contain repressed mRNP complexes (Northemann *et al.*, 1980). Whereas some cytoplasmic mRNP species, e.g. duck globin mRNP (Vincent *et al.*, 1980) or liver mRNP (Northemann *et al.*, 1980), are merely inactive in cell-free protein synthesis, other cytoplasmic mRNPs may be either inhibitory, e.g. brain (Mahony & Brown, 1979), or stimulatory, e.g. mRNPs from rabbit reticulocytes (Buhl *et al.*, 1981) and mouse sarcoma-ascites cells (Geoghegan *et al.*, 1979).

The range of effects on protein biosynthesis which have been observed with isolated mRNPs may be due to the presence of variable amounts of components with opposing activities as a result of differences in the isolation and purification procedures used in different laboratories. Thus, it appears that one or more of the low-M_r RNAs which have been detected in several mRNP preparations (Northemann *et al.*, 1980; Vincent *et al.*, 1980, 1981) may be inhibitory to cell-free protein synthesis, whereas a protein fraction from purified myosin was found to have a small but apparently significant stimulatory activity specifically for the translation of myosin mRNA in the wheat-germ cell-free system (Bester *et al.*, 1980). RNP complexes (10 S) devoid of mRNA but containing RNAs which inhibit cell-free protein synthesis have also been isolated from chick embryonic muscle cells (Bag *et al.*, 1980; Mukherjee & Sarkar, 1981; Sarkar *et al.*, 1981) and it is possible that *in vivo* they may become associated with mRNA-protein complexes.

Polysomal mRNPs

Eukaryotic mRNAs can be isolated from polysomes as RNP complexes having a buoyant density of approx. $1.4\ g \cdot ml^{-1}$. Such mRNP complexes have been obtained from many different animal cells (see Table 3 and the review by Preobrazhensky & Spirin, 1978) as well as plant cells (Ajtkhozhin & Akhanov, 1974; Pfisterer, 1978). Compared with the free cytoplasmic mRNA-protein complexes polysomal mRNP has a relatively simple structure, although there is little definite information about specific functions of the proteins in the translation of the messenger.

Structure of polysomal mRNP

Polyribosomes active in protein synthesis may be dissociated either in low or moderate salt by chelating Mg^{2+} with EDTA or in high salt (0.5 M) after releasing nascent polypeptide chains with puromycin. Dissociation with 2 M-urea has also been used (Rittschof & Traugh, 1980). These procedures lead to the release of mRNAs in the form of protein complexes which have buoyant densities of approx. 1.4–1.5 $g \cdot ml^{-1}$ and sedimentation coefficients generally 1.5–2 times those of the mRNA (Preobrazhensky & Spirin, 1978).

Electron microscopy of duck and rabbit polysomal globin mRNP has shown that the proteins are arranged at four to seven positions along the mRNA molecule (Dubochet *et al.*, 1973) and there have been many biochemical investigations of the proteins present in the polysomal mRNP complexes from different cells. There is now general agreement that an M_r 78000 protein is associated with the poly(A) tract of mRNA. This protein appears to be highly conserved in vertebrate species (Standart *et al.*, 1981) and it has been reported that HeLa cell poly(A) binding protein is antigenically related to the nuclear poly(A) polymerase from rat hepatoma tumour (Rose *et al.*, 1979).

Degradation of mRNP from Friend erythroleukaemia cells with T2 ribonuclease has demonstrated a repeating structure for the poly(A)-protein complex, fragments of 27 nucleotides being protected by the protein against digestion (Baer & Kornberg, 1980). Thus, multiple copies of the M_r 78000 protein are associated with the poly(A) tract and a typical mRNA with some 100–200 residues of adenylate probably contains four to eight copies of this protein.

The suggestion has been made that this protein may also bind to nucleotide sequences other than poly(A) (Goldenberg *et al.*, 1980) but it is now known that a second M_r 78000 protein, which has a different primary structure, is involved in binding to the non-poly(A) region (Setyono & Greenberg, 1981).

A major technical problem in investigating the structure of mRNA-protein complexes is the possibility that proteins may dissociate, migrate from their original locations or bind adventitiously during the isolation procedures. These factors may explain the lack of general agreement concerning the number of different mRNP-associated proteins except for the M_r 78000 poly(A)-binding protein and a second major protein of M_r 52000 (Table 2). Recently, proteins have been cross-linked to mRNA in polysomes by u.v. irradiation before dissociation of the mRNP complex, which ensures that there is no rearrangement of proteins during isolation (Greenberg, 1980). This technique has been applied to polysomal mRNP from mouse L cells, in which the poly(A) tract was shown to bind only the M_r 78000 protein, whereas the remaining nucleotide sequences of the mRNA interacted with six proteins of M_r 52000, 62000, 68000, 75000, 78000 [a second protein, distinct

Table 3. *Proteins present in mRNA–protein complexes released from eukaryotic polysomes*

Origin of polysomes	Method of isolating mRNP from polysomes	Proteins present in released polysomal mRNP ($10^{-3} \times M_r$)	References
Mouse Ehrlich ascites tumour cells	Release by EDTA	38, 40, 53, 56*, 67*, 71*, 77, 81*, 106, 131	Jeffery (1977)
Mouse L cells	U.v. irradiation of polysomes to crosslink proteins to RNA before release with EDTA	52, 62, 68, 75, 78*, 98	Greenberg (1980)
Rat liver	Poly(A)-protein particles, prepared by digestion of polysomes with RNAases A and T_1, were fractionated by sucrose gradient centrifugation	50*, 54*, 58*, 63*, 76*, 90*	Tomcsányi *et al.* (1981)
Rabbit reticulocytes	Release by 2 M-urea Release by EDTA	23 non-ribosomal proteins (23–175) 58*, 67*, 71*, 79*	Rittschof & Traugh (1980) Jeffery & Peters (1977)
Duck erythroblasts	Release by EDTA	One major protein (73*); 12 minor proteins (45–130)	Vincent *et al.* (1981)
Chick embryo muscle	Release by EDTA followed by affinity chromatography on oligo(dT)–cellulose	Two major proteins (52, 78*); three minor proteins (60*, 95, 110)	Jain & Sarkar (1979)
Human KB cells	Affinity chromatography on oligo(dT)–cellulose after cross-linking by u.v. irradiation of cells	52, 69, 73	Van Venrooij *et al.* (1981)
HeLa cells	Affinity chromatography on poly(U)–Sepharose of poly(A)-protein particles obtained by digestion of polysomes with RNAases A and T_1	62*, 74*	Kish & Pederson (1976)
Rainbow trout testis	Release by EDTA or puromycin in 0.5 M-KCl	Major proteins (73, 29); several minor proteins	Gedamu *et al.* (1977)

* Denotes protein bound to poly(A) tract.

from the poly(A) binding protein] and 98000. Interestingly, only three proteins (M_r 52000, 69000 and 73000) were found to be cross-linked to polysomal mRNA after the direct irradiation of intact human KB cells with u.v. light (van Venrooij *et al.*, 1981). On balance, therefore, it is likely that polysomal mRNA contains multiple copies of an M_r 78000 protein bound to the poly(A) tract, together with perhaps two or at most a few other proteins which are probably located elsewhere on the mRNA. There is evidence that these proteins are bound to specific RNA sequences (van Eekelen *et al.*, 1981*b*; Goldenberg *et al.*, 1979).

Do polysomal mRNP proteins have a function in protein biosynthesis?

Apart from one report that ribosomes bind mRNP better than they do mRNA (Cashion & Stanley, 1974) there is no direct evidence at present that mRNP proteins participate in the translation of the mRNA. It is also not clear whether the mRNP proteins remain bound to the mRNA during translation or, as has been suggested recently for the M_r 78000 poly(A)-binding protein (Greenberg, 1981), exchange with a pool of free cytoplasmic proteins at some stage during the ribosome cycle. In mouse plasmacytoma cells, phosphorylated proteins present in free mRNP do not accompany the mRNA sequences into polysomes during protein synthesis (Elkaim *et al.*, 1981). Apart from ribosomal proteins, phosphorylated proteins are probably initiation factors and it would not be surprising if these were either dissociated from the mRNA or dephosphorylated during formation of the initiation complex.

The evidence concerning the presence of the M_r 78000 poly(A)-binding protein in both polysomal and free cytoplasmic mRNP is contradictory. Thus, according to some reports the cytoplasmic mRNP in duck reticulocytes (Vincent *et al.*, 1981), rabbit reticulocytes and Ehrlich ascites tumour cells (van Venrooij *et al.*, 1977) lack this protein, but others indicate that it is present in both the polysomal and free cytoplasmic mRNP of chick embryonic muscle, rat liver, rabbit reticulocytes and Ehrlich ascites tumour cells (Jain *et al.*, 1979). In *Physarum polycephalum*, on the other hand, the poly(A)-binding protein seems to be restricted to the non-polysomal mRNP, possibly because polysomal mRNA in this organism has only very short (15 residues) poly(A) tracts (Adams *et al.*, 1980), and translation of free cytoplasmic mRNP is accompanied by poly(A) degradation and dissociation of the poly(A)–protein complex (Adams *et al.*, 1981). The poly(A)-binding protein has also been found to occur free in the cytoplasm of reticulocytes (Jeffery & Peters, 1977), HeLa cells (Setyono *et al.*, 1977), rat liver (Mazur & Schweiger, 1978) and *Artemia salina* embryos (Slegers *et al.*, 1981).

The suggestion that the M_r 78000 protein–poly(A) complex of polysomal mRNP is a dynamic structure, despite its resistance to extraction with high ionic strength buffers, has been made recently by Greenberg (1981), who observed that after inhibition of mRNA synthesis in mouse L cells by actinomycin D labelled M_r 78000 protein continued to be synthesized and attached to the polysomal mRNP. Whether this indicates a specific function in translation, however, remains uncertain since globin mRNA and mRNP are translated equally well in a cell-free system from mouse Krebs II ascites tumour cells (Ernst & Arnstein, 1975) and in the reticulocyte lysate (Thomas *et al.*, 1982). Moreover, the poly(A)-binding protein from globin mRNP appears to have no significant effect on the cell-free translation of globin mRNA (Fukami & Itano, 1976). As an alternative to a direct role in translation the poly(A)-binding protein may control the breakdown of eukaryotic polysomal mRNA by protecting the poly(A) sequence against nuclease degradation (Bergmann & Brawerman, 1977).

In order to trace the fate of the major mRNP proteins during translation in cell-free protein synthesis systems, we have recently labelled the protein moiety of polysomal globin mRNP from rabbit reticulocytes, either by direct iodination with ^{125}I or by reaction with the Bolton–Hunter reagent (M.S. Fenster & H.R.V. Arnstein, unpublished work). Using lactoperoxidase bound to Sepharose, iodination yielded globin mRNP in which both the M_r 52000 and M_r78000 proteins were highly labelled, as shown by SDS/polyacrylamide-gel electrophoresis (Fig. 4). In addition, high-M_r aggregates of labelled proteins were usually present in variable amounts, although in the best preparations only traces were found. Iodination by the lactoperoxidase method, and other similar procedures, markedly reduced the translational efficiency of the mRNP, but reaction with the Bolton–Hunter reagent was less deleterious.

Despite the relatively inefficient utilization of globin mRNP which had been iodinated by the lactoperoxidase method, translation of the ^{125}I-labelled mRNP in a normal reticulocyte lysate showed some incorporation of label into the 80 S ribosomes and polysomes, whereas there was no significant incorporation in a control experiment with ^{125}I-labelled mRNP fragments prepared by degradation of globin mRNP with micrococcal nuclease before iodination (Fig. 5). In another control experiment it was also shown that the observed incorporation was inhibited by cycloheximide or low temperature (Fig. 6). Taken together, these results suggest that the translation of mRNP and its incorporation into polysomes can take place without prior dissociation into the protein and mRNA moieties. We have not established, however, whether the M_r 52000 and M_r 78000 proteins are both retained. Moreover, the relatively inefficient incorporation of labelled protein into polysomes may indicate that some exchange between mRNP proteins and free cytoplasmic proteins can occur during protein synthesis. Further experiments are in progress to clarify these aspects.

Summary and Conclusions

A review of recent work on the structure of hnRNPs and mRNPs indicates that in general different proteins interact specifically with particular nucleotide sequences, although there is little detailed information. The variety of proteins present depends to some extent on the procedures used to isolate and purify the RNA–protein complexes. Relatively few proteins are present in the most highly purified preparations and the number of proteins cross-linked to the RNA by u.v. irradiation is also limited. This evidence suggests that only a few proteins are firmly bound in such a way that they interact directly with the RNA bases, whereas others may bind either to the phosphodiester backbone or to the core proteins of the primary RNA–protein complexes.

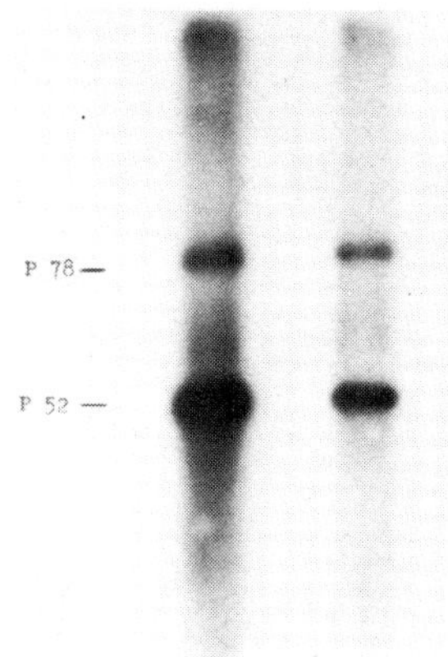

Fig. 4. *Preparation of mRNP labelled in the protein moiety by iodination*

Polysomal globin mRNP was isolated by dissociation of rabbit reticulocyte polysomes with EDTA and iodinated with ^{125}I by the lactoperoxidase method. The labelled proteins were analysed by SDS/polyacrylamide-gel electrophoresis showing radioactivity in the M_r 52000 (p 52) and M_r 78000 (p 78) mRNP proteins.

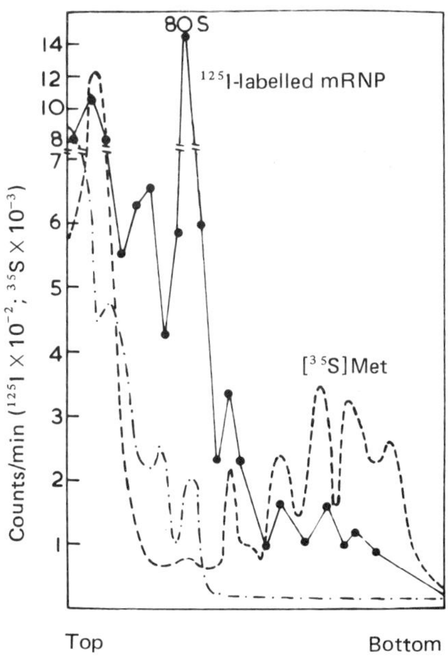

Fig. 5. *Utilization of ^{125}I-labelled globin mRNP for protein synthesis*

Globin mRNP was labelled with ^{125}I and added to a rabbit reticulocyte lysate (4 μg of mRNA/ml) containing [^{35}S]methionine and the usual components required for protein synthesis. After incubation at 30°C for 10 min, the lysate was analysed by fractionation in a sucrose gradient. – – –, [^{35}S]protein, showing nascent protein on polysomes; •, ^{125}I-labelled globin mRNP proteins; –.–.–, ^{125}I-labelled proteins using nuclease-treated globin mRNP.

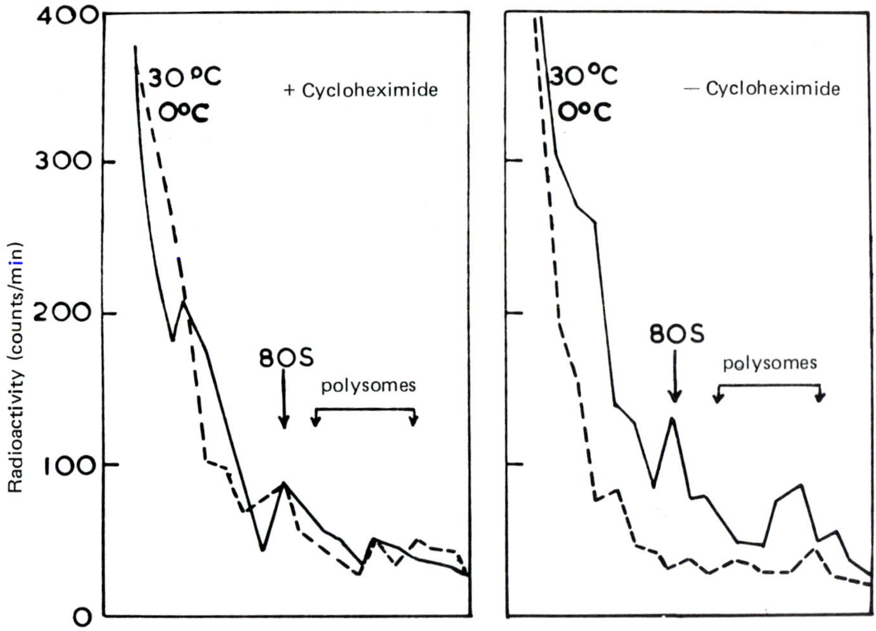

Fig. 6. *Inhibition of incorporation of* ^{125}*I-labelled globin mRNP into polysomes by cycloheximide*

Similar conditions as in Fig. 5 were used except that the labelled globin mRNP was used at 5.7 μg of mRNA/ml of lysate and an unincubated control was included. Where shown, cycloheximide (100 μg/ml) was added. —, 30°C; – – –, 0°C.

There now appears to be widespread agreement that nuclear hnRNP contains a set of proteins which differ from those associated with cytoplasmic mRNA, but there is little definite information about the function of any of these proteins. It is likely that some proteins act as translational repressors in the case of untranslatable or stored mRNP, although small inhibitory RNAs may also be involved. The fate of mRNP proteins during translation of the mRNA has not yet been fully elucidated and uncertainty remains about the possible function(s) of the major mRNP proteins in translation. With new approaches involving labelling of proteins in RNP complexes and the cross-linking of RNA and protein by u.v. irradiation further progress in solving some of these outstanding problems can be confidently predicted for the near future.

Support by a research grant from the Science and Engineering Research Council is gratefully acknowledged.

References

Adams, D.S., Noonan, D. & Jeffery, W.R. (1980) *Biochemistry* **19**, 1965–1970

Adams, D.S., Noonan, D. & Jeffery, W.R. (1981) *Proc. Natl. Acad. Sci. U.S.A.* 78, 1965–1970

Ajtkhozhin, M.A. & Akhanov, A.V. (1974) *FEBS Lett.* **41**, 275–279

Ajtkhozhin, M.A., Doschanov, Kh.J. & Akhanov, A.V. (1976) *FEBS Lett.* **66**, 124–126

Augenlicht, L.H. (1979) *Biochemistry* **18**, 3780–3786

Baer, B.W. & Kornberg, R.D. (1980) *Proc. Natl. Acad. Sci. U.S.A.* **77**, 1890–1892
Bag, J. & Sarkar, S. (1976) *J. Biol. Chem.* **251**, 7600–7609
Bag, J. & Sells, B.H. (1979) *Can. J. Biochem.* **57**, 1051–1057
Bag, J., Hubley, M. & Sells, B. (1980) *J. Biol. Chem.* **255**, 7055–7058
Bergmann, I.E. & Brawerman, G. (1977) *Biochemistry* **16**, 259–264
Bester, A.J., Durrheim, G., Kennedy, D.S. & Heywood, S.M. (1980) *Biochem. Biophys. Res. Commun.* **92**, 524–531
Beyer, A.L., Christensen, M.E., Walker, B.W. & LeStourgeon, W.M. (1977) *Cell* **11**, 127–138
Beyer, A.L., Miller, O.L. & McKnight, S.L. (1980) *Cell* **20**, 75–84
Beyer, A.L., Bouton, A.H. & Miller, O.L., Jr. (1981) *Cell* **26**, 155–165
Bhorjee, J.S. & Pederson, T. (1972) *Proc. Natl. Acad. Sci. U.S.A.* **69**, 3345–3349
Blanchard, J.-M., Brunel, C. & Jeanteur, P. (1977) *Eur. J. Biochem.* **79**, 117–131
Brunel, C., Widada, J.S., Lelay, M.-N., Jeanteur, P. & Liautard, J.-P. (1981) *Nucleic Acids Res.* **9**, 815–830
Buhl, W.-F., Ernst, H. & Hilse, K. (1981) *Biochem. Biophys. Res. Commun.* **99**, 1108–1116
Cashion, L.M. & Stanley, W.M., Jr. (1974) *Proc. Natl. Acad. Sci. U.S.A.* **71**, 436–440
Chisick, M.C., Brennessel, B.A. & Biswas, D.K. (1979) *Biochem. Biophys. Res. Commun.* **91**, 1109–1116
De Herdt, E., Slegers, H., Piot, E. & Kondo, M. (1979) *Nucleic Acids Res.* **7**, 1363–1373
Doetschman, T.C., Dym, H.P., Siegel, E.J. & Heywood, S.M. (1980) *Differentiation* **16**, 149–162
Dubochet, J., Morel, C., Lebleu, B. & Herzberg, M. (1973) *Eur. J. Biochem.* **36**, 465–472
Elkaim, R., Kempf, J. & Egly, J.M. (1981) *FEBS Lett.* **130**, 60–64
Ernst, V. & Arnstein, H.R.V. (1975) *Biochim. Biophys. Acta* **378**, 251–259
Fuchs, J.-P. & Jacob, M. (1979) *Biochemistry* **18**, 4202–4208
Fuchs, J.-P., Judes, C. & Jacob, M. (1980) *Biochemistry* **19**, 1087–1094
Fukami, H. & Itano, H.A. (1976) *Biochemistry* **15**, 3529–3535
Gedamu, L., Dixon, G.H. & Davies, P.L. (1977) *Biochemistry* **16**, 1383–1391
Geoghegan, T., Cereghini, S. & Brawerman, G. (1979) *Proc. Natl. Acad. Sci. U.S.A.* **76**, 5587–5591
Goldenberg, S. & Scherrer, K. (1981) *FEBS Lett.* **133**, 213–216
Goldenberg, S., Vincent, A. & Scherrer, K. (1979) *Nucleic Acids Res.* **6**, 2787–2797
Goldenberg, S., Vincent, A. & Scherrer, K. (1980) *Nucleic Acids Res.* **8**, 5057–5070
Greenberg, J.R. (1975) *J. Cell Biol.* **65**, 269–288
Greenberg, J.R. (1980) *Nucleic Acids Res.* **8**, 5685–5701
Greenberg, J.R. (1981) *Proc. Natl. Acad. Sci. U.S.A.* **78**, 2923–2926
Greenberg, J.R. & Setyono, B. (1981) *Biol. Cell* **41**, 67–78
Grosfeld, H. & Littauer, U.Z. (1975) *Biochem. Biophys. Res. Commun.* **67**, 176–181
Heinrich, P.C. & Northemann, W. (1981) *Mol. Biol. Rep.* **7**, 15–24
Heinrich, P.C., Gross, V., Northemann, W. & Scheurlen, M. (1978) *Rev. Physiol. Biochem. Pharmacol.* **81**, 101–134
Hendrick, D. & De Kloet, S.R. (1975) *Mol. Biol. Rep.* **2**, 225–232
Houssais, J.-F. (1977) *Mol. Biol. Rep.* **3**, 251–261
Infante, A.A. & Heilmann, L.J. (1981) *Biochemistry* **20**, 1–8
Ishikawa, K., Sato, T., Sato, S. & Ogata, K. (1974) *Biochim. Biophys. Acta* **353**, 420–437
Jain, S.K. & Sarkar, S. (1979) *Biochemistry* **18**, 745–753
Jain, S.K., Plaskal, M.G. & Sarkar, S. (1979) *FEBS Lett.* **97**, 84–90
Jeffery, W.R. (1977) *J. Biol. Chem.* **252**, 3525–3532
Jeffery, W.R. & Peters, C. (1977) *Mol. Biol. Rep.* **3**, 379–386
Jenkins, N.A., Kaumeyer, J.F., Young, E.M. & Raff, R.A. (1978) *Dev. Biol.* **63**, 279–298
Kinniburgh, A.J. & Martin, T.E. (1976) *Proc. Natl. Acad. Sci. U.S.A.* **73**, 2725–2729
Kish, V.M. & Pederson, T. (1975) *J. Mol. Biol.* **95**, 227–238
Kish, V.M. & Pederson, T. (1976) *J. Biol. Chem.* **251**, 5888–5894
Kumar, A. & Pederson, T. (1975) *J. Mol. Biol.* **96**, 353–365
Lang, V.W., Lerner, M.R., Steitz, J.A. & Flint, S.J. (1981) *Proc. Natl. Acad. Sci. U.S.A.* **78**, 1371–1375
Lerner, M.R., Boyle, J.A., Mount, S.M., Wolin, S.L. & Steitz, J.A. (1980) *Nature (London)* **283**, 220–224
Lukanidin, E.M., Zalmanzon, E.S., Komaromi, L., Samarina, O.P. & Georgiev, G.P. (1972) *Nature (London) New Biol.* **238**, 193–197
Mahony, J.B. & Brown, I.R. (1979) *J. Neurochem.* **33**, 1019–1030
Malcolm, D.B. & Sommerville, J. (1977) *J. Cell Sci.* **24**, 143–165
Martin, T., Billings, P., Levey, A., Ozarslan, S., Quinlan, T., Swift, H. & Urbas, L. (1973) *Cold*

Spring Harbor Symp. Quant. Biol. **38**, 921-932
Martin, T., Billings, P., Pullman, J., Stevens, B. & Kinniburgh, A. (1978) *Cold Spring Harbor Symp. Quant. Biol.* **42**, 899-909
Maundrell, K. & Scherrer, K. (1979) *Eur. J. Biochem.* **99**, 225-238
Maxwell, E.S., Maundrell, K., Puvion-Dutilleul, F. & Scherrer, K. (1981) *Eur. J. Biochem.* **113**, 223-247
Mayrand, S. & Pederson, T. (1981) *Proc. Natl. Acad. Sci. U.S.A.* **78**, 2208-2212
Mazur, G. & Schweiger, A. (1978) *Biochem. Biophys. Res. Commun.* **80**, 39-45
Moffett, R.B. & Webb, T.E. (1981) *Biochemistry* **20**, 3253-3262
Mukherjee, A.S. & Sarkar, S. (1981) *J. Biol. Chem.* **256**, 11301-11306
Murray, V. & Holliday, R. (1980) *FEBS Lett.* **106**, 5-7
Northemann, W., Schmelzer, E. & Heinrich, P.C. (1980) *Eur. J. Biochem.* **112**, 451-459
Obrig, T., Shen, A., Kwoka, S. & Chudyk, M.A. (1979) *Biochem. Biophys. Res. Commun.* **91**, 1062-1067
Parsons, J.T. & McCarty, K.S. (1968) *J. Biol. Chem.* **243**, 5377-5384
Pederson, T. (1974) *J. Mol. Biol.* **83**, 163-183
Pederson, T. & Davis, N.G. (1980) *J. Cell Biol.* **87**, 47-54
Pfisterer, J. (1978) *Z. Naturforsch.* **33c**, 359-362
Preobrazhensky, A.A. & Spirin, A.S. (1978) *Progr. Nucleic Acid Res. Mol. Biol.* **21**, 1-38
Quinlan, T.J., Billings, P.B. & Martin, T.E. (1974) *Proc. Natl. Acad. Sci. U.S.A.* **71**, 2632-2636
Quinlan, T.J., Kinniburgh, A.J. & Martin, T.E. (1977) *J. Biol. Chem.* **252**, 1156-1161
Rech, J., Brunel, C. & Jeanteur, P. (1979) *Biochem. Biophys. Res. Commun.* **88**, 422-427
Rittschof, D. & Traugh, J.A. (1980) *Eur. J. Biochem.* **115**, 45-52
Rogers, J. & Wall, R. (1980) *Proc. Natl. Acad. Sci. U.S.A.* **77**, 1877-1879
Rose, K.M., Jacob, S.T. & Kumar, A. (1979) *Nature (London)* **279**, 260-262
Roy, R.K., Lau, A.S., Munro, H.N., Baliga, B.S. & Sarkar, S. (1979) *Proc. Natl. Acad. Sci. U.S.A.* **76**, 1751-1755
Ruderman, J.V., Woodland, H.R. & Sturgess, E.A. (1979) *Dev. Biol.* **71**, 71-82
Samarina, O.P., Lukanidin, E.M., Molnar, J. & Georgiev, G.P. (1968) *J. Mol. Biol.* **33**, 251-263
Sarkar, S., Mukherjee, A.K. & Guha, C. (1981) *J. Biol. Chem.* **256**, 5077-5086
Setyono, B. & Greenberg, J.R. (1981) *Cell* **24**, 775-783
Setyono, B., Grossmann, M. & Liautard, J.-P. (1977) *Biochimie* **59**, 43-49
Slegers, H. & Kondo, M. (1977) *Nucleic Acids Res.* **4**, 625-639
Slegers, H., De Herdt, E. & Kondo, M. (1981) *Eur. J. Biochem.* **117**, 111-120
Standart, N., Vincent, A. & Scherrer, K. (1981) *FEBS Lett.* **135**, 56-60
Stevenin, J., Gallinaro-Matringe, H. & Jacob, M. (1977) *Mol. Biol. Rep.* **3**, 323-330
Thomas, N.S.B., Butcher, P.D., Kelly, J.M., Cox, R.A. & Arnstein, H.R.V. (1982) *Biochem. Soc. Trans.* **10**, 91-92
Thomas, P.S., Shepherd, J.H., Mulvihill, E.R. & Palmiter, R.D. (1981) *J. Mol. Biol.* **150**, 143-166
Tomcsányi, T., Komáromy, L. & Tigyi, A. (1981) *Eur. J. Biochem.* **114**, 421-428
van Eekelen, C.A.G., Mariman, E.C.M., Reinders, R.J. & van Venrooij, W.J. (1981*a*) *Eur. J. Biochem.* **119**, 461-467
van Eekelen, C.A., Rieman, T. & van Venrooij, W.J. (1981*b*) *FEBS Lett.* **130**, 223-226
van Venrooij, W.J. & Janssen, D.B. (1978) *Mol. Biol. Rep.* **4**, 3-8
van Venrooij, W.J., van Eekelen, C.A.G., Jansen, R.T.P. & Princen, J.M.G. (1977) *Nature (London)* **270**, 189-191
van Venrooij, W.J., Wagenmakers, A.J.M., van den Oetelaar, P. & Reinders, R.J. (1981) *Mol. Biol. Rep.* **7**, 93-99
Vincent, A., Civelli, O., Maundrell, K. & Scherrer, K. (1980) *Eur. J. Biochem.* **112**, 617-623
Vincent, A., Goldenberg, S. & Scherrer, K. (1981) *Eur. J. Biochem.* **114**, 179-193
Wahrman, M.Z. & Augenlicht, L.H. (1979) *Biochem. Biophys. Res. Commun.* **87**, 165-171
Woodland, H.R., Flynn, J.M. & Wyllie, A.J. (1979) *Cell* **18**, 165-171
Zieve, G.W. (1981) *Cell* **25**, 296-297

Biochem. Soc. Symp. **47**, 183–191
Printed in Great Britain

Mechanism of Protein Translocation across the Endoplasmic Reticulum

PETER WALTER and GÜNTER BLOBEL

Laboratory of Cell Biology, The Rockefeller University, New York, NY 10021, U.S.A.

Synopsis

This laboratory has recently formulated a general hypothesis on the mechanism of protein translocation across membranes, protein integration into membranes and subsequent protein traffic between membranes and membrane-bounded compartments [Blobel (1980) *Proc. Natl. Acad. Sci. U.S.A.* 77, 1496–1500]. It was hypothesized that the information for these processes, termed 'protein topogenesis', is encoded in discrete 'topogenic sequences' that constitute a permanent or transient part of the polypeptide chain. The information content of topogenic sequences would be decoded and processed by distinct effectors.

One specific example of intracellular protein topogenesis will be discussed in detail, namely translocation across and integration into the endoplasmic reticulum (ER) membrane. A protein (SRP, signal recognition protein) was extracted from the ER and purified to homogeneity [Walter & Blobel (1980) *Proc. Natl. Acad. Sci. U.S.A.* 77, 7112–7116]. Addition of this complex to extracted ER vesicles restores their ability not only to catalyse translocation of secretory and lysosomal proteins across the ER membrane but also to catalyse integration of a variety of membrane proteins into the ER membrane. SRP was shown to recognize the signal sequence of nascent secretory protein, thereby catalysing the specific binding of these polysomes to ER vesicles and initiating the translocation event. In the absence of ER vesicles, SRP reversibly arrests the elongation of secretory protein synthesis after its signal peptide and about 40 additional amino acids have been polymerized. This translation arrest can be released if ER vesicles are added, resulting in completed and efficiently translocated secretory protein.

A cell contains millions of protein molecules. These are steadily being synthesized and degraded. At homeostasis, a given species of protein is represented by a characteristic number of molecules that is kept constant within a narrow range. Very little is known about the cell's accounting procedures, i.e., how it balances and controls biosynthesis and biodegradation.

An important aspect of biosynthesis (Blobel, 1980) is the intracellular topology of proteins. Many protein species spend their entire life in the same compartment in which they are synthesized; others have to be translocated across the hydrophobic barrier of one, or in some cases, two distinct cellular membranes in order to reach the intracellular compartment or extracellular site where they exert their function.

Numerous protein species have to be integrated asymmetrically into distinct cellular membranes. For many proteins this requires partial translocation, i.e. selective transfer of one or several distinct hydrophilic or charged segments of the polypeptide chain across the hydrophobic barrier of one or two intracellular membranes.

Hypothetical models for intracellular protein translocation must deal with two essential tenets which appear to underly the observed phenomenology of this process. First, the permeability barrier of the membrane appears to be reversibly modified for the passage of each translocated polypeptide chain while being maintained for other solutes. Second, the species of protein to be translocated as well as the type of membrane across which a given protein is translocated are highly specific. Both of these tenets can be readily satisfied by postulating that protein translocation is a receptor-mediated process (Blobel, 1980) in which specificity is achieved by 'signal' sequences in the proteins to be translocated and by signal-sequence-specific translocation systems that are restricted in their location to distinct cellular membranes.

Until recently, the postulated recognition and translocation machineries (Blobel & Sabatini, 1971; Blobel & Dobberstein, 1975*a*; Blobel, 1980) remained largely undefined, so much so that they were deemed unnecessary (von Heijne & Blomberg, 1979; Wickner, 1979; Garnier *et al.*, 1980; Engelman & Steitz, 1981). Only after the development of an *in vitro* translocation system (Blobel & Dobberstein, 1975*b*) that was able to reproduce translocation of nascent secretory proteins across the ER membrane (isolated in form of closed microsomal vesicles) with apparent fidelity, did it become possible to assay and to characterize the ER's translocation activity *in vitro*. Two approaches were taken to dissect the membrane translocation activity: salt extraction (Warren & Dobberstein, 1978; Walter & Blobel, 1980) and limited proteolysis (Walter *et al.*, 1979; Meyer & Dobberstein, 1980*a*). Both approaches yielded membrane vesicles that were largely translocation-inactive; translocation activity could, however, be restored by re-addition of the salt or tryptic extract. These findings provided an assay for the purification of the active components of the salt extract (Walter & Blobel, 1980) and of the proteolytic extract (Meyer & Dobberstein, 1980*b*). The purified active component of the proteolytic extract consisted of an apparently single polypeptide chain of 60000 daltons (Meyer & Dobberstein, 1980*b*) that was shown to be generated by limited proteolysis from an integral membrane protein of 72000 daltons (Meyer *et al.*, 1982). In contrast, the purified active component of the salt-extract was shown to be a peripheral (completely salt-extractable) membrane protein sedimenting at 11S (approx. 250000 daltons) that consisted of six polypeptide chains which could not be separated from each other by a variety of non-denaturing procedures (Walter & Blobel, 1980). Both activities (the proteolytic fragment and the 11S salt-extracted protein) were required to reconstitute the translocation activity of vesicles that had been salt-extracted and proteolysed when assayed in the wheat germ cell-free translation system (P. Walter, unpublished work). Interestingly, the reticulocyte lysate translation system seems to contain an activity similar to that of the salt extract and thereby renders the reconstitution assays salt extract-independent (Walter *et al.*, 1979; Meyer & Dobberstein, 1980*a*).

The studies outlined below on the role of the 11S protein in the translocation process revealed that it is involved in the recognition of the signal sequence and therefore it was termed 'signal recognition protein' (SRP) (Walter *et al.*, 1981).

During the purification of SRP (Walter & Blobel, 1980), we noted that SRP inhibited protein synthesis in the wheat germ cell-free translation system if added in the absence of salt-extracted membrane. We decided to investigate the specificity of this phenomenon by comparing the effect of SRP on the translation of secretory versus cytoplasmic proteins. Rabbit reticulocyte RNA was chosen as a readily available mRNA pool containing species coding almost exclusively for the cytoplasmic proteins α- and β-globin. When SRP was added in increasing amounts to a translation mixture containing both globin and prolactin mRNAs (but no microsomal vesicles), a concentration-dependent inhibition of preprolactin synthesis was observed (Walter *et al.*, 1981). However, no significant effect on the amount of globin synthesized could be detected. At the highest concentration of SRP employed, preprolactin synthesis was essentially completely inhibited.

These findings indicated that SRP was able to affect the translation of specific mRNAs (i.e. inhibition of preprolactin synthesis and not of globin synthesis). On the premise that this effect of SRP on preprolactin synthesis was exerted via a direct interaction of SRP with polysomes synthesizing preprolactin, we attempted to demonstrate an affinity of SRP for these polysomes. To assemble preprolactin synthesizing polysomes, a staphylococcal nuclease-treated wheat germ cell-free translation system was used. If exogenous globin or prolactin mRNA was translated, polysomes were formed *in vitro*. Upon addition of radioactively labelled SRP an association of SRP with polysomes synthesizing preprolactin was observed, when the polysome profile was displayed on sucrose density gradients (Walter *et al.*, 1981). In contrast, little or no labelled SRP cosedimented with globin-synthesizing polysomes, even though these polysomes were present in approx. 4-fold higher amounts than preprolactin synthesizing polysomes. It should be noted that radiolabelled SRP was also found in the position of the monosomes whether or not either mRNA was being translated.

We have therefore demonstrated a specific interaction between SRP and polysomes synthesizing secretory protein. This highly specific interaction was detected by two independent assays: (i) SRP specifically inhibited the translation (in the wheat germ cell-free system) of bovine preprolactin (secretory protein) mRNA, in a concentration-dependent manner, but had no effect on the translation of rabbit globin (cytoplasmic protein) mRNA; (ii) ^{125}I-labelled SRP bound selectively and with high affinity (apparent $K_D < 8 \times 10^{-9}$M) to *in vitro* assembled polysomes synthesizing preprolactin and not to those synthesizing α- and β-globin. The translation-inhibitory effect of SRP was correlated with its selective polysome binding capacity. If the latter was impaired, the translation-inhibitory effect was also abolished.

Impairment of SRP's polysome binding capacity and, in parallel, impairment of its translation-inhibitory effect, was observed after modification of SRP by *N*-ethylmaleimide (Walter *et al.*, 1981; Jackson *et al.*, 1980). The relatively low affinity binding of SRP (apparent $K_D < 5 \times 10^{-5}$M) to monosomes (not engaged with mRNA in protein synthesis) was also completely abolished after modification of SRP by *N*-ethylmaleimide, strongly suggesting that this low-affinity binding to ribosomes is also a specific interaction and not nonspecific adsorption.

Impairment of high-affinity binding and abolition of the translation-inhibitory effect of SRP was also observed after modification of the preprolactin nascent chain. Incorporation of β-hydroxyleucine, an analogue of leucine, into nascent preprolactin chains of polysomes considerably reduced SRP's high-affinity binding to these polysomes, and, in parallel reduced its translation-inhibitory effect. The effect of β-hydroxyleucine could be abolished when incorporation took place in the presence of β-hydroxyleucine and L-leucine, arguing against the possibility that the observed effect of β-hydroxyleucine was the result of free β-hydroxyleucine rather than of incorporated β-hydroxyleucine residues. These data strongly suggested that it is the nascent polypeptide chain, and not mRNA, which is mediating the high-affinity binding of SRP to polysomes. Taken together with the recent demonstration by Hortin & Boime (1980) that replacement of leucine by β-hydroxyleucine abolishes translocation *in vitro* of secretory proteins that contain leucine-rich signal sequences [the signal peptide of bovine preprolactin contains eight leucine residues (Jackson & Blobel, 1980)], but not those that contain leucine-poor signal sequences, these data suggest that it is the signal sequence of the nascent chain that mediates SRP's high-affinity binding.

The specific binding of SRP to polysomes synthesizing secretory protein (Walter *et al*., 1981) suggested that SRP functions in the early steps of the overall protein translocation event, i.e. SRP would be required to recognize nascent polysomes and to mediate their specific binding to membranes. This binding has been postulated (Blobel & Dobberstein, 1975*a*) to occur early in the synthesis of the secretory proteins, presumably shortly after the signal sequence is synthesized (Boime *et al*., 1977; Rothman & Lodish, 1977). It was therefore of interest to investigate more closely the role of SRP in these early events of protein translocation, namely in the functional binding of the polysomes to membranes.

In the past, attempts have been made to study the nature of the ribosome-membrane junction of the rough ER by performing binding experiments with isolated ribosomes and microsomal vesicles (Borgese *et al*., 1977; Kreibich *et al*., 1978*a*,*b*; Hawkins & Freedman, 1979). However, the results of these studies could not correlate the observed binding event with the formation of a functional (with respect to translocation) ribosome–membrane junction. Binding studies of *in vivo* assembled 'detached' polysomes (polysomes isolated from rough microsomes after detergent solubilization) to microsomal membranes (Rayland *et al*., 1971) pose an artificial situation as well. These polysomes, containing chains at various stages of completion with the signal sequence already removed from many chains, do not exist in this form as 'free' polysomes in the cell.

An assay was therefore designed (Walter & Blobel, 1981*a*) which would permit to examine the binding of nascent polysomes to microsomal vesicles. This assay system offered the advantage of being able to correlate the binding of nascent, *in vitro* assembled polysomes to membranes with the actual translocation of the newly synthesized secretory protein. The assay therefore simulates, being based on ongoing protein synthesis, the dynamic conditions occurring *in vivo*. However, it poses limitations in that it is restricted to condition that are compatible with protein synthesis *in vitro*.

For technical reasons this assay was an indirect one; instead of measuring directly the fraction of nascent polysomes that bound to microsomal membranes, we determined the fraction of nascent polysomes that did not bind under given conditions. This was done by measuring the amount of mRNA that remained in the translation system following removal (by differential centrifugation) of mRNA that was assembled into membrane-bound polysomes.

Using this assay it was demonstrated (Walter & Blobel, 1981*a*) that microsomes that have been depleted of their translocation activity by salt extraction of SRP were unable to bind nascent preprolactin-synthesizing polysomes and that nascent polysome binding, as well as preprolactin translocation, could be restored only upon the re-addition of SRP. The binding of nascent preprolactin polysomes to microsomes is therefore dependent on SRP.

The overall translocation event can be viewed as occurring in steps, where the recognition of nascent polysomes synthesizing secretory protein and their binding to the microsomal membrane precede the physical translocation of the nascent polypeptide chain across the membrane. From the mentioned data we can then conclude that SRP is essential for recognition and binding to the microsomal membrane of nascent polysomes synthesizing secretory protein. Whenever we were able to interfere with the recognition by SRP of nascent *in vitro* assembled polysomes we also observed an interference with SRP-mediated binding of these polysomes to the microsomal membrane. For example, *N*-ethylmaleimide-modification of SRP, which had been shown to abolish both its low-affinity binding to the ribosome as well as its high-affinity binding to nascent polysomes synthesizing secretory protein (Walter *et al.*, 1981), also abolished SRP's capacity to bind these polysomes to the microsomal membrane. Likewise, modification of the nascent chain via incorporation of β-hydroxyleucine, previously shown to abolish the high-affinity binding of SRP to nascent polysomes synthesizing secretory protein (Walter *et al.*, 1981), also abolished binding of these polysomes to the microsomal membrane. Data which are discussed below (Walter & Blobel, 1981*b*) will demonstrate that SRP's recognition of specific polysomes and its ability to bind these polysomes to the microsomal membrane may be sequential events.

Considering the kinetics of the polysome binding, it was demonstrated (Walter & Blobel, 1981*a*) that at least the binding step of the translocation process is a strictly cotranslational event. The binding of polysomes is essentially completed after only 3–4 min, i.e. after at most half a preprolactin chain is synthesized in our *in vitro* system. Therefore, each prolactin molecule translocated has been completed on bound polysomes. Taken together, our data provide experimental support for the proposal (Blobel & Dobberstein, 1975*a*; Blobel *et al.*, 1979; Blobel, 1980) that protein translocation across the ER is indeed a receptor-mediated process and conclusively rule out an alternative proposal originally advanced by Bretscher (1973) and subsequently expanded upon by Wickner (1979) as well as by Engelman & Steitz (1981) that translocation across the lipid bilayer occurs spontaneously and does not require mediation by proteins. Our data also rule out translocation models (von Heijne & Blomberg, 1979; Garnier *et al.*, 1980; Steiner *et al.*, 1980; Di Rienzo *et al.*, 1978) that, although relying on the participation of proteins, postulate a primary interaction of the signal sequence (due to its hydrophobic nature) with the lipid bilayer rather than with protein. If this were the case, binding of nascent polysomes synthesizing secretory protein to salt-extracted microsomal vesicles should have taken place in the absence of SRP. It is thus clear that the initial events that lead to translocation and provide its specificity are, not protein–lipid interactions (signal sequence–lipid bilayer), but protein–protein interactions (signal sequence plus ribosome–SRP).

As mentioned above, SRP selectively inhibits translation of secretory proteins in the absence of microsomal membranes. We were able to demonstrate (Walter &

Blobel, 1981*b*) that prolactin mRNA translation could be arrested by SRP. This arrest occurs at a distinct point in preprolactin synthesis, resulting in the formation of a low molecular weight polypeptide that comprises the *N*-terminal portion of preprolactin. Its size of approx. 8000 daltons corresponds to about 70 amino acids polymerized. This translation arrest explains the observed (Walter *et al*., 1981) SRP-induced inhibition of preprolactin synthesis. The SRP-induced arrest of translation was reversible. Preprolactin synthesis continued when the translation arrest was released by the addition of salt-extracted microsomal membranes, leading to the formation of completed and efficiently translocated prolactin molecules.

To our knowledge, this is the first example of a site-specific modulation in the synthesis of specific proteins at the translational level. SRP was suggested above (Walter *et al*., 1981; Walter & Blobel, 1981*a*) to recognize polysomes synthesizing secretory protein by information contained in the nascent secretory protein. Since about 40 amino acids are required to 'span' the large ribosomal subunit (Malkin & Rich, 1967; Blobel & Sabatini, 1970) and since preprolactin chain elongation was arrested after about 70 amino acids were polymerized, an *N*-terminal segment comprising 30 residues would be exposed on the 'outside' of the ribosome. This exposed segment corresponds about to the size of the cleaved signal peptide of preprolactin (Jackson & Blobel, 1980). We therefore feel certain that the information required for recognition by SRP of a polysome synthesizing secretory protein is contained in the *N*-terminal signal peptide and/or closely adjacent sequences. The previously reported observation (Boime *et al*., 1977; Rothman & Lodish, 1977), namely that nascent chains of more than 80 residues could no longer be translocated, might be due to the inability of the longer chains to interact with SRP.

Because we have also been able to demonstrate an SH-group dependent interaction of SRP with ribosomes regardless of their state of biosynthetic activity (Walter *et al*., 1981) it seems likely that, upon recognition of a signal sequence and in the absence of microsomal membranes, SRP transmits the information to arrest protein synthesis through ribosomal components causing a site-specific elongation arrest.

SRP arrests the elongation of secretory proteins once their signal sequence (marking them as secretory proteins) is expressed outside the ribosome (Walter & Blobel, 1981*b*). This arrest can be fully released when microsomal membranes are added, thereby allowing the secretory protein to be completed and concomitantly translocated across the membrane. SRP may act to prevent the synthesis of this class of proteins until their sequestration out of their biosynthetic compartment (i.e. cytoplasm) is guaranteed. Teleologically this mode of action appears plausible if one considers that many secretory (and lysosomal) proteins may express enzymic activities which would be potentially harmful to the cell if fully synthesized in the cytoplasm. It should be noted, however, that the observed SRP-dependent elongation arrest might be an experimentally generated condition that may never or rarely arise in the intact cell, where ER membranes are always present.

Although we have only described experiments using prolactin as a model secretory protein and the two globin chains as a model cytoplasmic protein, we feel certain that the results obtained can be generalized to all proteins which utilize the translocation machinery of the ER. Although a systematic study with other proteins has not yet been carried out, several lines of evidence argue in favour of such a generalization: (i) the integration of a membrane protein, such as the δ-subunit of the acetylocholine receptor, into membranes is SRP-dependent (Anderson *et al.*, 1982); and (ii) the synthesis of the above mentioned integral membrane protein, as well as of the secretory proteins chicken lysozyme (P. Walter, unpublished work) and rat apolipoprotein AI (Stoffel *et al.*, 1981) and the lysosomal protein cathepsin D (A. Erickson, P. Walter & G. Blobel, unpublished work), was found to be arrested by SRP; in all cases arrest was released when microsomal vesicles were added.

Based on our findings (Walter & Blobel, 1980; Walter *et al.*, 1981; Walter & Blobel, 1981*a,b*), we would like to propose the scheme depicted in Fig. 1 as a model for the role of SRP in translocation of secretory and lysosomal proteins across and integration of membrane proteins into the membrane of the ER. SRP can be extracted from and rebound to microsomes. Data obtained following the distribution of SRP with a radioimmunoassay (P. Walter, unpublished work) suggest that SRP is in equilibrium (Fig. 1, A) between a membrane-bound and a free (soluble-cytoplasmic) form at salt concentrations approaching physiological conditions (150 mM-potassium acetate). Because of the observed low-affinity binding of SRP to monomeric non-translating ribosomes *in vitro* we would also like to propose an equilibrium between the putative free form of SRP and monomeric ribosomes (Fig. 1, B). Upon translation of a mRNA coding for a secretory protein (Fig. 1, C) the expression of the nascent secretory protein's signal sequence causes a 6000-fold enhancement in the apparent affinity of SRP for polysomes. Concomitantly, and presumably through the ribosome, SRP arrests the synthesis of the secretory protein (Fig. 1, D), preventing the completion of any secretory or presecretory protein in the cytoplasm. Only when membranes with 'translocation competent sites' [phospholipid vesicles and trypsinized microsomal vesicles have no effect (P. Walter, unpublished work)] are offered to this arrested polysome does it attach to the membranes (here by virtue of a putative ribosome and SRP receptor)

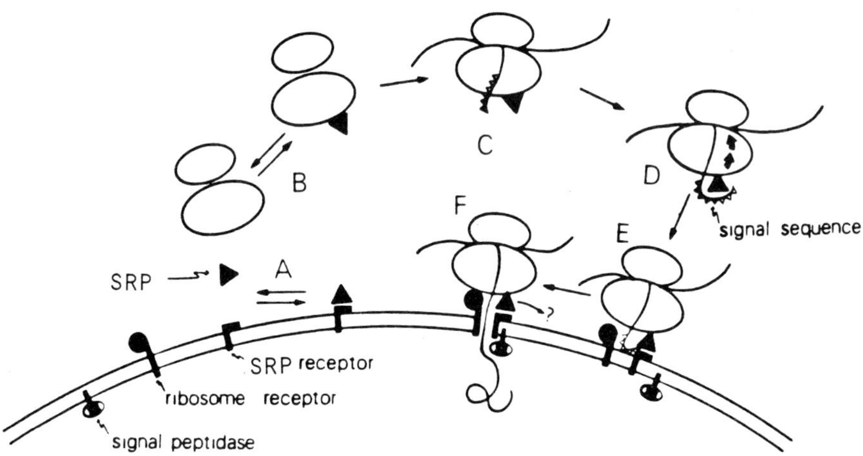

Fig. 1. *Model for the function of SRP in the protein translocation process*

For explanation see the text. Reproduced with the permission of the Rockefeller University Press from Walter & Blobel (1981*b*).

(Fig. 1, E). Preliminary studies using SRP as an affinity probe indicate that the 72000 dalton protein identified by Meyer *et al.* (1982) might function as a SRP-receptor (R. Gilmore, P. Walter & G. Blobel, unpublished work). This attachment results in the assembly of the functional translocation machinery which is here depicted as a pore-like structure. Synthesis of the secretory protein then continues, concomitant with its translocations across the membrane (Fig. 1, F).

Our choice of the term 'signal recognition protein' was made on the basis of the most striking property of this protein, namely its high-affinity binding to nascent polysomes containing ER-targeted signal sequences in their nascent chains (Blobel, 1980). Although we have not yet demonstrated a ribosome-independent and direct binding between such a signal peptide and SRP (and therefore have refrained from using the simpler term 'signal receptor') other lines of evidence suggest that such a direct interaction may indeed occur. For example, a signal peptide isolated from a tryptic digest of ovalbumin (Lingappa *et al.*, 1979) as well as a chemically synthesized peptide containing the signal peptide region of preproparathyroid hormone (Majzoub *et al.*, 1980) have been shown to compete when added to translocation system *in vitro* containing translocation-competent microsomal membranes. Likewise, preincubation of translocation-competent microsomal membranes with preproinsulin synthesized *in vitro* was shown to reduce the vesicles' translocation activity (Prehn *et al.*, 1981). However, despite our demonstration (Walter & Blobel, 1981*a*) that the microsomal membranes' ability to bind selectively nascent polysomes synthesizing secretory protein can be localized to SRP, the results of the above mentioned competition experiments constitute only indirect evidence for a direct signal–signal receptor interaction. They do not rule out the possibility that the high-affinity binding of SRP to nascent polysomes is due to a signal sequence-induced conformational change in the ribosome rather than the result of a direct binding of SRP to the signal sequence itself.

References

Anderson, D.J., Walter, P. & Blobel, G. (1982) *J. Cell Biol.* **93**, 501–506

Blobel, G. (1980) *Proc. Natl. Acad. Sci. U.S.A.* **77**, 1496–1500

Blobel, G. & Dobberstein, B. (1975*a*) *J. Cell Biol.* **67**, 835–851

Blobel, G. & Dobberstein, B. (1975*b*) *J. Cell Biol.* **67**, 852–862

Blobel, G. & Sabatini, D.D. (1970) *J. Cell Biol.* **45**, 130–145

Blobel, G. & Sabatini, D.D. (1971) *Biomembranes* (Manson, L.A., ed.), vol. 2, pp. 193–195

Blobel, G., Walter, P., Chang, C.N., Goldman, B.M., Erickson, A.H. & Lingappa, V.R. (1979) in *Symposium of the Society of Experimental Biology (Great Britain)*, (Hopkins, C.R. & Duncan, C.J., eds.), vol. 33, pp. 9–36, Cambridge University Press, London

Boime, I., Szczesna, E. & Smith, D. (1977) *Eur. J. Biochem.* **73**, 515–520

Borgese, N., Mok, W., Kreibich, G. & Sabatini, D.D. (1977) *J. Mol. Biol.* **88**, 559–580

Bretscher, M.S. (1973) *Science* **181**, 622–629

Di Rienzo, J.M., Nakamura, K. & Inouye, M. (1978) *Annu. Rev. Biochem.* **47**, 481–532

Engelman, D.M. & Steitz, T.A. (1981) *Cell* **23**, 411–422

Garnier, J., Gaye, P., Mercier, J.C. & Robson, B. (1980) *Biochimie* **62**, 231–239

Hawkins, H.C. & Freedman, R.B. (1979) *Biochim. Biophys. Acta* **558**, 85–98

Hortin, G. & Boime, I. (1980) *Proc. Natl. Acad. Sci. U.S.A.* **77**, 1356–1360

Jackson, R.C. & Blobel, G. (1980) *Ann. N.Y. Acad. Sci.* **343**, 392–404

Jackson, R.C., Walter, P. & Blobel, G. (1980) *Nature (London)* **284**, 174–176

Kreibich, G., Ulrich, B.C. & Sabatini, D.D. (1978*a*) *J. Cell Biol.* **77**, 464–487

Kreibich, G., Freienstein, C.M., Pereyra, P.N., Ulrich, B.C. & Sabatini, D.D. (1978*b*) *J. Cell Biol.* **77**, 488–505
Lingappa, V.R., Lingappa, J.R. & Blobel, G. (1979) *Nature (London)* **281**, 117–121
Majzoub, J.A., Rosenblatt, M., Fennick, B., Maunus, R., Kronenberg, H.M., Potts, J.T., Jr. & Habener, J.F. (1980) *J. Biol. Chem.* **265**, 11478–11483
Malkin, L.I. & Rich, A. (1967) *J. Mol. Biol.* **26**, 329–346
Meyer, D. & Dobberstein, B. (1980*a*) *J. Cell Biol.* **87**, 498–502
Meyer, D. & Dobberstein, B. (1980*b*) *J. Cell Biol.* **87**, 503–508
Meyer, D.I., Louvard, D. & Dobberstein, B. (1982) *J. Cell Biol.* **92**, 579–583
Prehn, S., Nuernberg, P. & Rapoport, T.A. (1981) *FEBS Lett.* **123**, 79–84
Rayland, W.L., Shires, T.K. & Pitot, H.C. (1971) *Biochem. J.* **121**, 271–278
Rothman, J.E. & Lodish, H.F. (1977) *Nature (London)* **269**, 775–789
Steiner, D.F., Quinn, P.S., Chan, S.J., Marsh, J. & Tager, H.S. (1980) *Ann. N.Y. Acad. Sci.* **343**, 1–16
Stoffel, W., Blobel, G. & Walter, P. (1981) *Eur. J. Biochem.* **120**, 519–522
von Heijne, G. & Blomberg, C. (1979) *Eur. J. Biochem.* **97**, 175–181
Walter, P. & Blobel, G. (1980) *Proc. Natl. Acad. Sci. U.S.A.* **77**, 7112–7116
Walter, P. & Blobel, G. (1981*a*) *J. Cell Biol.* **91**, 551–556
Walter, P. & Blobel, G. (1981*b*) *J. Cell Biol.* **91**, 557–561
Walter, P., Jackson, R.C., Marcus, M.M., Lingappa, V.R. & Blobel, G. (1979) *Proc. Natl. Acad. Sci. U.S.A.* **76**, 1795–1799
Walter, P., Ibrahimi, I. & Blobel, G. (1981) *J. Cell Biol.* **91**, 545–550
Warren, G. & Dobberstein, B. (1978) *Nature (London)* **273**, 569–571
Wickner, W. (1979) *Annu. Rev. Biochem.* **48**, 23–45

Biochem. Soc. Symp. 47, 193–209
Printed in Great Britain

Synthesis and Maturation of the Erythrocyte Anion Transport Protein – an Internal Sequence for Membrane Insertion

HARVEY F. LODISH and WILLIAM A. BRAELL

Department of Biology, Massachusetts Institute of Technology, Cambridge, MA 02139, U.S.A.

Synopsis

The biosynthesis of the erythrocyte anion transport glycoprotein, Band III (M_r 100000), is of interest, as its *N*-terminal half is hydrophilic and faces the cytoplasmic surface; the *C*-terminal half spans the phospholipid bilayer several times. Band III is synthesized by erythroid precursor cells obtained from the spleens of anaemic mice. Newly synthesized Band III was inserted into rough endoplasmic reticulum membranes with an asymmetric orientation which resembled that of mature Band III in erythrocyte membranes: the *N*-terminal portion of the molecule facing the cytoplasm. Newly made Band III contained a high-mannose asparagine-linked oligosaccharide, which was susceptible to cleavage by endoglycosidase H. During the next 20–30 min, this oligosaccharide was processed to a form resistant to endoglycosidase H degradation, presumably in the Golgi complex. The processed Band III was subsequently expressed on the cell surface, at about 30–45 min after synthesis.

To study the mechanism of insertion of Band III into microsomes, we used erythroid precursor cells from the spleens of anaemic mice as a source of messenger RNA for studies *in vitro* in the wheat germ and reticulocyte lysate cell-free system containing dog pancreatic microsomes. Immediately after synthesis, Band III was found to be inserted into microsomal membranes in its mature configuration, with the *N*-terminal portion exposed to the cytoplasm and its hydrophobic *C*-terminal portion spanning the lipid bilayer. The newly-synthesized Band III was also provided with a high-mannose asparagine-linked oligosaccharide. Band III was found to be inserted into dog pancreatic microsomes in a co-translational manner; in synchronized translation studies microsomes could be added as late as the time when the hydrophilic *N*-terminal half of the protein had been synthesized and still allow normal trans-membrane insertion and glycosylation. There is no cleavage of any *N*-terminal peptide during membrane insertion. In many respects, therefore, the biosynthesis of Band III resembles that of co-translationally-inserted proteins whose *N*-terminal portions are exposed on the exterior of the cell, like vesicular stomatitis virus glycoprotein, HLA-A antigens, and glycophorin. However, our results suggest that Band III contains a sequence near the middle of the protein which directs its insertion into endoplasmic reticulum membranes.

Introduction

The major integral membrane protein of the human erythrocyte is the anion transport protein, Band III (Ho & Guidotti, 1975; Steck, 1978; Cabantchik *et al.*, 1978). The *N*-terminal 40000–45000 daltons of the protein comprise a hydrophilic domain on the cytoplasmic face of the erythrocyte membrane (Steck *et al.*, 1976, 1978; Fukuda *et al.*, 1978), while the *C*-terminal 52000 daltons comprise a membrane-bound domain that spans the membranes two or more times (Drickamer, 1977; Steck *et al.*, 1978; Grinstein *et al.*, 1978; Rao & Reithmeier, 1979; Williams *et al.*, 1979). The *C*-terminal segment of Band III contains an asparagine-linked oligosaccharide with the repeating disaccharide ...Gal(β1→4)GlcNAc(β1→3)... (Tsuji *et al.*, 1980). The transmembrane configuration of Band III is thus quite different from that of well-studied membrane proteins like VSV glycoprotein or glycophorin (Fig. 1). There the *C*-terminal 20–30 amino acid residues project into the cytoplasm, while the *N*-terminal region is exposed on the extracytoplasmic face of the membrane (Katz *et al.*, 1977; Lodish *et al.*, 1981).

This difference in topography raises the question of whether the mechanism of biogenesis of Band III is at all related to that of proteins of the VSV G class. G protein is co-translationally inserted into the rough endoplasmic reticulum, where it receives two high-mannose asparagine-linked oligosaccharides (Lodish *et al.*, 1981; Rothman & Lodish, 1977). It then matures to the Golgi complex, where there occurs addition of fatty acids and also modifications of the oligosaccharides (Knipe *et al.*, 1977; Schmidt & Schlesinger, 1979; Katz & Lodish, 1979; Li *et al.*, 1978; Robbins *et al.*, 1977). Subsequently it arrives at the cell surface (Knipe *et al.*, 1977).

In this paper we show that membrane insertion of Band III is co-translational; in particular, microsomes can be added until the time when the nascent chain is half-complete and still permit subsequent insertion of the polypeptide. Thus, an internal sequence is used for membrane insertion. As we emphasize, insertion of Band III is distinct from all other glycoproteins studied to date, such as the VSV glycoprotein, and all secretory proteins, including ovalbumin, which utilize an *N*-terminal sequence for membrane insertion. However, we show that Band III follows a route of maturation to the cell surface which is similar in pathway and kinetics to that of co-translationally inserted proteins such as VSV G and HLA-A.

Details of all of the methods used in this work are fully presented in two recent publications (Braell & Lodish, 1981, 1982*a*).

Spleen Cells make Band III

Erythropoietic spleens from anaemic mice were excised, minced, aspirated, and forced through 35 μm Nitex cloth to yield a single cell suspension. This suspension was washed by centrifugation and labelled in culture for 1 h with [^{35}S]methionine. Band III was detected by immunoprecipitation. One antiserum was raised against total Band III (anti-Band III serum). This serum was specific for antigenic determinants on the hydrophilic domain of Band III. The other serum was raised against the *C*-terminal portion of Band III (anti-TR 52 serum), and was specific for the membrane-bound domain of Band III (Braell & Lodish, 1981). A detergent solution of a plasma membrane fraction from labelled spleen cells yielded a labelled protein

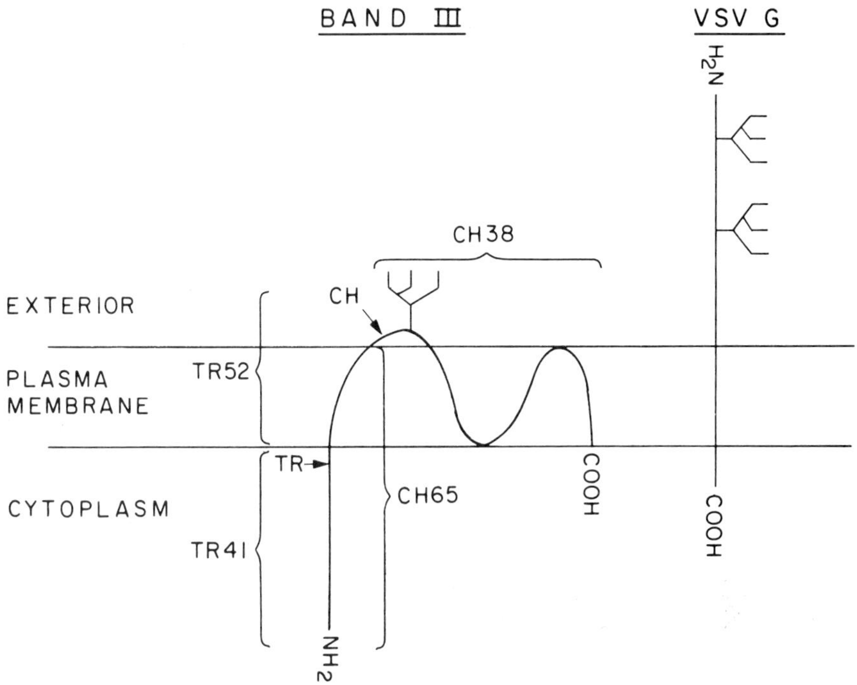

Fig. 1. *Configurations of Band III protein and VSV glycoprotein in the plasma membrane*

The configuration of Band III presented is that obtained from a compilation of structural information for the human erythrocyte protein (Drickamer, 1976, 1977, 1978; Steck *et al.*, 1976, 1978; Fukuda *et al.*, 1978; Rao & Reithmeir, 1979; Williams *et al.*, 1979). The sites of cleavage by chymotrypsin on the exterior and trypsin on the interior of the cells are indicated by CH and TR, respectively. The characteristic cleavage fragments CH 65 and CH 38 are produced by chymotrypsin digestion, while TR 52 and TR 41 are produced by trypsin digestion. The numbers associated with each fragment refer to the M_r of the fragment in kDa, as determined by ourselves for the mouse erythrocyte Band III protein in the Laemmli gel electrophoresis system. These M_r values are similar to those obtained by other investigators working with the human protein in other gel electrophoresis systems. The structure of the VSV G glycoprotein is based on information compiled in Lodish *et al.* (1981).

upon immunoprecipitation with the antiserum against erythrocyte Band III, which comigrated with authentic mouse Band III (Fig. 3). The methionine-labelled tryptic peptides of this protein matched those of authentic mouse Band III isolated from the erythrocytes of mice injected with [^{35}S]methionine (results not shown). Thus, these spleen cells were synthesizing Band III.

After labelling times as short as 2 min, newly-made Band III is found only in the membrane fraction of the cell (results not shown). To examine whether this protein has the same transmembrane orientation as in the erythrocyte plasma membrane, we studied the sensitivity to trypsin of newly-synthesized Band III in microsome preparations from labelled spleen cells. If the orientation of Band III in the microsomes is the same as that observed in the plasma membrane, with the *N*-terminal region facing the cytoplasmic surface, extravesicular trypsin should digest Band III to yield an M_r 52000 membrane-protected fragment, the TR 52 fragment (Fig. 1). This fragment was indeed generated (Fig. 4) as confirmed by analysis of its tryptic

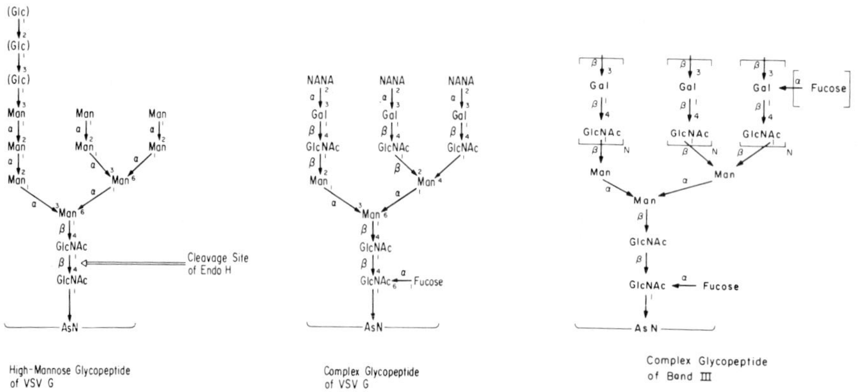

Fig. 2. *Structures of the asparagine-linked oligosaccharide (left to right) transferred to nascent VSV G protein, on mature VSV G protein, and on human Band III protein*

Taken from Tsuji *et al*. (1980) and Lodish *et al*. (1981).

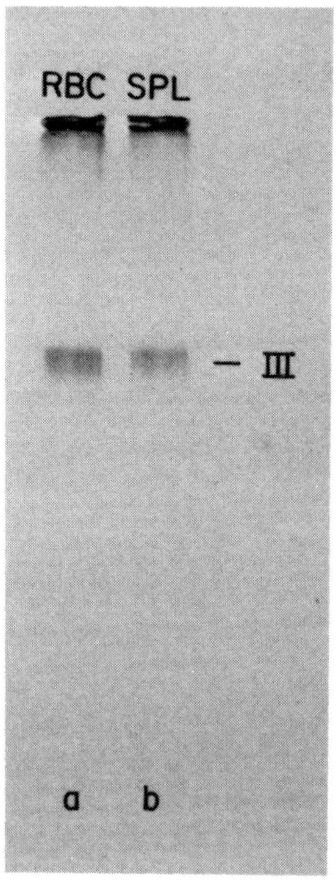

Fig. 3. *Immunoprecipitation of spleen cell Band III*

A detergent solution of plasma membranes prepared from spleen cells labelled in culture for 60 min with [^{35}S]methionine was immunoprecipitated with anti-Band III serum, and run in parallel on SDS/polyacrylamide-gel electrophoresis with labelled erythrocyte Band III immunoprecipitated with the same serum. Lane a, immunoprecipitation of mouse erythrocyte Band III; lane b, immunoprecipitation of mouse spleen cell Band III.

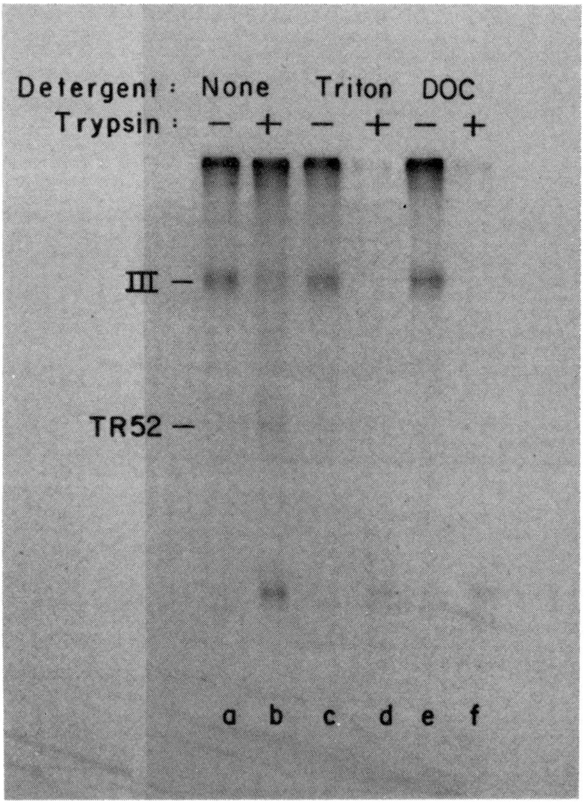

Fig. 4. *Trypsin sensitivity of spleen cell Band III in microsomes*

Spleen cells were labelled for 2 min with [^{35}S]methionine, rapidly washed, and microsomes were prepared. These microsomes were treated with or without 15 μg of TPCK-treated trypsin 1 ml, in the presence or absence of Triton X-100 or deoxycholate detergents, for 60 min at 0°C. Immunoprecipitation of Band III with anti-TR 52 serum (Braell & Lodish, 1981) was performed, and the samples were run in SDS/polyacrylamide-gel electrophoresis and fluorographed. Lane a, no detergent, no trypsin; lane b, no detergent plus trypsin; lane c, 0.5% Triton X-100, no trypsin; lane d, 0.5% Triton X-100 plus trypsin; lane e, 0.2% deoxycholate, no trypsin; lane f, 0.2% deoxycholate, plus trypsin.

peptides (results not shown). Moreover, the addition of nondenaturing detergents to microsomes during the trypsin digestion caused complete destruction of TR 52. This implies that the generation of this TR 52 fragment is dependent on the protection afforded by the permeability barrier of the microsomal lipid bilayer.

Newly made Band III is sensitive to endoglycosidase H; this implies that it contains an asparagine-linked high-mannose oligosaccharide (Fig. 5, lanes b and c). A study similar to that of Fig. 4, but performed with endoglycosidase H, demonstrates that the oligosaccharide of microsomal Band III resists the glycosidase unless the lipid bilayer is made permeable with detergents (Fig. 5, 5d–5g). Similar results were obtained by Hanover & Lennarz (1980) for other membrane and secreted proteins. This indicates that the location of the Band III oligosaccharide is within the lumen of the microsomal vesicle, in accord with the results for other glycoproteins.

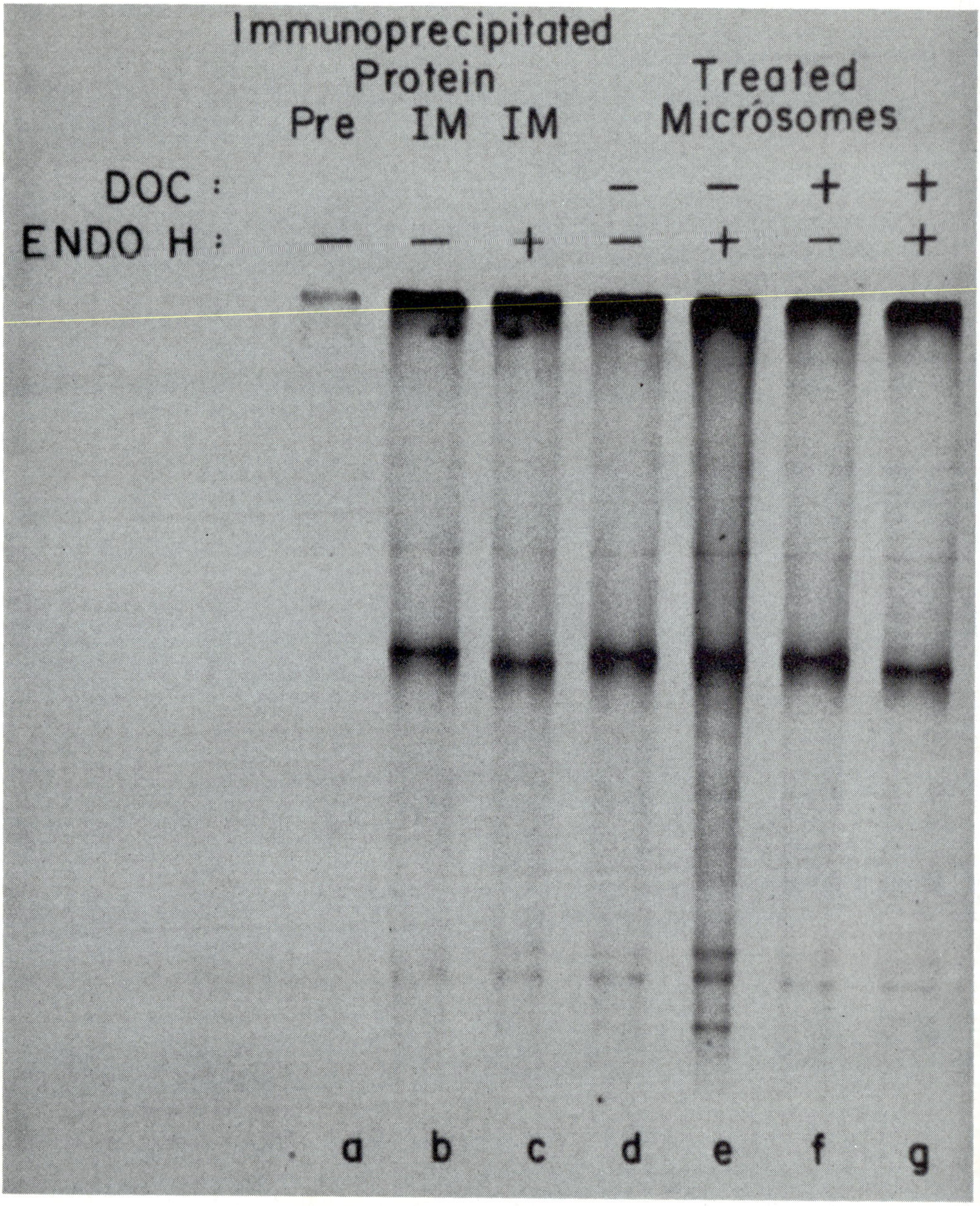

Fig. 5. *Endoglycosidase sensitivity of microsomal Band III*

A crude microsomal preparation of spleen cells labelled with [^{35}S] methionine for 2 min was digested for 60 min at 37°C with 10 μg of endoglycosidase H/ml in buffer with or without deoxycholate. Band III was immunoprecipitated with anti-Band III serum, and the samples run in SDS/polyacrylamide-gel electrophoresis and fluorographed. Lane a, immunoprecipitation of labelled microsomes with preimmune serum; lane b, immunoprecipitated spleen cell Band III incubated for 18 h at 37°C without glycosidase after immunoprecipitation; lane c, immunoprecipitated spleen cell Band III incubated for 18 h at 37°C with endoglycosidase H after immunoprecipitation; lanes d–g, immunoprecipitation was performed after glycosidase treatment: lane d, no detergent, microsomes incubated without glycosidase; lane e, no detergent, microsomes treated with endoglycosidase H; lane f, 0.2% deoxycholate, microsomes incubated without glycosidase; lane g, 0.2% deoxycholate, microsomes treated with endoglycosidase H.

We conclude that newly-made mouse Band III has about 45000 daltons of the *N*-terminal region extending into the cytoplasm, and about 52000 daltons of the membrane-bound *C*-terminal region spanning the microsomal membrane or within the lumen, and with its oligosaccharide sequestered within the lumen. The configuration of newly-made Band III in intracellular membranes is therefore like that of Band III on the mature erythrocyte membrane.

Band III is Made and Glycosylated in Cell-Free Systems

RNA was prepared from spleen cells of anaemic mice and translated in a reticulocyte lysate or wheat germ translation system. Band III was isolated from these reactions by immunoprecipitation, as it represents only about 10^{-4} of the total protein made by these cells. The predominant immunoprecipitated product co-migrated with authentic Band III, synthesized by intact spleen cells (Fig. 6). Tryptic peptide analysis (results not shown) established that this material indeed is Band III.

Band III isolated from cell-free translations in which dog pancreatic microsomes were present migrated more slowly on SDS gels than did Band III isolated from translations in which the microsomes were added only after translation had been halted with emetine (Fig. 6, lanes f and g). The difference in mobility between the two forms corresponds to about 2000 daltons, consistent with the presence of a high-mannose oligosaccharide on the slower migrating form of Band III. To demonstrate this point, treatment of this slower migrating form with endoglycosidase H converted it to the faster migrating form (Fig. 6, lanes c and d); the latter form is resistant to endoglycosidase H (results not shown). This is evidence that Band III is glycosylated by dog pancreatic microsomes in a co-translational manner.

When the slower migrating form of Band III, produced by cell-free translation in the presence of microsomes, was treated with endoglycosidase H, it co-migrated with the faster migrating form which did not insert into the membrane. This and other results suggests that there is no peptide cleaved from Band III during insertion into membranes. In particular, the cell-free product (made in the absence of microsomes) co-migrated with the Band III made by spleen cells and treated with endoglycosidase H (Braell & Lodish, 1982*a*).

Band III Synthesized *in vitro* is Inserted Normally into Microsomal Membranes

To demonstrate this point, Band III was synthesized in cell-free reactions containing dog pancreatic microsomes. Proteinase digestion of reaction products yields the characteristic TR 52 fragment, which has a similar mobility to the homologous fragment from erythrocyte membranes (Fig. 7, lanes i and m). Partial proteolytic digestion of the erythrocyte TR 52 and the TR 52 derived from the cell-free translation product yielded the same fragments (results not shown), establishing that they are indeed the same. No TR 52 was obtained from the faster moving form of Band III synthesized in the absence of microsomes (Fig. 7, lane j). Also, addition of Triton X 100 during the trypsin digestion resulted in destruction of TR 52 (lane k), indicating that the resistance of the *C*-terminal TR 52 segment of Band III to cytoplasmic proteinase is due to the permeability barrier of the microsomal membrane. We conclude that the slower migrating glycosylated form of

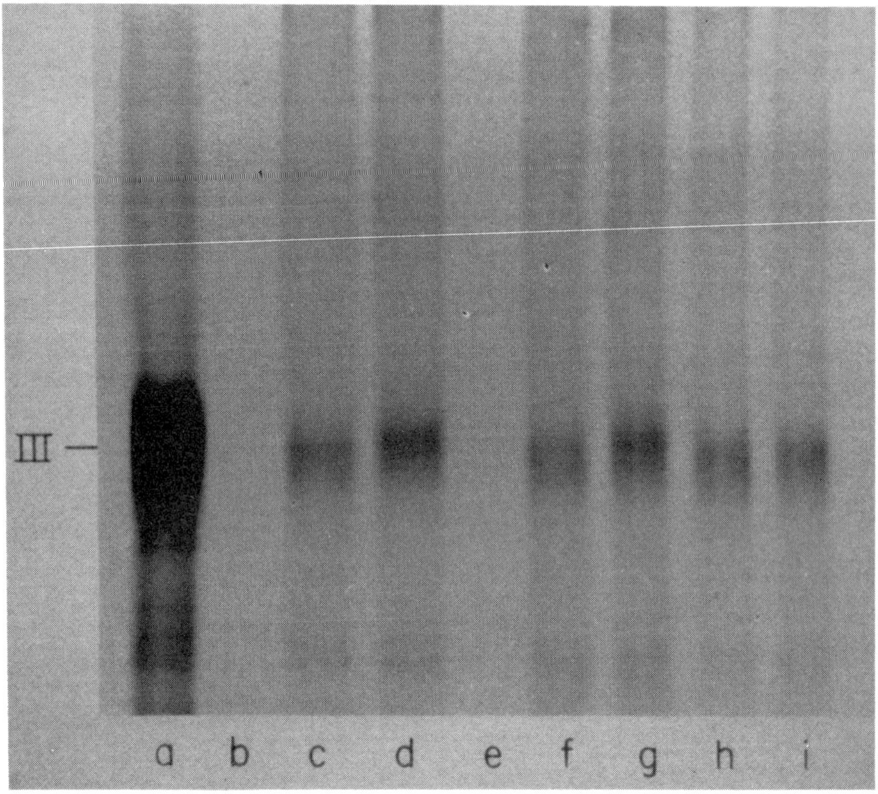

Fig. 6. *Cell-free synthesis of spleen cell Band III*

Spleen cell RNA was prepared by the guanidinium thiocyanate method (Braell & Lodish, 1982*a*) and translated in the wheat germ system at 25°C in the presence or absence of dog pancreatic microsomes, or in the presence of microsomes treated in ways known to destroy their ability to insert VSV G type proteins (heating at 60°C for 10 min, or treatment for 30 min at 25°C with 1.5 mM-*N*-ethylmaleimide). The samples translated in the absence of microsomes were inhibited by 100 μg of emetine/ml, and then incubated for 30 min at 25°C with dog pancreatic microsomes. Band III was immunoprecipitated from the reactions with preimmune or anti-Band III serum. Some samples translated with membranes were digested with or without endoglycosidase H after immunoprecipitation. Lane a contained immunoprecipitated spleen cell Band III. Lanes c and d contained cell-free synthesized Band III treated with (lane c) or without (lane d) endoglycosidase H after immunoprecipitation. Lanes f, g, h, and i contained Band III immunoprecipitated with anti-Band III serum from reactions in which membranes were present only after translation was inhibited (lane f), membranes were present always (lane g), membranes treated with *N*-ethylmaleimide were added (lane h), and membranes treated at 60°C were added (lane i). Lane b was an immunoprecipitation with anti-Band III serum from a reaction with microsomes but lacking spleen cell RNA. Lane e was an immunoprecipitation with preimmune serum from a reaction with both RNA and microsomes present.

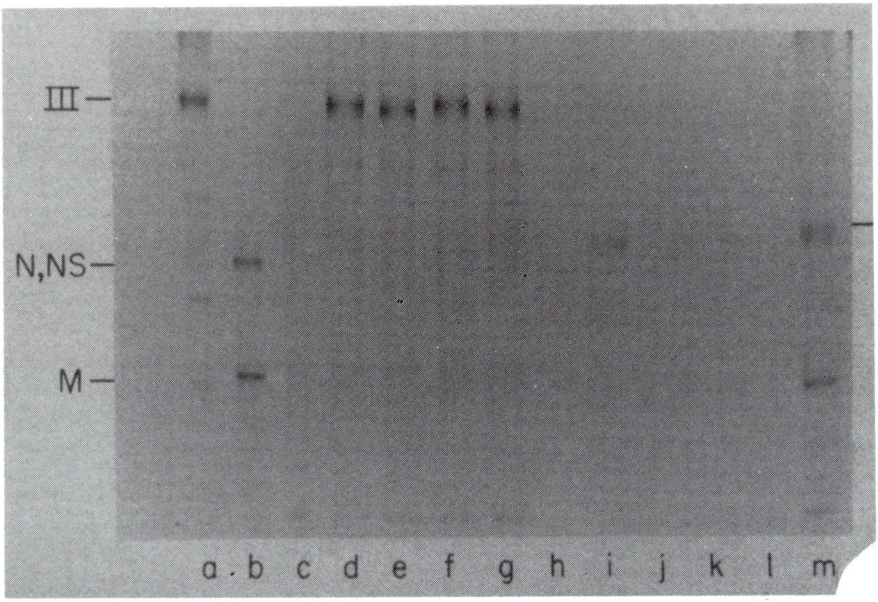

Fig. 7. *Trypsin protection of cell-free Band III*

Spleen cell RNA was translated in the wheat germ system with dog pancreatic microsomes as described in Fig. 6. Reactions were split and incubated with or without 20 μg of TPCK-treated trypsin/ml at 0°C for 60 min, in the presence or absence of 0.5% Triton X-100 detergent. Digestion was halted with soya bean trypsin inhibitor, and Band III was immunoprecipitated with preimmune or anti-Band III sera. Lane a contains labelled erythrocyte membrane proteins. Lane b contains VSV proteins N, NS, and M. Lanes c and h contain precipitates with pre-immune serum from a reaction with microsomes, either untreated (lane c) or trypsinized after translation was halted (lane h). Lanes d and i contain the immunoprecipitates from reactions containing microsomes, either untreated (lane d) or treated with trypsin after ending translation (lane i). Lanes e and j contain the immunoprecipitates from reactions incubated with microsomes added only after translation was halted with emetine, undigested (lane e) or subsequently trypsinized (lane j). Lanes f and k were like lanes d and i, respectively, except that 0.5% Triton X-100 was present during the 1 h of digestion at 0°C. Lanes g and l were like lanes e and j, respectively, except that 0.5% Triton X-100 was present during the 1 h of digestion at 0°C. Lane m contains labelled trypsin-digested erythrocyte membranes. Note that the fragment comigrating with TR 52 is only produced by the digestion of the sample where membranes were present during translation.

Band III synthesized *in vitro* was inserted properly into the microsomal membrane, providing the membranes are present during protein synthesis.

Band III is Co-translationally Inserted into Microsomes

This determination can be made by means of a synchronized translation experiment. In such an experiment, initiation of protein synthesis is allowed for a brief period, after which initiation is specifically inhibited, and the ribosomes undergoing translation of the messenger are allowed to traverse the message synchronously. The addition of dog pancreatic microsomes at various stages of growth of the nascent chain, followed by completion of the protein in the presence of the membranes, will show when during elongation such membranes must be present if insertion is to occur.

We have performed this experiment on Band III synthesized in the reticulocyte lysate. Synchronization was effected using 4.5 mM-7-methylguanosine 5′-phosphate and 1.5 μM-edeine; this combination of inhibitors, originally used by Scheele *et al.* (1980), inhibited initiation to greater than 99% in our hands (results not shown). The degree of completion of Band III was monitored by delivery at various times of aliquots of the synchronized reaction mix into immunoprecipitation buffer containing emetine and sodium EDTA, which inhibited further peptide chain elongation during the subsequent immunoprecipitation.

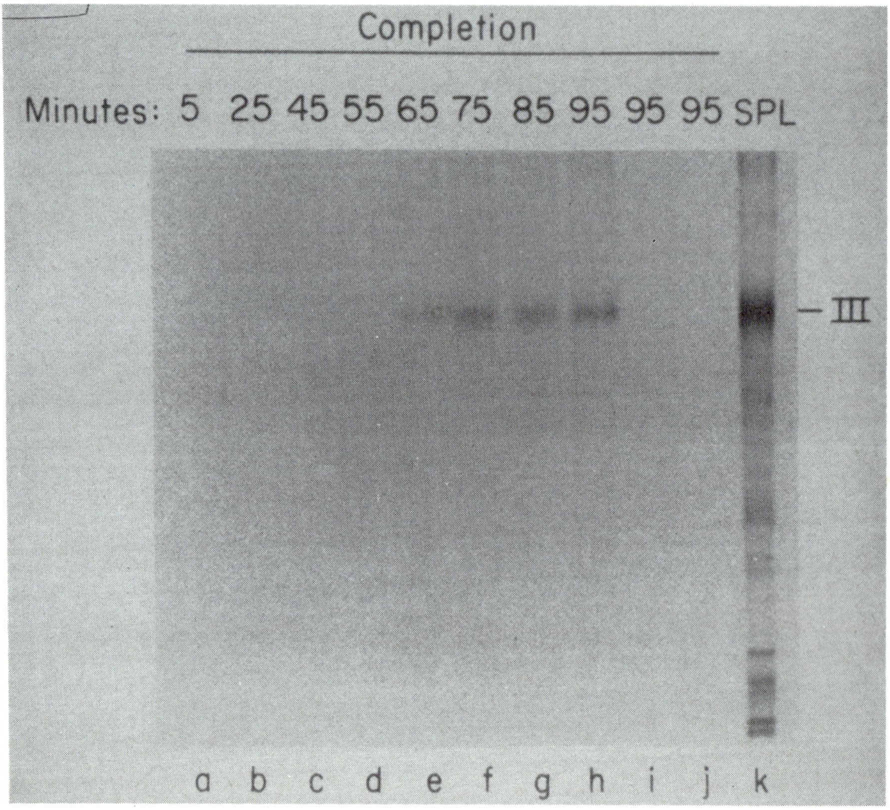

Fig. 8. *Synchronized completion of cell-free Band III*

Spleen cell RNA was translated at 22°C in the reticulocyte lysate system, with 4.5 mM-7-methylguanosine 5′-phosphate and 1.5 μM-edeine added after 5 min to inhibit initiation. Starting from the time of addition of inhibitor, aliquots of the reaction were delivered into immunoprecipitation buffer containing 100 μg of emetine/ml and sodium EDTA, which inhibited elongation totally. Band III was immunoprecipitated with anti-TR 52 serum from these samples, and analysed by SDS/polyacrylamide-gel electrophoresis and fluorography. The times shown are elapsed times from the beginning of translation of the RNA. Lanes a–h respectively show the amount of completed Band III after 5, 25, 45, 55, 65, 75, 85, and 95 min of translation. Lane i shows the result of a second anti-TR 52 immunoprecipitation of the supernatant of the 95 min sample. Lane j shows the result of a preimmune precipitation of a sample incubated for 95 min. Lane k has immunoprecipitated spleen cell Band III. Quantification of completion is shown in Fig. 11.

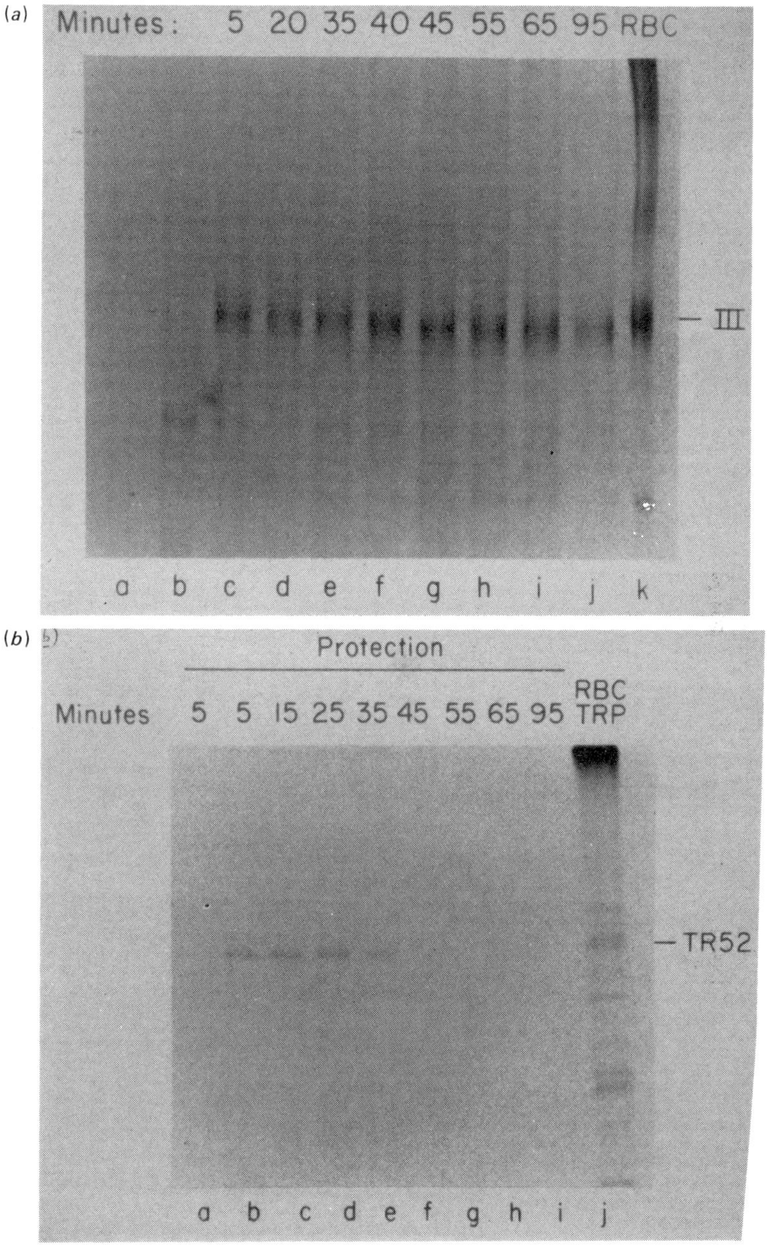

Fig. 9. *Synchronized glycosylation (a) and trypsin protection (b) of cell-free Band III*

(*a*) Aliquots of the reaction were added at various times to 6 A_{260} units of dog pancreatic microsomes/ml and the translation contined at 22°C for the remainder of 100 min from the start of translation. Band III was immunoprecipitated with anti-TR 52 serum and analysed by SDS/polyacrylamide-gel electrophoresis and fluorography, the gels being run three times longer than usual for extra resolution. Lane a shows a preimmune precipitation from a sample with membranes added at 5 min; lane b shows the result of a second immunoprecipitation from the 5 min membrane addition of lane c. Lanes c–j, respectively, show the immunoprecipitations from samples with membranes added at 5, 20, 35, 40, 45, 55, 65, and 95 min after the start of translation. Lane k is erythrocyte Band III. (*b*) Samples were translated and membranes added

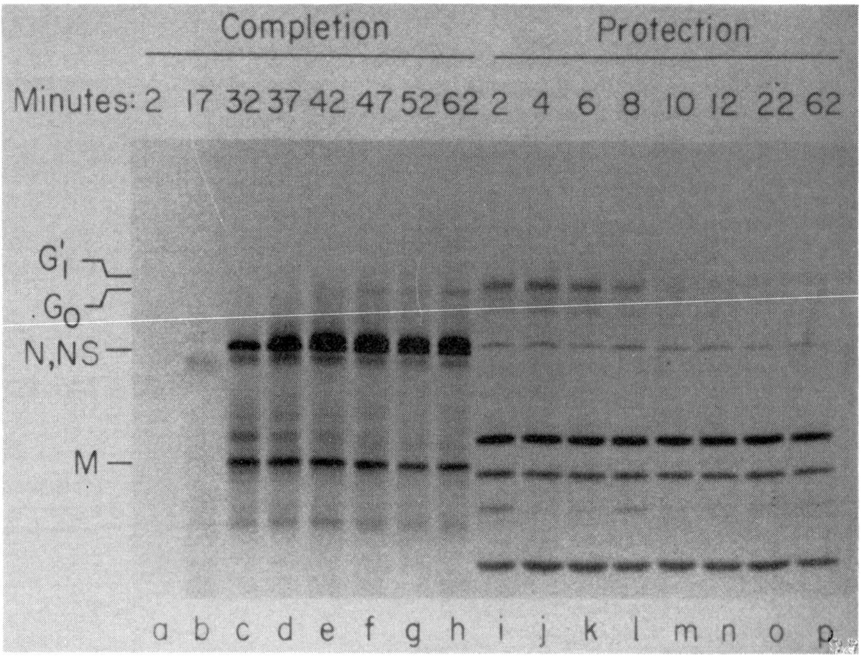

Fig. 10. *Completion and protection of VSV G protein*

As an internal control, VSV messenger RNA was added to the reactions described in parts (*a*) and (*b*) of Fig. 9 at 2 min before the addition of initiation inhibitors. The amount of VSV RNA (240 μg total RNA/ml) was well below that which would interfere with Band III insertion, but still could be directly observed. Completion was monitored by addition of aliquots of the reaction at various times to SDS gel sample buffer; protection was monitored by digestion of samples with trypsin prior to sample buffer addition. Samples were analysed by SDS/polyacrylamide-gel electrophoresis and fluorography. Lanes a–h, respectively, show the degree of completion of VSVG protein at 2, 17, 32, 37, 42, 47, 52 and 62 min after VSV RNA addition. Lanes i–p, respectively, show the degree of trypsin protection when membranes were added at 2, 4, 6, 8, 10, 12, 22 and 62 min after translation. Quantification of VSV G protein completion and protection are shown in Fig. 11.

Under the conditions of translation in the reticulocyte lysate system, some 60–65 min were required for completion of the Band III polypeptide (Figs. 8 and 11). Based on a value of 895 amino acid residues, obtained by dividing the M_r of 100000 by the average amino acid M_r of 112, the average rate of elongation for Band III was about 14 residues/min.

The latest time at which membranes could effect insertion was found by adding microsomes to aliquots of the synchronized reaction at various times, and allowing read-off to finish. Each aliquot was either immunoprecipitated directly and analysed

---

as described in part (*a*), and treated with trypsin at 0°C for 60 min prior to immunoprecipitation, to determine the degree of proteinase protection. Digested samples were immunoprecipitated with anti-TR 52 serum and analysed by SDS/polyacrylamide-gel electrophoresis and fluorography. Lane a shows a preimmune precipitation of a sample with membranes added at 5 min. Lanes b–i show the immunoprecipitations from samples with membranes added at 5, 15, 25, 35, 45, 55, 65, and 95 min after the start of translation. Lane j shows erythrocyte TR 52. Quantification of the proteinase protection of Band III is shown in Fig. 11.

in order to determine the state of glycosylation, or digested with trypsin and then immunoprecipitated in order to determine the insertion into microsomes.

As is shown in Fig. 9(b), microsomes could be added as late as 35–40 min after the start of synthesis, and still allow normal trans-membrane insertion of the subsequently completed polypeptide. Also, if membranes were added before 35–40 min the subsequently completed Band III migrated on SDS gels with a slower mobility than did Band III completed after a later addition of membranes (Fig. 9a). Most probably, this was a shift at this time from synthesis of the slower migrating glycosylated Band III to synthesis of unglycosylated Band III, although the quantification was more ambiguous than for the protection of Band III from proteinase. Thus, the signal directing the insertion of Band III appears to occur prior to residue 550 in the protein, some 55–60% of the distance from the *N*-terminus of Band III (Fig. 11).

As an internal standard, VSV messenger RNA was translated simultaneously with spleen cell RNA in the same synchronized reaction. VSV G protein (570 amino acids) was completed in about 43 min, yielding a rate of elongation of about 13 amino acid residues/min. VSV G was protected from trypsin digestion by its insertion into microsomes only if membranes were added within about 8 min (Figs.

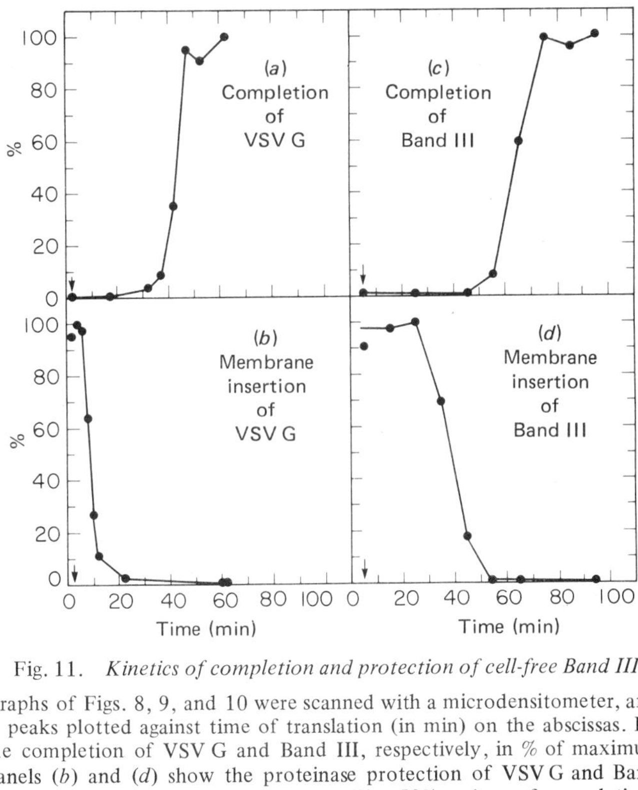

Fig. 11. *Kinetics of completion and protection of cell-free Band III*

The fluorographs of Figs. 8, 9, and 10 were scanned with a microdensitometer, and the area of the relevent peaks plotted against time of translation (in min) on the abscissas. Panels (*a*) and (*c*) show the completion of VSV G and Band III, respectively, in % of maximum, along the ordinate. Panels (*b*) and (*d*) show the proteinase protection of VSV G and Band III, respectively, in % of maximum, along the ordinate. The 50% points of completion and loss of protection were taken in each case as the mean values of these parameters.

10 and 11). This places the signal sequence of VSV G prior to residue 100, confirming earlier results in the wheat germ cell-free system for VSV G (Rothman & Lodish, 1977).

Maturation of Band III in Intact Spleen Cells

During a chase period, pulse-labelled Band III, synthesized in intact spleen cells, was observed gradually to become resistant to endoglycosidase H, with a half-life of about 30 min (Fig. 12). This result establishes, first, that the shift in mobility observed with the pulse-labelled Band III is not due to nonspecific proteolytic degradation. Second, the processing steps which remove mannose from the high-mannose oligosaccharide and thereby confer resistance to endoglycosidase H have been associated with enzyme activities found in the Golgi complex (Li *et al.*, 1978; Tabas & Kornfeld, 1979; Bretz *et al.*, 1980). These results imply that Band III

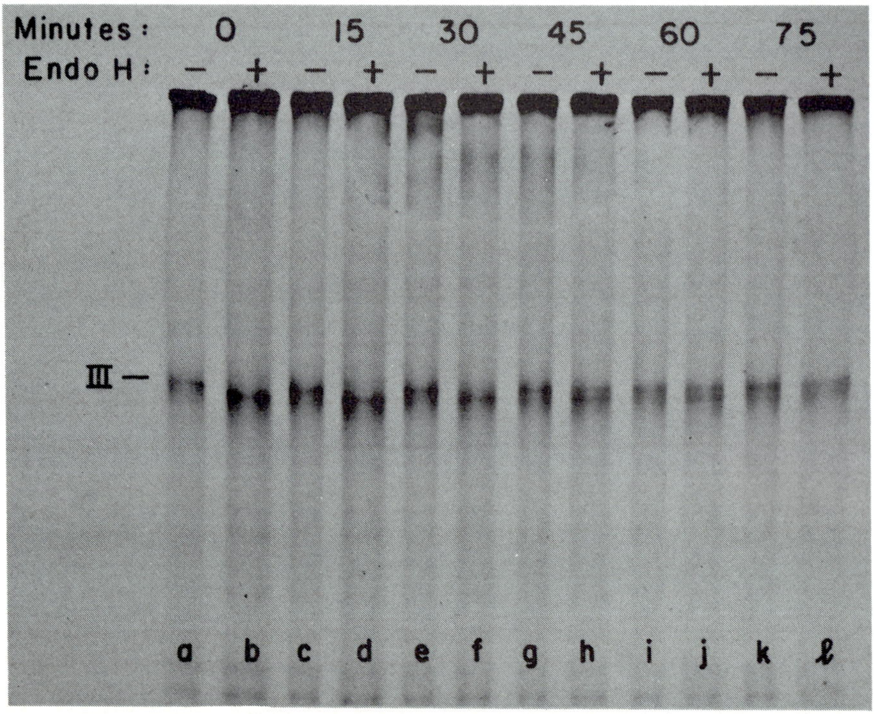

Fig. 12. *Sensitivity of newly-made spleen cell Band III to endoglycosidase H*

Spleen cells were labelled with [^{35}S]methionine for 5 min, and chased with excess unlabelled methionine. Aliquots of cells were removed, lysed, and immunoprecipitated with anti-Band III antiserum. The immunoprecipitates were split and digested for 18 h at 37°C with or without endoglycosidase H. The digests were run in SDS/polyacrylamide-gel electrophoresis and fluorography was performed. Lane a, 0 min of chase, no enzyme; lane b, 0 min of chase, glycosidase-treated; lane c, 15 min of chase, no enzyme; lane d, 15 min of chase, glycosidase-treated; lane e, 30 min of chase, no enzyme; lane f, 30 min of chase, glycosidase-treated; lane g, 45 min of chase, no enzyme; lane h, 45 min of chase, glycosidase-treated; lane i, 60 min of chase, no enzyme; lane j, 60 min of chase, glycosidase-treated; lane k, 75 min of chase; no enzyme; lane l, 75 min of chase, glycosidase-treated.

receives a high-mannose oligosaccharide in the endoplasmic reticulum, and progresses through the Golgi complex. It appears on the cell surface about 10 min later (Braell & Lodish, 1981).

Discussion

Taken together, the above data demonstrate that Band III is co-translationally inserted into microsomes, as such microsomes must be present before the protein is about 60% complete if insertion is to occur at all. Moreover, the insertion of Band III is not prevented until microsomes are added after the protein is about half complete, although in the same reaction the insertion of VSV G protein is prevented if microsomes are added after the first 100 amino acid residues are synthesized. These data suggest that Band III may have an internal signal sequence, which directs the insertion into microsomes, mid-way through the 895 amino acid polypeptide.

Synchronized translation experiments have been performed with several proteins, including membrane proteins like VSV G and Sindbis pE_2, pancreatic secretory proteins, and the peculiar case of ovalbumin (Rothman & Lodish, 1977; Garoff *et al.*, 1978; Lingappa *et al.*, 1979; Scheele *et al.*, 1980). In all of these proteins, the *N*-terminal segment is transferred to the extracytoplasmic side of the endoplasmic reticulum membrane. In all cases, insertion is prevented if more than about 100 amino acid residues are synthesized before the addition of membranes. This constitutes good evidence for the existence of an *N*-terminal signal sequence directing the insertion, which must act rapidly after its emergence from the ribosome or else become impotent. One report (Lingappa *et al.*, 1979) indicates that ovalbumin utilizes an internal signal sequence, near residue 250, for membrane transport. However, re-evaluation of these synchrony experiments indicate that the critical sequence for membrane insertion is within the *N*-terminal 150 residues (Braell & Lodish, 1982*b*). This is in agreement with other recent results (Meek *et al.*, 1980; R.L. Meek, K.A. Walsh & R.D. Palmiter, personal communication) which indicate that nascent ovalbumin chains of 120 residues are already found inserted into the endoplasmic reticulum membrane.

Rigorously, in the case of Band III it is inappropriate to claim that the actual signal for membrane insertion is congruent with the end of the longest nascent chain which can interact positively with the microsomes, at residue 450–500. This critical point is that at which the actual signal can no longer bring about insertion into membranes, and the essential signal sequence theoretically could reside anywhere within the previously-synthesized 55000 daltons. It is by analogy to other signal sequences that we suggest that only a few thousand daltons of excess amino acid sequence are required to abolish the signal's effectiveness. Under such an assumption, the signal for Band III would reside roughly halfway through the polypeptide. It is reasonable to make this assumption, however, as the hydrophilic *N*-terminal domain of Band III removed by trypsin is not membrane-bound and there is no evidence which suggests that that segment actually crosses the membrane. Thus, the signal sequence, which would presumably be inserted in the membrane, does not seem to reside in this 45000 dalton *N*-terminal domain. Logically, this localizes the signal sequence of Band III to a region 45000–60000 daltons from the *N*-terminus, making it a truly internal signal sequence.

Because Band III spans the membrane multiple times, it is possible that more than one insertion signal resides in the protein. These later signals would not be detected in our experiment if the failure of the first hydrophobic domain to insert prevented the insertion of subsequent domains. Such signals also could work independently, if one assumes the anti-TR 52 serum is specific only for determinants on the most *N*-terminal domain of the *C*-terminal TR 52 fragment. If this domain were not inserted, its destruction by proteinase would leave the later domains of Band III undetectable in our experiment, whether or not they were inserted into the membrane. The synchronization experiment would then localize only the most *N*-terminal of the insertion signals.

We do not know whether the mechanism by which the *C*-terminal region of Band III inserts into the membrane is at all similar to that by which the *N*-terminal of VSV G and secretory proteins cross the membrane. Undoubtedly, however, the mechanism for transfer of a high-mannose oligosaccharide from a dolichol carrier to the nascent chain is universal. The dog pancreatic microsomes are apparently able to provide Band III with a single high-mannose oligosaccharide, similar to that which it acquires upon insertion into spleen cell membranes. The transferase which accomplishes this has been shown only to glycosylate the denatured nascent polypeptide chain of VSV G (Rothman *et al.*, 1978), implying that Band III receives its high-mannose oligosaccharide during translation.

Whether or not Band III, like Sindbis pE_2 or ovalbumin, lacks a cleaved *N*-terminal peptide, it differs from all other proteins synthesized on the rough endoplasmic reticulum which have been studied to date in that it may utilize an internal amino acid sequence for membrane insertion. We have proposed one model for such a process (Lodish *et al.*, 1981), as have others (Blobel, 1980; Engelman & Steitz, 1981), but the exact details of this remain to be elucidated.

However, in its maturation to the cell surface, newly-made Band III apparently passes through the Golgi complex, where its oligosaccharide is processed to an endoglycosidase H-resistant form in a manner reminiscent of the maturation of the sugars of VSV G and glycophorin. Newly-synthesized Band III begins appearing on the cell surface about 20 min after its synthesis, a time span for maturation similar to that observed for VSV G protein and other glycoproteins of similar structure (Lodish *et al.*, 1981).

Acknowledgements

This research was supported by grants AI-08814, AM-15322, AM-15929 and AM-27375 from the National Institutes of Health. W.A.B. was supported by a Training Grant from the National Institutes of Health. We thank Miriam Boucher for preparing the manuscript.

References

Blobel, G. (1980) *Proc. Natl. Acad. Sci. U.S.A.* **77**, 1496–1500
Braell, W.A. & Lodish, H.F. (1981) *J. Biol. Chem.* **256**, 11337–11344
Braell, W.A. & Lodish, H.F. (1982*a*) *Cell* **28**, 23–31
Braell, W.A. & Lodish, H.F. (1982*b*) *J. Biol. Chem.*, in the press
Bretz, R., Bretz, H. & Palade, G.E. (1980) *J. Cell Biol.* **84**, 87–101
Cabantchik, Z.I., Knauf, P.A. & Rothstein, A. (1978) *Biochim. Biophys. Acta* **515**, 239–302
Drickamer, L.K. (1976) *J. Biol. Chem.* **251**, 5115–5123

Drickamer, L.K. (1977) *J. Biol. Chem.* **252**, 6909–6917
Drickamer, L.K. (1978) *J. Biol. Chem.* **253**, 7242–7248
Engelman, D.M. & Steitz, T.A. (1981) *Cell* **23**, 411–422
Fukuda, M., Eshdat, Y., Tarone, G. & Marchesi, V.T. (1978) *J. Biol. Chem.* **253**, 2419–2428
Garoff, H., Simons, K. & Dobberstein, B. (1978) *J. Mol. Biol.* **124**, 587–600
Grinstein, S., Ship, S. & Rothstein, A. (1978) *Biochim. Biophys. Acta* **507**, 294–304
Hanover, J.A. & Lennarz, W.J. (1980) *J. Biol. Chem.* **255**, 3600-3604
Ho, M.K. & Guidotti, G. (1975) *J. Biol. Chem.* **250**, 675–683
Katz, F.N. & Lodish, H.F. (1979) *J. Cell Biol.* **80**, 416–426
Katz, F.N., Rothman, J.E., Knipe, D.M. & Lodish, H.F. (1977) *J. Supramol. Struct.* **7**, 353–370
Knipe, D., Lodish, H.F. & Baltimore, D. (1977) *J. Virol.* **21**, 1121–1127
Li, E., Tabas, I. & Kornfeld, S. (1978) *J. Biol. Chem.* **253**, 7762–7770
Lingappa, V.R., Lingappa, J.R. & Blobel, G. (1979) *Nature (London)* **281**, 117–121
Lodish, H.F., Braell, W.A., Schwartz, A.L., Strous, G.J.A.M. & Zilberstein, A. (1981) *Int. Rev. Cytol. Suppl.* **12**, 247–307
Meek, R.L., Walsh, K.A. & Palmiter, R.D. (1980) *Fed. Proc. Fed. Am. Soc. Exp. Biol.* **39**, 1867
Rao, A. & Reithmeier, R.A.F. (1979) *J. Biol. Chem.* **254**, 6144–6150
Robbins, P.W., Hubbard, S.C., Turco, S.J. & Wirth, D.F. (1977) *Cell* **12**, 893–900
Rothman, J.E. & Lodish, H.F. (1977) *Nature (London)* **269**, 775–780
Rothman, J.E., Katz, F.N. & Lodish, H.F. (1978) *Cell* **15**, 1447–1454
Scheele, G., Jacoby, R. & Carne, T. (1980) *J. Cell Biol.* **87**, 611–628
Schmidt, M.F.G. & Schlesinger, M. (1979) *Cell* **17**, 813–819
Steck, T.L. (1978) *J. Supramol. Struct.* **8**, 311–324
Steck, T.L., Ramos, B. & Strapazon, E. (1976) *Biochemistry* **15**, 1154–1161
Steck, T.L., Koziarz, J.J., Singh, M.K., Reddy, G. & Kohler, H. (1978) *Biochemistry* **17**, 1216–1222
Tabas, I. & Kornfeld, S. (1979) *J. Biol. Chem.* **254**, 11655–11663
Tsuji, T., Irimura, T. & Osawa, T. (1980) *Biochem. J.* **187**, 677–686
Williams, D.G., Jenkins, R.E. & Tanner, M.J.A. (1979) *Biochem. J.* **181**, 477–493

Subject Index

Anti-(cap binding protein) (anti-CBP), inhibition of protein crosslinking to mRNA and translation, 145–161
Asparagine-linked oligosaccharide in Band III protein, 193-209

BHK cells, cytoskeleton and M_r 50000 cap binding protein, 145–161
Bacteria, properties of ribosomes, 11–48
Band III (erythrocyte anion transport protein), 193–209

Cap binding protein
  association with cytoskeleton in BHK cells, 145–161
  eukaryotic mRNA function, 129–143
Caps, see 5′-terminal caps
Charging, 1-9
Chloroplasts, properties of ribosomes, 11–48
Co-translational membrane insertion of erythrocyte anion transport protein, 193–209
Crosslinking
  intra-RNA in ribosomal subunits, 49–60
  rRNA–protein in ribosomal subunits, 49–60
Cytoplasm, properties of ribosomes, 11–48
Cytoskeleton, association with M_r 50000 cap binding protein, 145–161

Deadenylylation of mRNA, 61–78

Elongation factors EF-Ts and EF-Tu, kinetics of ribosome structure, 1–9
Endoplasmic reticulum, mechanism of protein translocation, 183–191
12,13-Epoxytrichothecenes, 79–93
Erythrocyte anion transport protein, synthesis and maturation, 193–209
Escherichia coli
  expression of IF3 gene, 95–112
  kinetics of ribosome structure, 1–9
  rRNA structure, 11–48
  secondary structure of rRNA, 49–60
  structure and expression of IF3 gene, 95–112

Gene dosage, effect on IF3 synthesis, 95–112
Gene expression and mRNP complexes, 163–182
Glycosylation of band III in cell-free systems, 193-209

Heterogeneous nuclear ribonucleoprotein (hnRNP), 163–182
High-mannose oligosaccharide, processing in anion transport protein, 193–209
β-Hydroxyleucine, effect on translation of preprolactin, 183–191

Immunofluorescence microscopy, cap binding protein and cytoskeleton, 145–161
Initiation factor IF3
  infC gene, 95–112
  purification from overproducing strains, 95–112
  structure and expression of gene in *E. coli*, 95–112
Initiation factor IF3 (*infC*) gene
  localization and expression, 95–112
  nucleotide sequence, 95–112
Initiation of protein synthesis, 95–112, 129–143, 145–161
Initiation site selection in eukaryotic mRNA, 113–128
Initiator codon, recognition by eukaryotic ribosomes, 113–128

Kinetics, ribosome structure, 1–9

Magic spot, kinetics of ribosome structure, 1–9
Maturation of erythrocyte anion transport protein, 193–209
Membrane insertion of erythrocyte anion transport protein, 193–209
Messenger RNA (mRNA)
  5′-terminal caps and cap-binding proteins, 129–143
  deadenylylation, 61–78
  effect of secondary structure on ribosome transit, 61–78

Messenger RNA (mRNA) – *contd*
  inhibition of protein cross-linking and translation by anti-CBP, 145–161
  methods of structure analysis, 61–78
  modification of primary transcripts, 129–143
  preparation of 5′-labelled, 61–78
  recognition of initiator codon by ribosomes, 113–128
  secondary structure of eukaryotic, 61–78
  translation mechanism in prokaryotes, 95–112
Mesenger ribonucleoprotein (mRNP) complexes in gene expression, 163–182
Microsomal membranes, insertion of band III, 193–209
Mitochondria, properties of ribosomes, 11–48

Nuclear pre-mRNP (hnRNP), 163–183
Nucleotide sequence
  5′-end of L-rRNA, 11–48
  flanking AUG triplets in eukaryotic mRNAs, 113–128
  gene for IF3, 95–112
  rRNA, 11–48

Pausing of ribosomes, 61–78
Peptidyltransferase centre as target for 12,13-epoxytrichothecenes, 79–93
Phenylalanyl-tRNA synthetase, 95–112
Polymix, 1–9
Polyribosomes, *see* polysomes
Polysomes
  association with signal recognition protein, 183–191
  binding of anti-CBP antibody, 145–161
  binding to endoplasmic reticulum, 183–191
  messenger ribonucleoprotein complexes, 163–182
Pre-messenger ribonucleoprotein (pre-mRNP), 163–182
Pre-ribosomal RNA (pre-rRNA), processing of transcript, 79–93
Preprolactin, effect of signal recognition peptide on synthesis, 183–191
Protein
  cap-binding, 129–143, 145–161
  components of ribonucleoproteins, 163–182
  erythrocyte anion transport, 193–209
  inhibition of crosslinking to mRNA by anti-CBP, 145–161
  lysosomal, 183–191
  membrane, 183–191
  ribosomal, *see* ribosomal protein
  secretory, 183–191
  signal recognition, 183–191
  topogenesis, 183–191
  translocation across the endoplasmic reticulum, 183–191

RNA
  messenger, *see* messenger RNA
  ribosomal, *see* ribosomal RNA
Regulation
  expression of IF3 (*infC*) gene, 95–112
  transcriptional initiation in eukaryotes, 113–128
Restriction endonuclease mapping of the *infC* gene, 95–112
Ribonucleoprotein (RNP), 163–182
Ribosomal RNA (rRNA)
  5.8 S, 11–48
  crosslinking in ribosomal subunits, 49–60
  cytoplasmic, 11–48
  determination of secondary structure, 49–60
  eukaryotic ribosomes, 11–48
  organization within functional ribosomes, 11–48
  organization within ribosomal subunits, 49–60
  primary structure, 11–48
  secondary structure, 11–48, 49–60
Ribosomal RNA (rRNA) genes, spacer sequences, 11–48
Ribosomal proteins, differences in antibiotic-resistant yeast strains, 79–93
Ribosomal subunits, migration on mRNA, 113–128
Ribosomes
  binding to mRNA, 113–128
  biogenesis of yeast, 79–93
  comparison of properties of various types, 11–48
  effect of mRNA secondary structure on transit, 61–78
  kinetic view of structure, 1–9
  migration of subunits on mRNA, 113–128
  organization of ribosomal RNA in subunits, 49–60

Ribosomes – *contd*
recognition of initiator codon by eukaryotic, 113–128
structure of eukaryotic, 11–48
structure of prokaryotic, 1–9
structure of yeast, 79–93

Scanning mechanism of initiation in eukaryotes, 113–128
Secondary structure
computer-aided studies of RNA, 61–78
determination for *E. coli* rRNA, 49–60
eukaryotic mRNA, 61–78
ribosomal RNA, 11–48, 49–60
Secretory proteins, effect of signal recognition peptide on synthesis, 183–191
Signal recognition protein, mechanism of protein translocation, 183–191
Spacer sequences in rRNA genes, 11–48
Stop signal, migration of ribosomal subunits, 113–128
Stringent response and the kinetics of ribosome structure, 1–9
Synthesis of erythrocyte anion transport protein, 193–209

Temperature-sensitive (*ts*) mutants of yeast, ribosomal proteins, 79–93
5′-Terminal caps
binding proteins, 129–143, 145–161
eukaryotic mRNA function, 129–143
inhibition of initiation by analogues, 129–143
removal before mRNA secondary structure analysis, 61–78
synthesis and transcription initiation, 129–143
Threonyl-tRNA synthetase, 95–112
Transcription
5′-terminal cap synthesis and initiation of, 129–143
cluster of genes in the *infC* region, 95–112
Translation
in vitro, 1–9
inhibition by antibody to cap binding protein, 129–143
inhibition of initiation by cap analogues, 129–143
mechanism in prokaryotes, 95–112
Trichodermin, yeast mutants resistant to, 79–93

Yeast, biogenesis of ribosomes, 79–93
Yeast (*Kluyveromyces*), ribosomes of antibiotic-resistant strains, 79–93
Yeast (*Saccharomyces*)
antibiotic-resistant mutants, 79–93
temperature-sensitive mutants, 79–93
ribosomes of antibiotic-resistant strains, 79–93